T0206205

3D SPECTROSCOPY IN ASTRONOMY

Simultaneously storing both spectral and spatial information, 3D spectroscopy offers a new way to tackle astrophysical problems, and opens up new lines of research. Since its inception in the eighties and early nineties, research in this field has grown enormously. Large telescopes all around the world are now equipped with integral field units, and two instruments of the James Webb Space Telescope will have integral field spectroscopic capabilities. Nowadays, more effort is dedicated to refining techniques for reducing, analysing and interpreting the data obtained with 3D spectrographs.

Containing lectures from the seventeenth Winter School of the Canary Islands Astrophysics Institute, this book explores new 3D spectroscopy techniques and data. A broad and balanced presentation of research in this field, it introduces astronomers to a new generation of instruments, widening the appeal of integral field spectroscopy and helping it become a powerful tool in tackling astrophysical problems.

Canary Islands Winter School of Astrophysics

Volume XVII

Editor in Chief
F. Sánchez, *Instituto de Astrofísica de Canarias*

Volumes in this series

I. Solar Physics
II. Physical and Observational Cosmology
III. Star Formation in Stellar Systems
IV. Infrared Astronomy
V. The Formation of Galaxies
VI. The Structure of the Sun
VII. Instrumentation for Large Telescopes: a Course for Astronomers
VIII. Stellar Astrophysics for the Local Group: a First Step to the Universe
IX. Astrophysics with Large Databases in the Internet Age
X. Globular Clusters
XI. Galaxies at High Redshift
XII. Astrophysical Spectropolarimetry
XIII. Cosmochemistry: the Melting Pot of Elements
XIV. Dark Matter and Dark Energy in the Universe
XV. Payload and Mission Definition in Space Sciences
XVI. Extrasolar Planets
XVII. 3D Spectroscopy in Astronomy
XVIII. The Emission-Line Universe

Participants of the XVII Canary Islands Winter School of Astrophysics

3D SPECTROSCOPY IN ASTRONOMY

XVII Canary Island Winter School of Astrophysics

Edited by

EVENCIO MEDIAVILLA, SANTIAGO ARRIBAS, MARTIN ROTH, JORDI CEPA-NOGUÉ and FRANCISCO SÁNCHEZ

Instituto de Astrofísica de Canarias, Tenerife

CAMBRIDGE UNIVERSITY PRESS
Cambridge, New York, Melbourne, Madrid, Cape Town,
Singapore, São Paulo, Delhi, Tokyo, Mexico City

Cambridge University Press
The Edinburgh Building, Cambridge CB2 8RU, UK

Published in the United States of America by Cambridge University Press, New York

www.cambridge.org
Information on this title: www.cambridge.org/9781107403475

First published 2010
First paperback edition 2011

A catalogue record for this publication is available from the British Library

Library of Congress Cataloguing in Publication data
3D spectroscopy in astronomy / [edited by] Evencio Mediavilla . . . [et al.].
p. cm. – (Canary Islands Winter School on astrophysics ; v. 17)
Three-D spectroscopy in astronomy
Includes bibliographical references.
ISBN 978-0-521-89541-5 (hardback)
1. Astronomical spectroscopy – Congresses. 2. Integral field spectroscopy – Congresses.
3. Three-dimensional imaging in astronomy – Congresses. I. Mediavilla, E. II. Title. III. Series.
QB465.A16 2010
522′.67 – dc22 2009031003

ISBN 978-0-521-89541-5 Hardback
ISBN 978-1-107-40347-5 Paperback

Contents

Contributors

MATTHEW A. BERSHADY, University of Wisconsin, Madison, USA

LUIS COLINA, Departamento de Astrofísica, Molecular e Infrarroja, Instituto de Estructura de la Materia, CSIC, Spain

FRANK EISENHAUER, Max-Planck-Institut für Extraterrestrische Physik, Germany

PIERRE FERRUIT, Observatoire de Lyon, CRAL, Université de Lyon, France

BEGOÑA GARCÍA LORENZO, Instituto de Astrofísica de Canarias, Spain

ARLETTE PÉCONTAL-ROUSSET, Observatoire de Lyon, CRAL, Université de Lyon, France

MARTIN ROTH, Astrophysikalisches Institut, Potsdam, Germany

SEBASTIÁN F. SÁNCHEZ, Centro Astronómico Hispano-Alemán de Calar Alto, Spain

JAMES E.H. TURNER, Gemini Observatory, Chile

Participants

Ahumada, Andrea Verónica	Observatorio Astronómico de Córdoba (Argentina)
Antichi, Jacopo	Universita di Padova (Italy)
Brown, Warren	Smithsonian Astrophysical Observatory (USA)
Cardiel, Nicolas	Centro Astronómico Hispano Alemán (Spain)
Cardwell, Andrew	Open University/Isaac Newton Group (UK/Spain)
Castillo-Morales, África	Universidad Complutense de Madrid (Spain)
Cervantes Rodríguez, José L.	Instituto de Astrofísica de Canarias (Spain)
Charcos Llorens, Miguel	University of Florida (USA)
Christensen, Lise	Astrophysikalisches Institut Postdam (Germany)
de Hoyos Fernández de Cordova, Carlos	Universidad Autónoma de Madrid (Spain)
de Lorenzo-Cáceres Rodríguez, Adriana	Instituto de Astrofísica de Canarias (Spain)
Díaz, Cristina	Universidad Complutense de Madrid (Spain)
Díaz, Rubén Joaquín	Observatorio Astronómico de Córdoba (Argentina)
Fernández Ontiveros, Juan A.	Instituto de Astrofísica de Canarias (Spain)
Ferrand, Mickaël	CNRS Lab. Sciences de l'Image, de l'Informatique et de la Télédétection (France)
Ferrero, Patrizia	Thueringer Landssternwarte Tautenburg (Germany)
Flores-Estrella, Roberto Atahualpa	Universidad Nacional Autónoma Mexico (Mexico)
Floyd, David	Space Telescope Science Institute (USA)
Fuentes Carrera, Isaura	Inst. Astronomía, Geofísica y C. Atmosféricas (Brazil)
García-Benito, Rubén	Universidad Autónoma de Madrid (Spain)
García-Marín, Macarena	Instituto de la Materia-CSIC (Spain)
Gavrilovic, Natasa	Astronomical Observatory Belgrade, Serbia and Montenegro
Germeroth, Andre	University of Heidelberg (Germany)
Gieles, Mark	Utrecht University (The Netherlands)
Gilli, Gabriela	Instituto de Astrofísica de Canarias (Spain)
Gómez, Matías	Universidad de Concepción (Chile)
Gómez-Cambronero Alvarez, Pedro	GRANTECAN (Spain)
González Pérez, José M.	Instituto de Astrofísica de Canarias (Spain)
Goodsall, Tim	University of Oxford (UK)
Gössl, Claus A.	Ludwig-Maximilians-Universitaet Muenchen (Germany)
Gustafsson, Maiken	University of Aarhus (Denmark)
Hernández Peralta, Humberto	Instituto de Astrofísica de Canarias (Spain)
Herrmann, Kimberly	Pennsylvania State University (USA)
Houghton, Ryan	University of Oxford (UK)

Jahnke, Knud	Max Planck Institute for Astronomy (Germany)
Jalobeanu, André	CNRS Lab. Sciences de l'Image, de l'Informatique et de la Télédétection (France)
Kehrig, Carolina	Instituto de Astrofísica de Andalucía (Spain)
Kelz, Andreas	Astrophysikalisches Institut Postdam (Germany)
Kjaer, Karina	European Southern Observatory (Germany)
Koehler, Ralf	Max-Planck-Institut für Extraterrestrische Physik (Germany)
Kristensen, Egstrom	Observatoire de Paris-Meudon (France)
Lagos Lizana, Patricio Andrés	Observatorio Nacional ON/MCT (Brazil)
Lara López, Martiza Arlene	Instituto de Astrofísica de Canarias (Spain)
López Martín, Dr. Luis	Instituto de Astrofísica de Canarias (Spain)
Lorente Espin, Oscar	Universidad Politécnica de Cataluña (Spain)
Lyubenova, Mariya	European Southern Observatory (Germany)
Maier, Millicent	University of Oxford (UK)
Martín Gordón, David	Instituto de Astrofísica de Andalucía (Spain)
Moiseev, Alexei	Special Astrophysical Observatory (Russia)
Monreal Ibero, Ana	Astrophysikalisches Institut Postdam (Germany)
Mora, Marcelo	European Southern Observatory (Germany)
Nowak, Nina	Max-Planck-Ins Extraterrestrische Physik (Germany)
Ocvirk, Pierre	University of Central Lancashire (UK)
Parker, Richard John	University of Sheffield/ING (UK/Spain)
Pastorini, Guia	Universita di Florence (Italy)
Prez Gallego, Jorge	University of Florida (USA)
Prez-Montero, Enrique	Universidad Autónoma de Madrid (Spain)
Povic, Mirjana	Instituto de Astrofísica de Canarias (Spain)
Ragaini, Silvia	Universita di Padova (Italy)
Ramirez Ballinas, Isidro	Universidad Nacional Autónoma Mexico (Mexico)
Rebaza Castillo, Ovidio	Instituto de Astrofísica de Andalucía (Spain)
Rix, Samantha Anne	Isaac Newton Group Telescopes (Spain)
Scheepmaker, Remco	Astronomical Institute Utrecht (The Netherlands)
Smirnova, Aleksandrina	Special Astrophysical Observatory (Russia)
Stoklasova, Ivana	Observatoire de Lyon (France)
Tamburro, Domenico	Max-Planck-Institute for Astronomy (Germany)
Trachternach, Clemens	Ruhr-Universität Bochum (Germany)
Valdivielso Casas, Luisa	Instituto de Astrofísica de Canarias (Spain)
van Eymeren, Janine	Ruhr-Universität Bochum (Germany)
Vidrih, Simon	University of Cambridge (UK)
Wang, Wei	Peking University (China)

Preface

3D spectroscopy has a relatively short history. Most of the present instrument concepts were developed in the 1980s and early 1990s. During those pioneering years a great deal of work was done in optical labs in an attempt to understand how the optical fibres, micro-lenses and image slicers behave. Only a few groups (often formed by one or two people) worked on this topic. Communications were not very good, which explains why virtually all the groups decided to refer to this technique by a different name. So we ended up with 'spectral imaging', 'bidimensional spectroscopy', 'integral field spectroscopy', 'two-dimensional spectroscopy', '3D spectroscopy', etc.

During those years it was more than doubtful whether this technique was going to be useful at all. In fact, it looked like a kind of curiosity of limited practical interest to astronomy. However, in the 1990s the first scientific results were obtained and they inmediately produced a change of perception.

In the last few years investment in this type of instrumentation has been enormous. Large telescopes all around the world are now equipped with integral field units. Two instruments of the future James Webb Space Telescope will also have integral field spectroscopic capabilities, etc. Instead of being based in the optical lab trying to characterize optical fibres or micro-lenses, more effort is dedicated nowadays to refining techniques for reducing, analysing and interpreting the data obtained with a new generation of 3D spectrographs. Clearly, we are in a wholly different phase; by attending the lectures and viewing the posters of our Winter School, it is clear that 3D spectroscopy has truly arrived on the scene.

This Winter School is particularly timely. It is aimed at instructing a new generation of astronomers in a new generation of instruments. This will probably be the first and last school on this topic. In a few years' time, organizing a similar school would seem like organizing a school today on 'imaging'. However, the string of recent scientific conferences on 3D spectroscopy and the over-subscription to this school would seem to indicate that this is the right topic at this time.

The Instituto de Astrofísica de Canarias (IAC) is specially pleased to be organizing a Winter School on the topic of 3D spectroscopy because it has contributed to this field since its inception. More than 15 years ago a small group of researchers at the IAC started working in this field. They set up from scratch an optical fibre lab at the institute and developed the first experimental integral field systems for the telescopes on La Palma. That was just the beginning of the IAC's contribution to the development and diffusion of this technique, a contribution that continues today to extend on several fronts, including the organization of this workshop in collaboration with the European Euro3D network.

The aim of this Winter School is to widen the appeal of integral field spectroscopy beyond the limited community of experts and to help it become a powerful tool in the hands of a new generation of astronomers for tackling new (and old) astrophysical problems. We thank all the participants, lecturers and students for their valuable contributions to this objective.

The Editors

Acknowledgements

The editors want to express their warmest gratitude to the lecturers for their efforts in preparing their classes and the chapters of this book. We also wish to thank our efficient secretaries: Nieves Villoslada and Lourdes Gonzalez, and many staff members of the IAC: Jess Burgos, Begoña López Betancor, Carmen del Puerto, Ramón Castro, Terry Mahoney, Miguel Briganti and the technicians of Servicios Informaticos Comunes del IAC. We acknowledge Ismael M. Delgado for the technical edition of this book for Cambridge University Press.

Finally, we also wish to thank the Astrophysical Institute Potsdam and the European Research Training Network Euro3D for their support to the finances, and the organization and the following institutions for their economical support: The European Commission, the Spanish Ministry of Education and Science, Iberia, the local governments (*cabildos*) of the islands of Tenerife and La Palma, and the Puerto de La Cruz Town Council.

Abbreviations

2dF	The Two Degree Field system
2D-FIS	2-Dimensional Fiber ISIS System
AAT	Anglo-Australian Telescope
ACS	The Advance Camera for Surveys
AOB	Adaptive Optics Bonnette
APO	Apache Point Observatory
APOGEE	Apache Point Observatory Galactic Evolution Experiment
ARGUS	
ARIES	Arizona infrared imager and echelle spectrograph
CAHA	Centro Astronómico Hispano Alemán
CCD	charge coupled device
CFHT	Canada–France–Hawaii Telescope
CIGALE	Scanning Perot–Fabry Interferometer, comparable with the UK TAURUS system, developed by Marseille Observatory to work at the Cassegrain focus of the 3.6 m CFH Telescope
CIRPASS	Cambridge Infrared Panoramic Survey Spectrograph
COHSI	Cambridge OH Suppression Instrument
EFOSC	ESO Faint Object Spectrograph and Camera
ESI	Echellette Spectrograph and Imager
ESO	European Southern Observatory
F2T2	Flamingos-2 Tandem Tunable Filter
FaNTOmM	Fabry–Perot of New Technology for the Observatoire du mont Megantic
FGS-TF	Tunable Filter
FISICA	The Florida Image Slicer for Infrared Cosmology and Astrophysics
FITS	Flexible Image Transport System
FLAMES	Fibre Large Array Multi Element Spectrograph
FLAMES/GIRAFFE	FLAMES Intermediate Resolution Spectrometer
FLAMINGOS	The Facility Near-Infrared Wide-field Imager and Multi-Object Spectrograph for Gemini
FORS	Focal Reducer and low dispersion Spectrograph
FWHM	full-width-at-half-maximum
GMOS	Gemini Multi-Object Spectrograph
GNIRS	Gemini Near Infrared Spectrograph
GriF	Grating Infrared Fabry–Perot
GTC	Gran Telescopio CANARIAS
HARPS	High Accuracy Radial Velocity Planet Searcher
HET	Hobby–Eberly Telescope
HEXAFLEX	Hexagonal Lattice Fiber Linked Experimental
HEXAFLEX-II	Hexagonal Lattice Fiber Linked Experimental II
HIFI	Hawaii Imaging Fabry–Perot Interferometer
HIRES	High Resolution Echelle Spectrometer
HST	Hubble Space Telescope

IAU	International Astronomical Union
ICA	Instituto de Astrofísica de Canarias
IDL	Interactive Data Language
IFS	Integral Field Spectroscopy
IFU	Integral Field Unit
IMACS-IFU	Inamori Magellan Cassegrain Spectrograph–Integral Field Unit
IRAF	Image Reduction and Analysis Facility
ISIS	Intermediate-Dispersion Spectroscopic and Imaging System
JWST	James Webb Space Telescope
KMOS	K-Band Multi-Object Spectrometer
KPNO	Kitt Peak National Observatory
MACAO	Multi-Application Curvature Adaptive Optics
MCAO	Multi-Conjugate Adaptive Optics
MIDAS	Mid-Infrared Asteroid Spectroscopy
MIRI	Mid-InfraRed Instrument
MOS	Multi-Object Imaging Spectrograph
MPE 3D	Max Planck Institut für Extraterrestrische Physik 3D
MPFS	Multi-Pupil Field Spectrograph
MUSE	Multi-Unit Spectroscopic Explorer
NaCo	short for NAOS-CONICA (Nasmyth Adaptive Optics System – Near-Infrared Imager and Spectrograph)
NIC-FPS	Near-Infrared Camera and Fabry–Perot Spectrometer
NICMOS	Near Infrared Camera and Multi-Object Spectrometer
NIFS	Near-Infrared Integral Field Spectrograph
NIRSpec	Near-Infrared Spectrograph
NOAO	National Optical Astronomy Observatory
NOT	Nordic Optical Telescope
OAN-SPM	National Astronomical Observatory at San Pedro Martir
OASIS	Optically Adaptive System for Imaging Spectroscopy
OHP	Observatoire de Haute-Provence
OMM	Observatoire du mont Megantic
ORM	Observatorio del Roque de los Muchachos
OSIRIS	Optical System for Imaging and low/intermediate-Resolution Integrated Spectroscopy
PACS	Photoconductor Array Camera and Spectrograph
PMAS	Potsdam Multiaperture Spectrophotometer
PPAK	PMAS fiber pack
PSF	Point Spread Function
PUMA	UNAM (Universidad Nacional Autónoma de Mexico) Scanning Fabry–Perot Interferometer
PUMILA	Near-infrared Spectrograph for the Kinematic Study of the Interstellar Medium
PYTHEAS	Prisme Interferomètre Trames de lentilles pour l'Holometrie, et l'Endoscopie des Astres et des Sources
QSO	quasi-stellar object

RFP	Rutgers Fabry–Perot
RSS	Robert Stobie Spectograph
RSS	Row Stacked Spectra
RTN	Research Training Network
SALT	Southern African Large Telescope
SAO	Smithsonian Astronomical Observatory
SAURON	Spectrographic Areal Unit for Research on Optical Nebulae
SHM	shared memory server
SILFID	Spectrographe Integral Linarisation par Fibres de l'Image Directe
SINFONI	Spectrograph for Integral Field Observations in the Near Infrared
SIS	Stabilized Imager and Spectrometer
SMM	Submillimetre galaxy
SNIFS	SuperNovae Integral Field Spectrograph
SPIFFI	Spectrometer for Infrared Faint Field Imaging
SWIFT	Short Wavelength Integral Field Spectrograph
TF	Tully–Fisher
TIGER	Traitement Intégral des Galaxies par l'Etude de leurs Raies
TTF	TAURUS Tunable Filter
UFTI	UKIRT Fast-Track Imager
UIST	UKIRT Imager-Spectrometer
UKIRT	United Kingdom Infrared Telescope
VIMOS	Visible Imaging Multi-Object Spectrograph
VIRUS	Visible Integral-Field Replicable Unit Spectrograph
VLT	Very Large Telescope
VPH	Volume Phase Holographic
WFPC	Wide-Field Planetary Camera
WHAM	Wisconsin H-alpha Mapper
WHT	William Herschel Telescope
WiFeS	Wide-Field Spectrograph
WIYN	Wisconsin–Indiana–Yale
WYFOS	Wide-Field Optic Spectrograph

1. Introductory review and technical approaches

MARTIN ROTH

1.1 Preface

The topic of the XVII IAC Winter School is '3D Spectroscopy': a powerful astronomical observing technique, which has been in use since the early stages of the first prototype instruments about a quarter of a century ago. However, this technique is still not considered a standard common user tool among most present-day astronomers.

3D Spectroscopy (hereafter '3D') is also called 'integral field spectroscopy' (IFS), sometimes 'two-dimensional' or even 'area' spectroscopy, and commonly also 'three-dimensional' spectroscopy; in other areas outside astronomy it is called 'hyperspectral imaging', and so forth. It is already this diversity in the nomenclature that perhaps reflects the level of confusion. For practical reasons, the organizers of this Winter School and the Euro3D network (which will be introduced below) have adopted the terminology '3D', which is intuitively descriptive, but, as a caveat early on, is conceptually misleading if we restrict our imagination to the popular picture of the 'datacube' (Figure 1.1). Although this term will commonly be used throughout this book, we need to point out for the reasons given later in the first chapter that the idealized picture of an orthogonal cube with two spatial, and one wavelength, coordinate(s) is inappropriate in the most general case.

Whatever the terminology, it is the aim of this Winter School to help alleviate the apparent lack of insight into 3D instrumentation, its use for astronomical observations, the complex problems of data reduction and analysis, and to spread knowledge among a significant number of international young researchers at the beginning of their career. The training of early stage researchers has always been the rationale in the tradition of the well-established series of Instituto de Astrofísica de Canarias (IAC) Winter Schools, but this is the first time in which a consortium of 11 leading European research institutes appears as co-organizer, represented by the 'Euro3D' Research Training Network (RTN) – which may well be recognized as the significant impact expected from this school. Euro3D was funded by the European Commission (EC) from July 2002 to December 2005 under Framework Programme 5 (Walsh and Roth, 2002). The RTN participants are as follows:

- Astrophysikalisches Institut Potsdam (Germany, coordinator)
- IoA University of Cambridge (UK)
- Durham University (UK)
- MPE Garching (Germany)
- European Southern Observatory (International Organization)
- Sterrewacht Leiden (The Netherlands)
- Observatoire de Lyon (France)
- CNRS Marseille (France)
- IFCTR Milano (Italy)
- Observatoire de Paris, Meudon (France)
- IAC Tenerife (Spain).

At the time of planning the RTN in 2001, these groups represented practically the entire expertise in 3D instrumentation in Europe. As can be seen from a compilation of existing instruments and those under development at the time (Table 1.1), significant capital

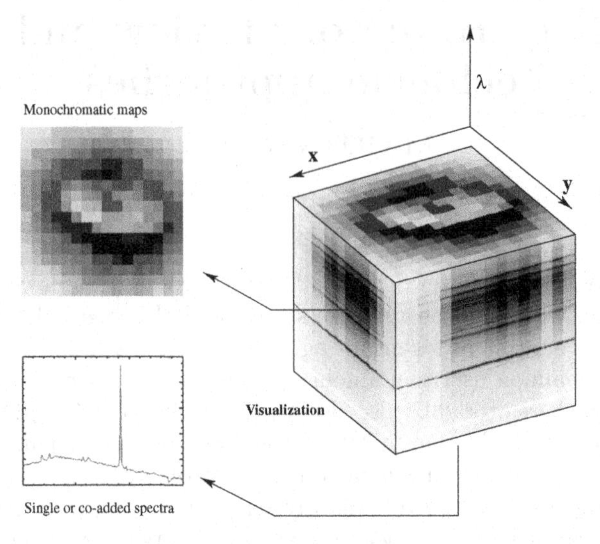

FIGURE 1.1. Schematic representation of a datacube, produced as the result of 3D observations. It can be interpreted as a stack of quasi-monochromatic images or, alternatively, as an assembly of n × m spectra.

FIGURE 1.2. Participants of the Euro3D Kickoff Meeting, held at IAC, Tenerife, 2–4 July 2002.

investments have been realized at almost all major observatories, with a clear technological lead in Europe. Disappointingly, however, the expertise within the user community for the reception of the data from such instruments was found to be very limited. It was confined almost entirely to members of the groups that have built the 3D instruments.

Therefore, and in response to the EC requirement to demonstrate 'the scientific, technological, or socio-economic reasons for carrying out research in the field covered by the network', the project objectives of the Euro3D consortium were identified as follows:

- to promote 3D technique with selected science projects
- to develop new observing and data reduction techniques
- to provide training and education of students and young researchers
- to popularize 3D technique in various research fields
- to provide user support and general information

TABLE 1.1. Compilation of 3D instrumentation on some major telescopes (as of 2001). Notes: Bold face: 3D instruments built by RTN members. (•): facilities accessible through Euro3D. 1): mobile instrument.

Telescope	Diameter (m)	Access Euro3D	3D Instrument	$\Delta\lambda$ (μm)	Maximum FoV arcsec^2	Maximum resolution $\lambda/\Delta\lambda$	Status
NGST	8.0	•	**IFMOS**				proposed
HST	2.4	•	–				
Keck	2 × 10		OSIRIS	0.9–2.5	3 × 6	4000	planned
GTC	10	•	**ATLANTIS**	1–2.4	8 × 8	5000	planned
HET	10		–				
LBT	2 × 8.4	•	**PMAS** 1)	0.35–1.0	16 × 16	2200	private
SUBARU	8.3		Kyoto-3D				
GEMINI	2 × 8.1	•	**GMOS**	0.4–1.0	50 × 450	10000	common user
		•	**GNIRS**	1–5	20 × 20	5000	common user
		•	**CIRPASS**	0.9–1.8	35 arcsec^2	3000	common user
		•	**NIFS**	1–2.5	3 × 3	5300	planned
VLT	4 × 8.1	•	**VIMOS**	0.37–1	54 × 54	2000	common user
		•	**NIRMOS**	1–1.8	28 × 28	2500	common user
		•	**SINFONI**	1–2.5	8 × 8	4500	common user
		•	**FLAMES**	0.37–0.9	11.5 × 7.3	25000	common user
Magellan	2 × 6.5	•	**IMACS**	0.4–0.9	50 × 50	5000	common user
Selentchuk	6.0	•	MPFS	0.39–0.9	16 × 16	3000	common user
Hale	5.0		PIFS	1–5	5.4 × 9.6	3000	experimental
WHT	4.2	•	**INTEGRAL**	0.39–1.0		1800	common user
		•	**SAURON**	0.47–0.54	39.5 × 32.9	2300	private
		•	**TEIFU**	0.47–0.54	4.1 × 5.9		private
CTIO	4.0		–				
KPNO	4.0		–				
AAT	3.9	•	**SPIRAL**	0.48–1.0	22 × 11	11000	'expert' user
CFHT	3.6	•	**TIGER**	0.4–1.0	12 × 9	2000	decommissioned
		•	**OASIS**	0.4–1.0	15 × 12	3000	common user
ESO 3.6 m	3.6	•	–				
WYIN	3.5		DENSEPAK	0.38–1.0	90 fibers	22000	common user
Calar Alto	3.5	•	**PMAS** 1)	0.35–1.0	16 × 16	2200	private
NTT	3.5	•	–				
TNG	3.5	•	GOHSS	0.8–1.8		3000	planned
		•	**MPE-3D** 1)	1.5–2.2	6.4 × 6.4	2100	private

FIGURE 1.3. The conceptually problematic picture of a 'datacube'

- to provide feedback from observers to technology
- to identify technology transfer opportunities.

Organizing the XVII IAC Winter School is an important contribution to its major goals and appropriately concludes the activities at the end of the lifetime of the RTN. It is hoped that the participants of the Winter School will benefit from the collective expertise presented in the lectures for the rest of their future scientific careers; furthermore that 3D Spectroscopy shall experience more widespread use in the community to further exploit and advance its unique potential for astronomy.

1.2 Introductory review

1.2.1 *Conceptual outline*

Principle of operation and terminology

We define integral field (3D) spectroscopy (IFS) as an astronomical observing method that creates in a single exposure spectra of (typically many) spatial elements ('spaxels') simultaneously over a two-dimensional field-of-view (FoV) on the sky. Owing to this sampling method, each spaxel can be associated with its individual spectrum. Once all of the spectra have been extracted from the detector frame in the process of data reduction, it is possible to reconstruct maps at arbitrary wavelengths. For instruments with an orthonormal spatial sampling geometry, the spectra can be arranged on the computer to form a three-dimensional array, which is most commonly called a 'datacube'. Datacubes are also well known as the natural data product in radio astronomy. Note, however, that there are many integral field spectrographs which do not sample the sky in an orthonormal system. In this case the term datacube is misleading. We shall see further below that atmospheric effects in the optical also make the term inappropriate in the most general case.

Historically, various designations have been invoked to describe integral field spectroscopy (area/imaging/2D/3D/three-dimensional spectroscopy, etc.). As spectroscopy involves one dimension per se (the wavelength scale), two-dimensional spectroscopy would arguably also describe the situation correctly. However, the datacube picture and the perhaps more explicit label '3D' have conventionally led to the now accepted terminology of three-dimensional spectroscopy, which is adopted throughout this book.

Instruments that create three-dimensional datasets in the sense described above, however not simultaneously but rather in some process of sequential data acquisition (scanning), e.g. tuneable filter (Fabry–Perot) instruments, scanning long slits, etc., are not strictly 3D spectrographs according to our definition. Because of similarities in the

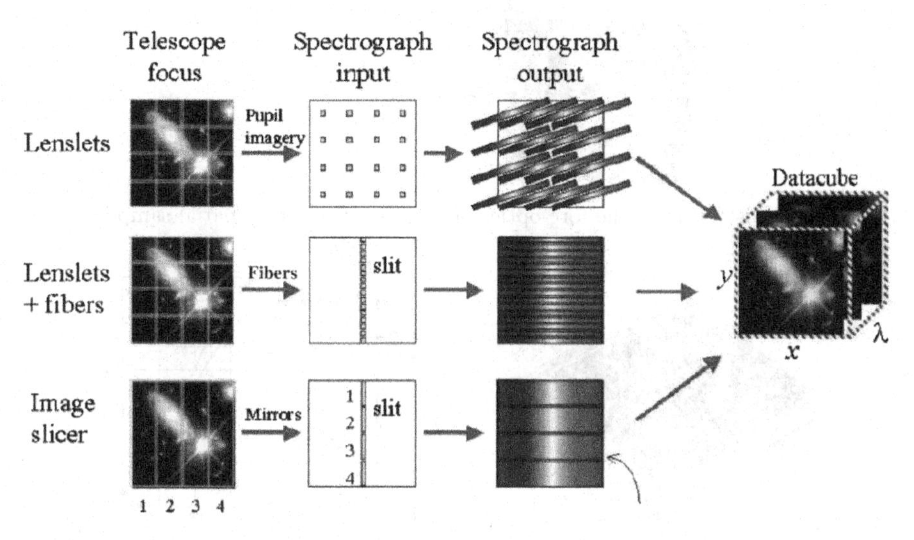

FIGURE 1.4. The three major principles of operation of present-day IFUs (Source: J. Allington-Smith).

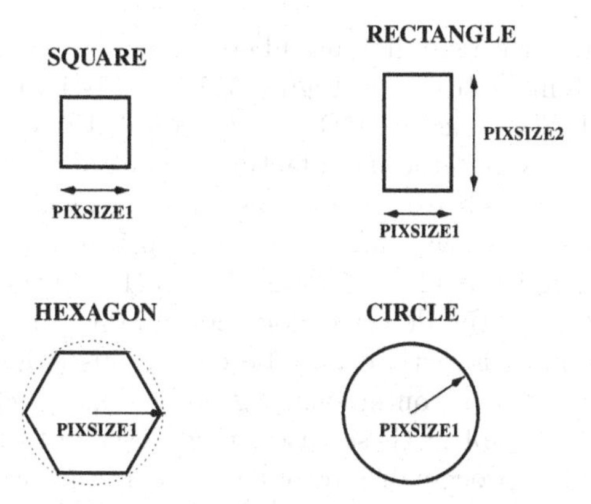

FIGURE 1.5. Different types of spaxel geometries: square, rectangular, hexagonal, bare fiber (circular).

analysis and interpretation of their data, however, such instruments will also be mentioned briefly below.

Methods of image dissection: spatial sampling

Integral field spectrographs have been built based on different methods of dissecting the FoV into spaxels, e.g. optical fiber bundles, lens arrays, optical fibers coupled to lens arrays, or slicers (Figure 1.4). We will discuss these methods further below.

The term 'spaxel' was introduced by the Euro3D consortium in order to distinguish spatial elements in the image plane of the telescope from pixels, which are the spatial elements in the image plane of the detector (Kissler-Patig *et al.*, 2004). The optical elements that accomplish the sampling of the sky are often called 'integral field units' (IFUs), and thus IFS is also sometimes called 'IFU spectroscopy'. Spaxels can have different shapes and sizes, depending on instrumental details and the type of IFU (Figure 1.5).

It is necessary to point out that, contrary to the persuasive implication of the datacube picture, IFUs do not necessarily sample the sky on a regular grid, e.g. fiber bundles, where

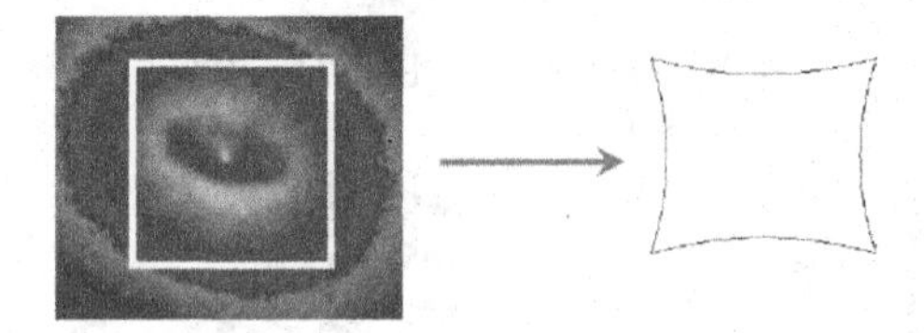

FIGURE 1.6. Field distortion in the fore-optics of an IFU affects the spatial sampling.

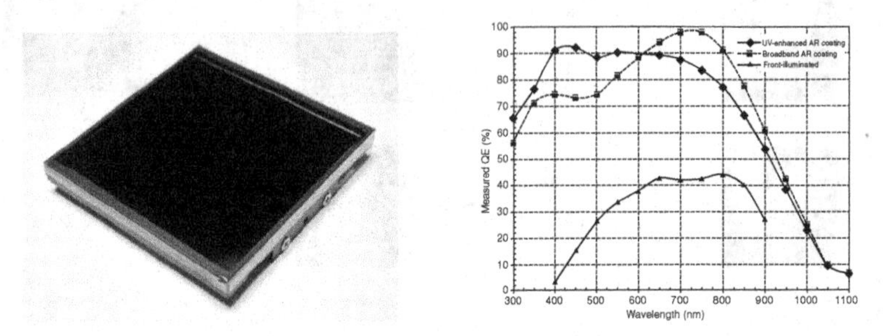

FIGURE 1.7. Fairchild CCD 4K × 4K, 15 μm pixels (left). The QE of modern CCDs can be made quite high, very nearly approaching 100% (right).

due to the manufacturing process individual fibers cannot be arranged to arbitrary precision (for a real fiber bundle IFU, see Figure 1.23). This is another reason why the popular description of IFS as 'like a CCD [charge coupled device] with a spectrum in each pixel' is misleading. Even if the manufacture of an IFU allows a perfectly regular sampling pattern to be created, e.g. in the case of a hexagonal lens array, the sampling is not necessarily orthonormal. Moreover, real optical systems create aberrations and, sometimes, non-negligible field distortions (Figure 1.6), which also means that the set of spectra extracted from the detector does not sample an orthogonal FoV on the sky. Furthermore, the sampling method may be contiguous (e.g. lens array) with a fill factor very close to unity, or non-contiguous (e.g. fiber bundle) with a fill factor of significantly less than 100%. In all of these cases, it is possible to reconstruct maps at a given wavelength through a process of interpolation and, repeating this procedure over all wavelengths, to convert the result into a datacube. Note, however, that interpolation often produces artifacts and generally involves loss of information.

Detectors and spectrographs

In order to better understand the potential and natural constraints of 3D instrumentation, it is useful to briefly review detectors, since they ultimately limit the total number of spatial and spectral resolution elements, hence FoV and wavelength coverage. Depending on the type of application, one would normally try to maximize the number of detector pixels in order to maximize one or both of these parameters. A selection of modern large area detectors is illustrated below. For visual wavelengths in the interval 0.35–1.0 μm, CCDs are nowadays the most common detectors, featuring high quantum efficiency (QE) and low read-out noise. Single chips with pixel resolutions of 4096 × 4096 (Figure 1.7) presently represent the state-of-the-art, although much larger CCDs are under development. In order to obtain even larger pixel numbers, buttable CCDs can be arranged to form mosaic configurations similar to modern wide-field imaging cameras (MegaCAM, OmegaCam, etc.), e.g. the detector for the IMACS IFU (Inamori Magellan Cassegrain Spectrograph – Integral Field Unit), which is built as a mosaic of 2 × 3 2K × 4K CCD chips. Image-intensified photon counting detectors that were still common in the 1980s

FIGURE 1.8. Rockwell 2K × 2K Infrared Array (left), 4K × 4K mosaic (right).

are no longer popular, except for read-noise limited applications with very low photon count rates. An interesting new development is the photon-counting LLCCD, which uses an on-chip avalanche gain technology for low noise at high frame rates (Jerram *et al.*, 2001).

In the near infrared (NIR) wavelength regime there has been enormous progress in detector technologies, providing large area infrared (IR) arrays with pixel resolutions of up to 2K × 2K and excellent electro-optical performance, and even larger arrays being under development. A mosaic of 2 × 2 such buttable arrays is shown in Figure 1.8. While a strong driver for this evolution has come from the development of the next generation space telescope – the James Web Space Telescope (JWST) – it is also the demand for detectors from NIR instrumentation for ground-based astronomy that has made NIR astronomy a rapidly growing field.

The number of spectra delivered by present-day 3D spectrographs is typically in the range of several hundreds up to a few thousand: Visible Imaging Multi-Object Spectrograph (VIMOS) IFU: 6400 (Le Fevre *et al.*, 2003). The corresponding number of spaxels is disappointingly small if one compares with direct imaging detectors: even the simple first generation commercial CCD imagers were offering, for example, $384 \times 576 \approx 221$ Kpixels: a factor of 34 larger than the number of spaxels of the largest IFU to date. So what is it that limits the size of an integral field spectrograph in terms of number of spatial and spectral elements?

First of all, IFS is obviously more complex than direct imaging: for any spatial element, a full spectrum must be accommodated. If n, m, k are the number of spatial resolution elements in x and y, and of the spectral resolution elements, respectively, then the total number N of required detector pixels is $N = n \times m \times k$. In comparison with direct imaging where $k = 1$, IFS requires at least a factor of k more detector pixels. Clearly, instrumentation for IFS is expensive in terms of detector space.

Another important factor is the spectrograph optical system. The input focal ratio as determined by the telescope and any fore-optics, the magnification from the spectrograph slit to the detector pixel size, as well as the total length of slit determine the spectrograph parallel beam size and thus the overall size of the optical system.

For a large number of spectra, a wide-field optical system is typically required. Such optical systems are challenging in terms of technical feasibility (controlling optical aberrations), and cost. As an example, Figure 1.9 shows the layout of the PMAS (Potsdam Multiaperture Spectrophotometer) spectrograph optics, which is optimized to cover the wavelength range 350–900 nm (Roth *et al.*, 2005). It is worthwhile noting that the lens sizes are 200 mm in diameter. While this size represents the state-of-the-art for modern

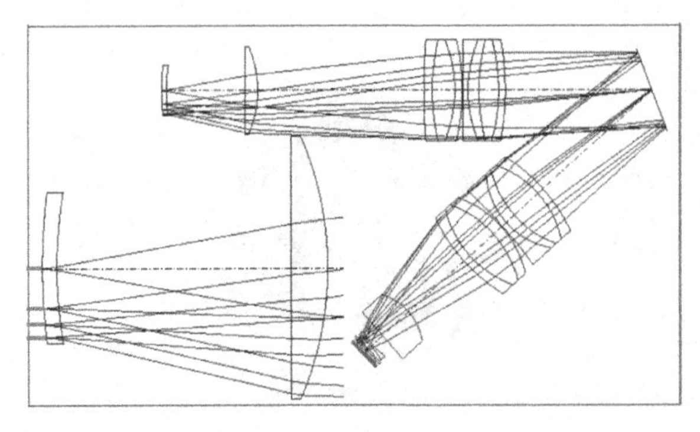

FIGURE 1.9. PMAS fiber spectrograph as an example for an IFS optical system. Left: a cross section, showing the layout of collimator, grating and camera. The parallel beam measures 150 mm in diameter. The magnified detail to the lower left illustrates the fiber interface. Right: photograph of the camera, which has a total mass of 60 kg.

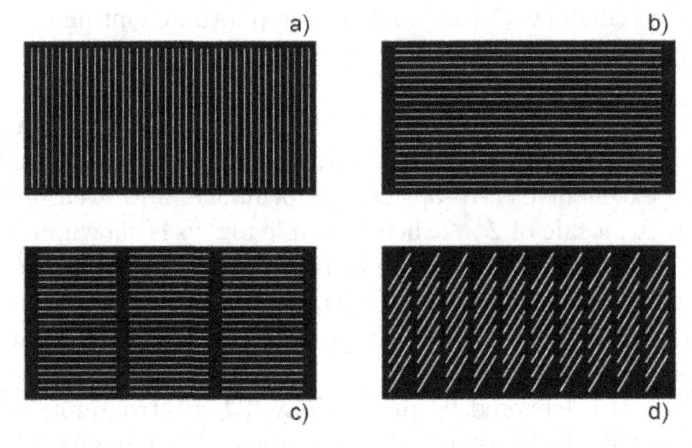

FIGURE 1.10. Different ways of allocating spectra on a detector: a) maximizing number of spaxels with full wavelength coverage, b) maximizing number of spectral pixels, c) maximizing number of spaxels with several banks of spectra, d) diagonal, maximizing number of spaxels at expense of wavelength coverage.

refractive optical systems for spectroscopy, the apochromatic correction from the ultraviolet (UV) to the NIR is unusual and quite demanding. The cost of such a system is high, mainly because of the use of expensive materials, e.g. CaF_2, which are required for high throughput and the large wavelength coverage. Instruments that are not designed for general-purpose operation but rather for a specific science case can often be built simpler and with less effort.

Given the total number of spaxels of the IFU, the critical question of how to allocate the corresponding spectra on the detector arises. Obviously, the total number of detector pixels is related to the number of spatial and spectral resolution elements like:

$$\Sigma_{Pixels} \geq N_{Spaxels} \times N_{Spectra} \tag{1.1}$$

In most conventional 3D spectrograph designs the detector space is not contiguously filled with spectra, basically in order to avoid cross-talk between adjacent spectra and to facilitate data reduction. Novel slicer designs like, for example, SINFONI (Eisenhauer *et al.*, 2003) or the OSIRIS lens array IFU (Larkin *et al.*, 2006) have, however, demonstrated that this is not a necessary condition. Figure 1.10 shows several basic ways of arranging the spectra. Generally, a trade-off needs to be made between maximizing the number of spaxels, or the number of spectral pixels.

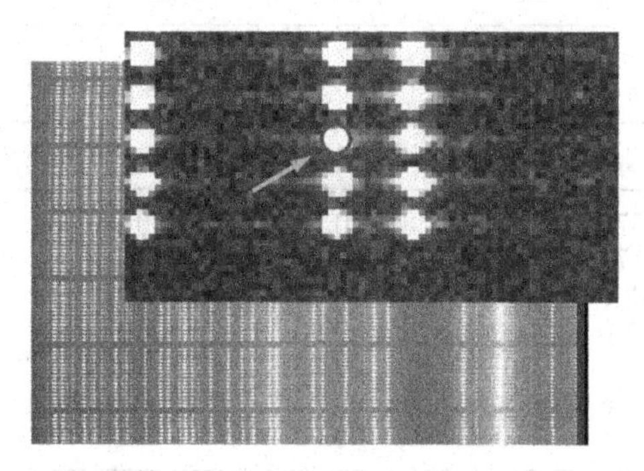

FIGURE 1.11. Typical CCD frame of an arc lamp exposure from a fiber bundle IFU. The direction of dispersion is horizontal. The magnified insert demonstrates how the instrumental profile is given by the FWHM (full-width-at-half-maximum) of the projected fiber size on the detector (≈ 4 pixels).

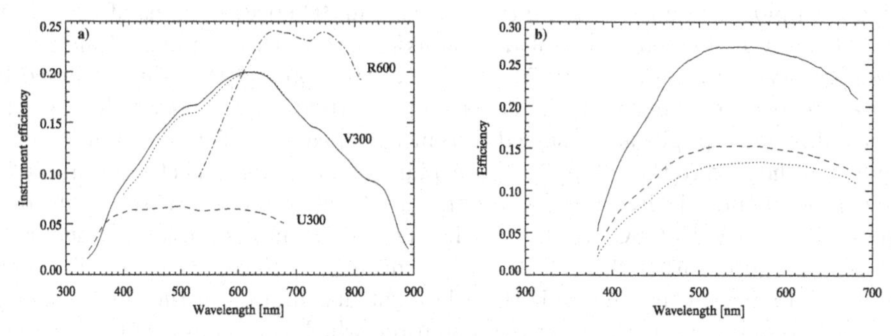

FIGURE 1.12. Examples for instrumental efficiency as a function of wavelength for different gratings. a) PMAS with lens array IFU, b) PMAS with PPak IFU.

Spectral resolution and wavelength coverage

The spectral resolution in most 3D instruments does not reach the diffraction limit. It is rather determined by the projected (pseudo-)slit width on the detector. Figure 1.11 illustrates this situation with an example dataset, which was obtained with a fiber-based 3D spectrograph.

As mentioned before, the wavelength coverage is normally tuned to the application and scientific scope of an instrument. Two extreme examples on 4 m class telescopes would be:

- SAURON (WHT) [Spectrographic Areal Unit for Research on Optical Nebulae (William Herschel Telescope)], which has a wavelength coverage of 54 nm in the operational wavelength interval of 450–1100 nm; and
- PMAS (CAHA 3.5 m) [Potsdam Multiaperture Spectrophotometer], with a wavelength coverage of up to 680 nm in the interval of 350–900 nm. Instrumental efficiency curves for the latter instrument are given in Figure 1.12.

Coupling the IFU to the spectrograph

Although this is a somewhat technical detail, it is worth pointing out the fundamental difference between two different methods of coupling the IFU to the spectrograph optical system. One method is the one that has conventionally been used for normal slit-based spectrographs, namely to image the (dispersed) telescope focal plane (= slit) onto the detector. This technique is normally used for the 'image slicer' type of 3D spectrograph.

TABLE 1.2. Examples for 3 types of 3D spectrographs

	$N_{spectra}$	FoV [″]	scale [″]	R	λ range/$\Delta\lambda$ [nm]
SAURON LR	1577	41 × 33	0.94	1200	450–700/54
SAURON HR	1577	11 × 9	0.27	1400	450–700/54
VIMOS LR	6400	54 × 54	0.67	220	370–1000/400
VIMOS HR	1600	13 × 13	0.33	2700	420–870/240
		8 × 8	0.25		
SINFONI	2048	3 × 3	0.10	1500–4000	1050–2450
		0.8 × 0.8	0.025		

Another method has emerged with the introduction of optical fibers in that the image information at the telescope focal plane is scrambled, and thus the one-to-one correspondence between points on the image plane and on the detector plane is lost. This principle is employed in an even more rigorous way in 3D spectrographs of the TIGER-type (TIGER = Traitement Intégral des Galaxies par l'Etude de leurs Raies; named after the prototype; Bacon *et al.*, 1995), where a simple fore-optics system followed by a lens array creates a tiny image of the entrance pupil (telescope aperture) for each lens. These so-called micropupils are designed to typically measure a few tens of micrometers in diameter. When fed to the collimator input plane of an imaging spectrograph, the family of micropupils practically operates like the apertures of a conventional multi-object spectrograph (e.g. the EFOSC-type), with the one significant distinction, namely that it is the telescope pupil rather than the telescope image that forms the multi-slit configuration. The important consequence is that the light distribution of the slit illumination no longer matters in terms of wavelength definition, which is an important argument for precision spectroscopy. The principle is not entirely new, but was first employed using a 'Fabry lens' for photoelectric photometers in order to eliminate the photometric noise introduced by seeing (Michlovic, 1972). Another variant of the TIGER micropupil technique is employed in fiber-optical 3D spectrographs, where the fiber bundle is connected to a lens array.

Examples of real IFUs

For the purpose of making the techniques introduced in the preceding sections more imaginable, three instruments are presented, which are meant to be representative for the (1) fiber-optical, (2) micropupil lens array (TIGER), and (3) slicer types of 3D spectrographs, namely the VIMOS (Visible Imaging Multi-Object Spectrograph; Le Fevre *et al.*, 2003), SAURON (Bacon *et al.*, 2001), and SINFONI instruments (Spectrograph for Integral Field Observations in the Near Infrared; Eisenhauer *et al.*, 2005). Some important parameters (number of spectra, FoV, spaxel size, spectral resolving power, wavelength coverage) are listed in Table 1.2 (LR/HR stand for low resolution and high resolution modes, respectively). Figure 1.13 shows examples of what the spectra look like on raw CCD frames for these three instruments. The VIMOS frame shows a science exposure from a stellar field, so some spectra are illuminated well, while most others contain only sky. Note that the direction of dispersion is oriented horizontally, and that the frame contains 4 banks of spectra that are placed as columns next to each other, with a gap between bank numbers 1 and 2. Note also that there are several artifacts as a result from zero and higher order contamination. The SAURON example shows a continuum flat-field exposure, which exhibits the characteristic pattern of the diagonal arrangement of the total 1577 spectra, whose density is so high that individual spectra are difficult to

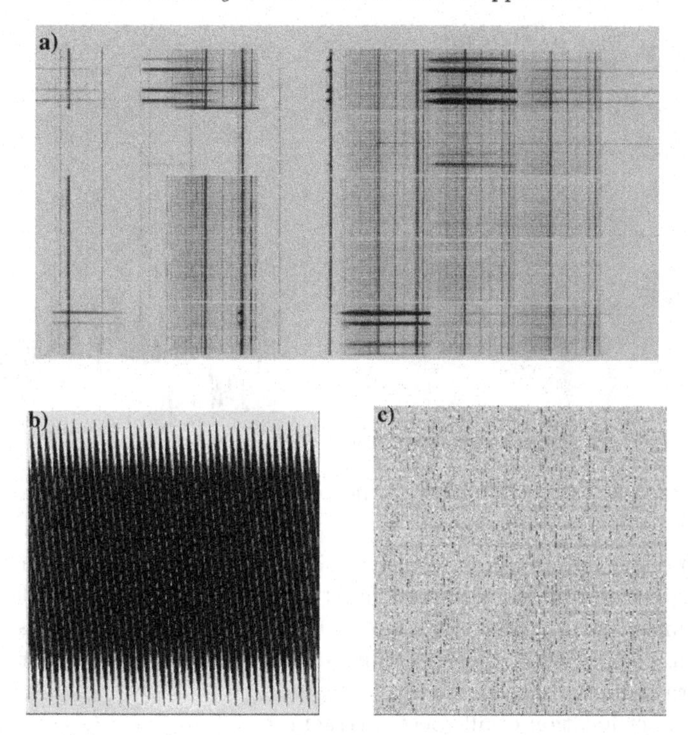

FIGURE 1.13. a) Raw detector frames from VIMOS IFU; b) SAURON; c) SPIFFI. See text for explanation.

discern. Finally, the SINFONI example illustrates the appearance of slicer data, which at first glance looks like noise. Closer inspection reveals the staggered slit arrangement as a vertical pattern right from the center. It is obvious that this IFU type produces the highest density of spectra on the detector, making it a good choice for IFUs requiring a large FoV and many spatial elements.

Extracting the spectra

Keeping in mind the different characteristic appearances of IFU types, it is useful to address the critical step of data reduction, which is required to produce data ready for analysis (a comprehensive discussion is presented by Turner in chapter 2 of this volume). Unlike direct imaging data – from which the observer can generally recover some fundamental properties – 3D spectroscopy produces raw data that are very difficult to understand, let alone interpret even only qualitatively. In order to do this, it is first necessary to extract the spectra from the raw detector frame. Figure 1.14 illustrates this step with an example from the MPFS instrument (Multi Pupil Field Spectrograph; Si'lchenko and Afanasiev, 2000). The CCD frame to the left is a continuum flat-field exposure, showing 16 groups of 16 spectra each, the dispersion direction being horizontal (left to right). The horizontal black line in the center indicates the spectrum, whose intensity with wavelength is plotted to the upper right. The combined effects of instrumental transmission, vignetting, CCD QE and the grating blaze function lead to a pronounced drop of recorded intensity from right ('red' wavelengths) to left ('blue').

Similarly, the plot to the lower right shows a vertical profile (column) near the center. As the CCD has 1024×1024 pixels; the total of 256 spectra is distributed such that each spectrum takes up 4 rows of pixels. Since the FWHM of a spectrum perpendicular to

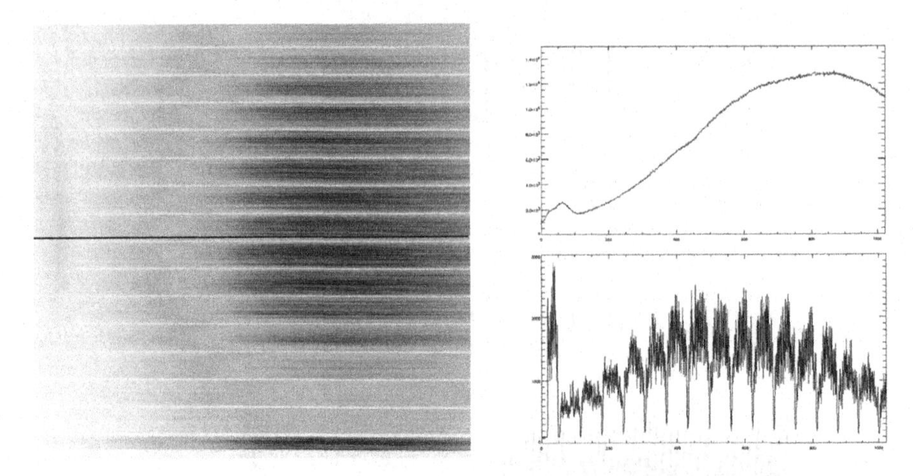

FIGURE 1.14. Example: extraction of MPFS spectra from raw CCD frame.

the direction of dispersion is approximately 2 pixels, and given the more or less Gaussian profile of the spectrograph point-spread function (PSF), there is considerable overlap between adjacent spectra. The average inter-spectra minimum intensity is $\approx 65\%$ of the average peak intensity, while the spectrum-to-spectrum intensity variation is $> 20\%$. Ideally, the data reduction algorithm must:

(1) find the exact location of all spectra (tracing);
(2) extract the flux per wavelength bin for each spectrum;
(3) compensate for different amount of cross-talk introduced by the overlap of spectra;
(4) apply a 'fiber-flat-field' correction to compensate for the variation of throughput over different fibers; and
(5) rebin all spectra to a common wavelength scale, before the set of spectra can be used to reconstruct maps, etc.

All or at least part of these steps are actually necessary at the telescope during observing if the observer wishes to find out whether the IFU was centered correctly on target, whether the expected signal-to-noise (S/N) has been reached, and whether other indicators of data quality are met. Note that the data reduction process is a non-trivial exercise, which is one of the reasons why it took quite a long while for 3D spectroscopy to develop from the stage of prototypes to the break-through as common user technique.

Data representation, data formats

As already pointed out in Section 1.2.1, the tempting concept of the 'datacube' may be misleading, which is particularly true when considering the representation of 3D data in digital form. The generic result of data reduction is a set of spectra, associated with a corresponding set of spaxel positions. The spaxel coordinate system may or may not be orthonormal, but in the most general case it is not. It is only through the process of interpolation in the spatial coordinate system that arbitrary IFU geometries are converted to orthonormal, i.e. a datacube compatible form (atmospheric refraction not considered; see Section 1.2.2 below). Interpolation, however, inevitably incurs loss of information. This is why the Euro3D consortium has introduced and promotes a special data format for transportation of reduced 3D data which is different from the seemingly simple application of the standard FITS (flexible image transport system) NAXIS = 3 format, which has in fact emerged from radio astronomy, where datacubes are a natural data product (Wells *et al.*, 1981). The Euro3D data format (Kissler-Patig *et al.*, 2004) avoids

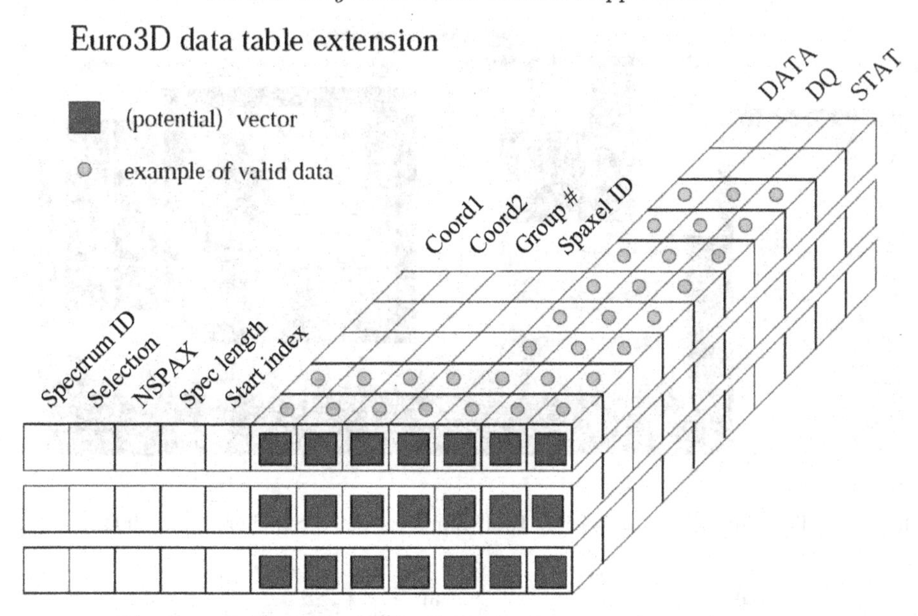

FIGURE 1.15. The Euro3D data format (from Kissler-Patig *et al.*, 2004).

this latter step of interpolation, merely assuming that the basic steps of data reduction have been applied to remove the instrumental signature, but otherwise presenting the data as a set of spectra with corresponding positions on the sky. Figure 1.15 illustrates the spaxel-oriented approach of the Euro3D (E3D) FITS data format.

Data analysis

3D data analysis comprises the inspection of data quality and the determination of scientifically useful physical quantities. It needs to be clearly distinguished from data reduction, which should rather be considered instrumental, and as such (almost) part of the data acquisition process. Some conventional goals of 3D data analysis are: reconstruction of maps at certain wavelengths, line-ratio maps, velocity fields, co-adding/binning of spectra, sky/background subtraction, spectrophotometry, etc. These topics will be addressed extensively later in this volume. At this point it is, however, appropriate to point out that there is a considerable level of complexity involved if one compares these with well-established astronomical observing techniques. Although significant progress has been made over the years, it seems fair to state that 3D spectroscopy as a general technique is still in its infancy. In an attempt to popularize 3D spectroscopy in the astronomical user community, the Euro3D consortium has made an effort to develop concepts to facilitate the process of generating science results from 3D data. A key instrument to this end is a 3D visualization tool, which can be described as a general graphical user interface to 3D data. The E3D programme (Figure 1.16) as a result of this effort, developed by S. Sánchez, is extensively described in this book.

1.2.2 *Comparison with classical techniques*

3D spectroscopy – as opposed to conventional slit spectroscopy – has a number of unique advantages, which can be summarized as follows, and which will be described in more detail below:

- a posteriori advantage, pointing;
- absence of slit effects;

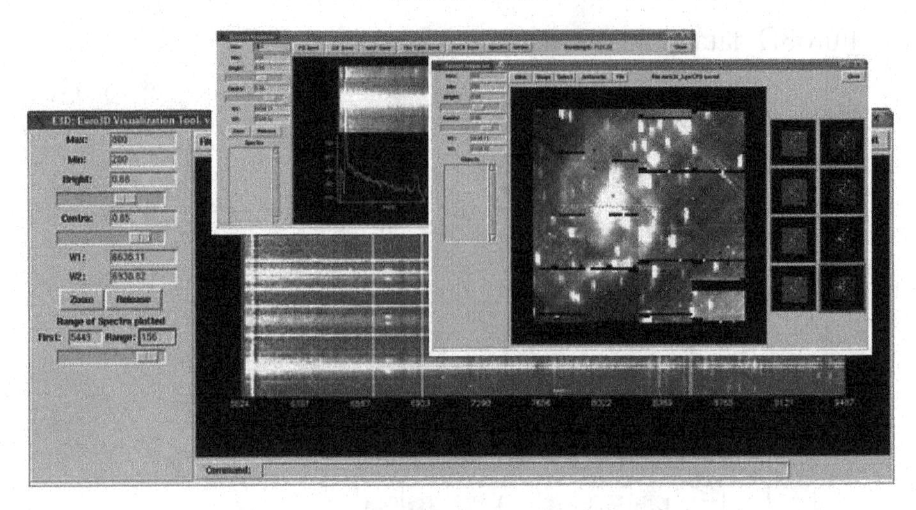

FIGURE 1.16. The Euro3D Visualization Tool (Sánchez, 2004; Sánchez *et al.*, 2004a).

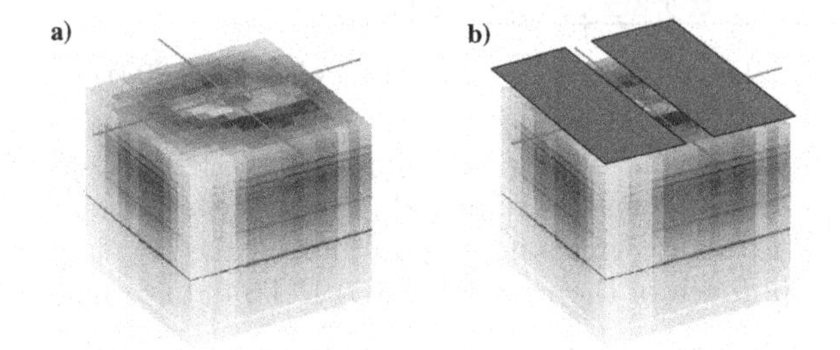

FIGURE 1.17. a) '*A posteriori*' advantage of 3D; b) unlike in slit spectroscopy, it is not necessary to center the object exactly within the aperture of the instrument.

- atmospheric refraction;
- spatial binning, low surface brightness;
- differential spectrophotometry;
- crowded field 3D spectroscopy;
- ultra-deep faint object 3D spectroscopy.

A posteriori advantage, pointing

This advantage is based on the fact that an IFU within its useful FoV has no physical aperture, so that there is no need to accurately center a given target of interest. The importance of this property has been highlighted in, for example, papers of Bacon *et al.* (1995) on the double nucleus of M31, or Mediavilla and Arribas (1993) on the offset AGN in NGC 3227, where the physically important feature (peculiar kinematics, emission line region) were in fact not coinciding with the intensity peak in broad-band light, which on a standard telescope acquisition camera would normally tend to suggest where to center the slit.

Slit effects

Slit effects are essentially light losses caused by the finite width of a spectrograph slit. The severeness of the effect is a function of seeing FWHM in relation to the slit width (Jacoby and Kaler, 1993), and a function of wavelength. Bacon *et al.* (1995) have pointed

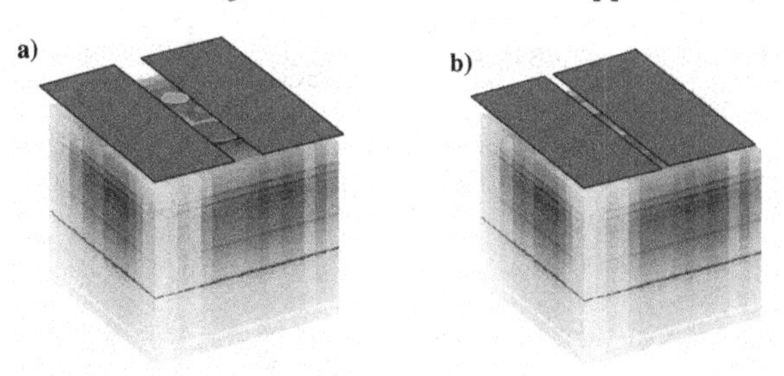

FIGURE 1.18. Slit effects: a) a wide slit collects a large fraction of target flux; however, it also collects a large amount of background light. b) A narrow slit minimizes the background contribution (and also increases spectral resolution), but at the same time reduces the recorded target flux. No such effect occurs in an IFU.

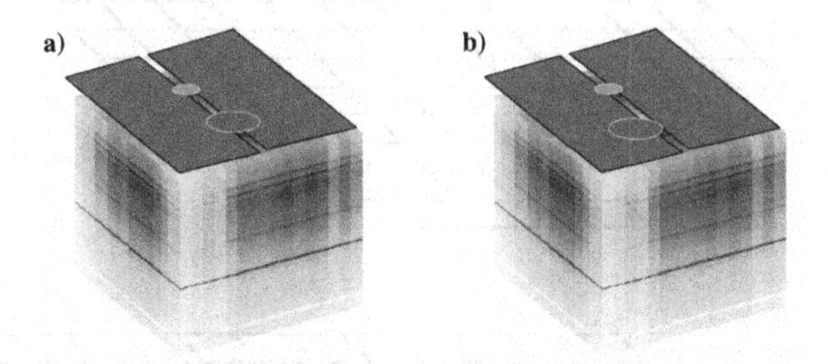

FIGURE 1.19. a) In slit spectroscopy, the effect of atmospheric refraction can be minimized by aligning the spectrograph slit with the atmospheric spectrum (parallactic angle). b) Because the seeing FWHM varies with wavelength, flux calibration is affected by systematic errors; however, this is not so for IFUs.

out that a surface brightness gradient across the slit in the direction of dispersion causes a bias in the instrumental profile, which in turn translates into an error on the wavelength scale. 3D spectrographs based on the micropupil principle are totally unaffected by this problem, which is why these instruments are superior tools to measure subtle velocity field variations on small scales, e.g. in the nuclear regions of galaxies.

Atmospheric refraction

Atmospheric refraction is due to dispersion in the atmosphere, which causes a different angle of deflection as a function of wavelength for rays at oblique angles of incidence, i.e. for objects at airmass >1. For point source images, the effect is equivalent to creating an extended spectrum instead of a well-defined spot in the focal plane of the telescope. The spectrum is extended in the direction of the parallactic angle, i.e. a great circle through the zenith and the target, as projected on the sky. Atmospheric refraction is a function of air temperature, pressure and relative humidity. The effect is described quantitatively by Filippenko (1982), who advised that a spectrograph should be placed such that the slit orientation is along the parallactic angle or, in other words, that all light of the atmospheric spectrum falls into the slit aperture. Since, however, seeing is a function of wavelength, different fractions of total stellar flux are recorded along the atmospheric spectrum. Due to this fact, spectrophotometric calibrations by means of a flux standard are affected by systematic errors whenever the standard is observed at an airmass that is different from the airmass of the object. None of these problems is an issue for 3D spectroscopy, at least whenever an IFU with a fill factor close to unity is employed. Contrary to a slit, the effect of atmospheric refraction translates merely into

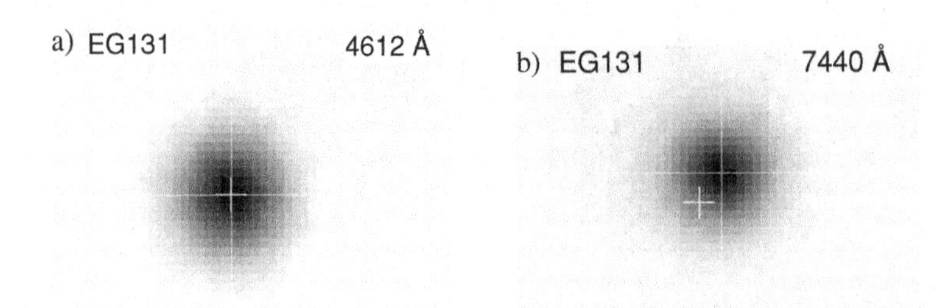

FIGURE 1.20. Maps of standard star exposure at two different wavelengths: a) blue; b) red. The offset due to atmospheric refraction and the wavelength dependence of the seeing FWHM are clearly visible. Data obtained with the GMOS-IFU at GEMINI-S (courtesy of J. Turner).

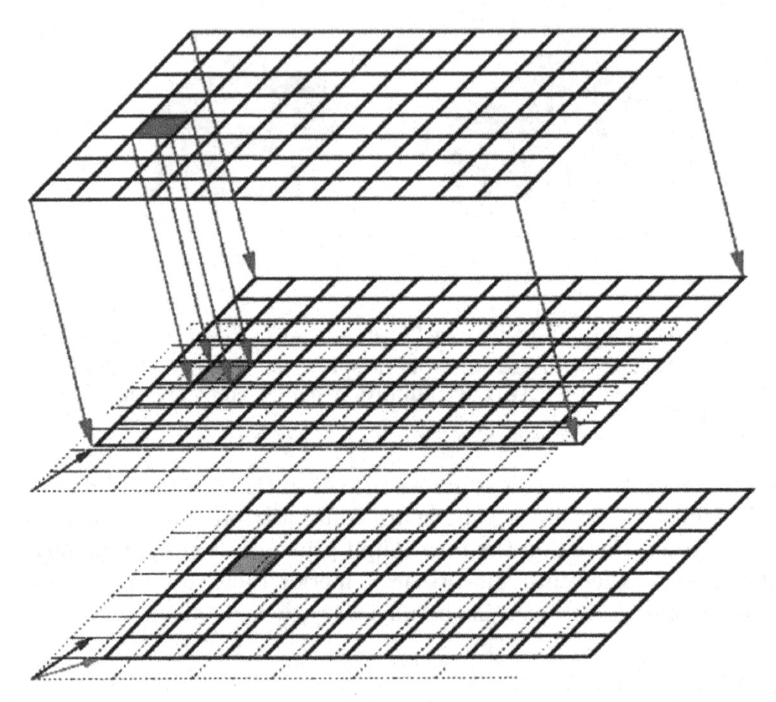

FIGURE 1.21. Atmospheric refraction in the datacube picture: two maps at two different wavelengths are shown as orthonormal grids. Owing to atmospheric refraction, the centroid of a stellar image experiences a translation from one wavelength to another. Note that, in the most general case, two exposures obtained at different times (and, therefore, at different airmass and hour angles) are affected by different translation vectors. Hence, the combination of two datasets requires a transformation to a common coordinate system, i.e. a correction of the translation due to refraction.

a shift in x,y as a function of wavelength in the familiar picture of a datacube; however, no flux is lost due to light falling outside an aperture (Figure 1.20 and Figure 1.21). The effect on 3D spectroscopy was first described by Arribas *et al.* (1995) and Emsellem *et al.* (1996). Note that it is atmospheric refraction that renders the rigorous definition of a datacube inappropriate: Figure 1.21 suggests a data rhomboid rather than a datacube. Because the atmospheric translation vector is never the same for a series of exposures taken at different times, the combination of datasets for the purpose of improving S/N, is a non-trivial exercise, involving rebinning and dithering techniques.

Spatial binning

From direct imaging data it is known that poor S/N in regions of low surface brightness can be improved by binning pixels into superpixels, which contain more signal and, hence, better S/N. Whenever it is justified to sacrifice high spatial resolution, one can trade spatial resolution for S/N. The same method can be applied to 3D spectroscopy by binning spaxels in regions of low surface brightness, where no dramatic variations are expected to play a role. The gain scales with the square root of the number of spaxels which are co-added into one superspaxel. This technique has been refined to what is called 'adaptive binning' (Cappellari and Copin, 2003), which is an algorithm that automatically generates co-added groups of spaxels through the use of a Voronoi tessellation. The regions are shaped by design such as to share the same level of S/N.

Differential spectrophotometry

As outlined already above (under the headings of *Slit effects* and *Atmospheric refraction*), 3D spectroscopy is, in principle, the ideal method to perform spectrophotometry. Spectrophotometry, however, is severely compromised under non-photometric observing conditions. Provided the observer is in the position to find a secondary or tertiary flux standard near the position of the target, one can apply the technique of 'differential spectrophotometry', analogous to the development of differential photometry (Barwig *et al.*, 1987). Using this technique, it is possible to continue observing even through thin clouds and to compensate for transmission variations through the simultaneous observation of the standard star. Novel 3D spectrographs with large FoV, or instruments with deployable IFUs, or 3D spectrographs incorporating a direct imager for field acquisition and guiding are ideally suited to adopt this technique.

Crowded field 3D spectroscopy

Spectroscopy of individual sources embedded in densely populated stellar fields, or in bright nebular emission, or in a combination of both is a notoriously difficult undertaking, where the precise subtraction of the background component is often a challenge. Estimating the background at the position of a target is particularly difficult for slit spectroscopy, where there is only information in one dimension (= along the slit) about the two-dimensional surface brightness distribution, which is often quite variable in the spatial domain, as well as with wavelength. Building on the established techniques of PSF-fitting CCD photometry, it has been shown that the same methods applied to datacubes yield superior results when trying to disentangle point sources from a bright background (Roth *et al.*, 2004b). In practical terms, this is accomplished by processing all monochromatic 'slices' of a datacube one by one. The PSF fitting technique can be even further enhanced in the presence of high resolution images (e.g. from space), where the centroids of point sources can be used to iteratively deconvolve the blended features contained in a datacube (Hook and Lucy, 1993; Becker *et al.*, 2004).

Ultra-deep faint object 3D spectroscopy

Arguably among the most spectacular data products from observational Astronomy are the Hubble Deep Field images and subsequent similar results, providing the deepest views into the Universe ever obtained (Williams *et al.*, 1996). These very deep images with total exposure times of on the order of ≈ 100 hours have only been possible owing to the linearity and excellent quality of CCD images, which were resampled and stacked using the sophisticated 'drizzle' algorithm (Fruchter and Hook, 2002). To perform similar deep observations with 3D spectroscopy was first suggested by Roth *et al.* (2000), and proposed

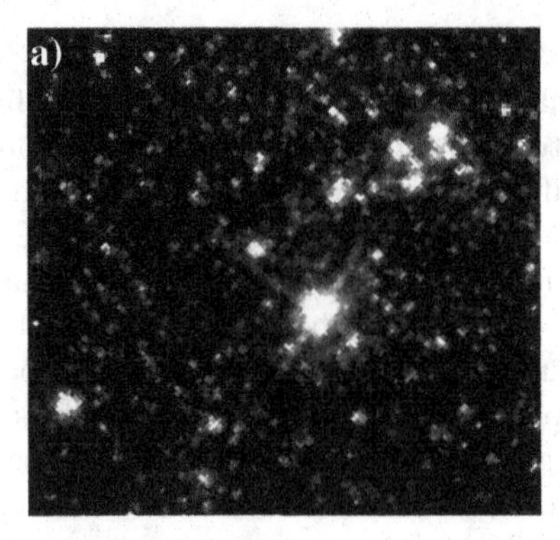

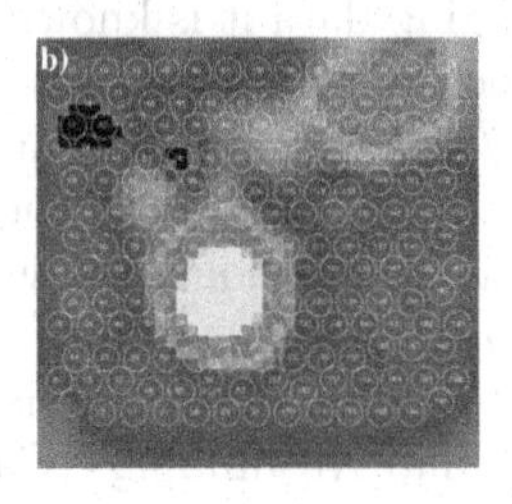

FIGURE 1.22. Crowded field 3D spectroscopy: the LBV candidate star B416 in M33. a) Hubble Space Telescope (HST) field, used for deconvolution technique; b) reconstructed map as inferred from INTEGRAL data. See Fabrika *et al.* (2005).

for an actual second generation very large telescope (VLT) instrument by Bacon *et al.* (2002). It is worth mentioning that the data reduction and stacking process of ground-based 3D data are significantly more complex than the same process for direct images from space. Note in particular the complication of different amounts of atmospheric refraction for datasets obtained at different times (see Figure 1.21). However, with the validation of concepts for faint object 3D spectroscopy with real observations and stacking data from many hours of exposure time (e.g. Bower *et al.*, 2004; Roth *et al.*, 2004b), the development of MUSE (Multi Unit Spectroscopic Explorer, a dedicated faint object 3D spectrograph for the European Southern Observatory – VLT) is now underway (Bacon *et al.*, 2004). The deep surveys planned with this instrument are expected to discover faint Lyα emitters within the 1 arcmin FoV at unprecedented flux levels down to $\approx 3 \times 10^{-19}\,\mathrm{erg\,s^{-1}\,cm^{-2}}$.

1.3 A brief history of 3D spectroscopy

Since the early 1980s, the desire to simultaneously acquire spectral and spatial information in a single exposure motivated various groups to develop integral field spectrographs for astronomical telescopes. According to the definition given in Section 1.2.1, the goal was to obtain three-dimensional datasets (datacubes), which should be generated from a single exposure under identical observing conditions, contrary to datacubes that are assembled through a scanning process involving multiple exposures. As an example for the latter method, Wilkinson *et al.* (1986) investigated the kinematics in the Cen A galaxy (NGC 5128) with a datacube generated from a series of offset long-slit exposures. They used the RGO Spectrograph at the 3.9 m Anglo-Australian Telescope (AAT) with a V1200 grating, a slit width of 0.3 mm, and the Image Photo Counting System (IPCS) as detector. Along the slit, the sampling was 2.2″ per spatial element over a total length of 61 elements. With a total exposure time of 13,327 s for the entire datacube, 71 telescope offset positions in increments of 2.4″ were needed. The spectral domain was sampled with a total of 1650 increments of 0.5 Å. Resulting from these observations, the velocity field over the face of the galaxy was derived and conclusions drawn concerning the kinematics of the stellar component with regard to the warped dust lane in this nearest large elliptical galaxy, which has just experienced a major merger event. The example demonstrated vividly the need for spatially resolved area spectroscopy, which indeed could be fulfilled with the long-slit scanning technique. There were other successful applications for different science problems – e.g. the study of planetary nebulae (Balick *et al.*,

1987) – but it was soon realized that the scanning techniques have their shortcomings. Not only was the sequential data-taking process cumbersome and inefficient, but it was also sensitive to variable atmospheric conditions. For example, an identical PSF throughout the FoV was found to be important for measurements of the two-dimensional kinematics in the nuclei of galaxies, requiring high spatial resolution, as pointed out by Bacon *et al.* (1995). Therefore, one would ideally like to obtain the entire datacube from a single exposure.

However, before the time was ripe for true integral field spectroscopy, another scanning method was introduced which did indeed provide a stable PSF for any map derived from the datacube (albeit not necessarily identical from wavelength to wavelength step), namely the imaging Fabry–Perot (FP) interferometer. In fact, applications of this technique go back to the early 1970s, e.g. when Tully (1974a–c) used the Maryland photographic Fabry–Perot Interferometer at the KPNO (Kitt Peak National Observatory) 2.1 m Telescope to determine the Hα velocity field in M51. For this work, photographic plates with IIa-O emulsion, and a two-stage Carnegie image intensifier with an S20 photocathode were employed, requiring typical exposure times of 0.25–2 hours plate^{-1} (10 h total). The remarkable result was the kinematic mapping of the galaxy over a FoV as large as 12 arcmin in diameter with a spatial resolution of 2″, and a spectral resolution of 0.5 Å, corresponding to a velocity resolution of 27 km s^{-1}, over a free spectral range of 4.2 Å. The sensitivity limit of these observations was 10^{-15} erg s^{-1} cm^{-2} arcsec^{-2}, and radial velocities could be measured wherever the Hα intensity exceeded 10^{-14} erg s^{-1} cm^{-2} arcsec^{-2}. Tully was able to fit a three-component model of the mass distribution, comprising disk, bulge and the nucleus. Subsequently, FP tunable filter instruments became more and more popular, e.g. the famous TAURUS and TAURUS-2 at WHT and AAT, respectively (Atherton *et al.*, 1981, 1982; Bland-Hawthorn and Taylor, 1994), HIFI at CFHT (Canada–France–Hawaii Telescope), and the prime focus FP at the Calar Alto 3.5 m (Meisenheimer and Hippelein, 1992). The major advantage of these instruments was a large FoV, high spectral resolution and large throughput. Several disadvantages – such as ghost images, stray light, interference artifacts, problems with flux calibration, the narrow useful wavelength range for a given etalon, and a relatively complicated data reduction process – have prevented these systems from becoming standard instruments for the common observer. Also, FP imagers are not true 3D spectrographs in the sense that they require scanning.

By the time that FP instruments had their technological breakthrough and the concept of IFS was more commonly appreciated, optical fibers had been discovered as a means of increasing the multiplex of spectroscopy towards making large spectroscopic surveys feasible: The first astronomical spectrum through a fiber was taken in 1978, and the first fiber-optical multi-object spectroscopy was performed using the Steward Observatory 2.3 m telescope just one year later (Hill *et al.*, 1980; see also a review by Hill, 1988). Almost at the same time Vanderriest (1980) used the design flexibility offered by optical waveguides to build the first fiber bundle prototype, which was installed at the Mauna Kea 2.2 m telescope. It comprised a total of 169 fibers, covering a hexagonal footprint of 27″ × 43″ with a fill factor of 75% and a projected fiber diameter of 0.42″ on the sky. Thirty-six additional fibers, arranged in four blocks away from the central FoV, were foreseen for sky background subtraction. While this setup was coupled to an existing, however ill-adapted, photographic image-tube spectrograph, in its early form it was already incorporating all major features that were later on included in subsequent dedicated fiber bundle IFS. Because of the technical constraints of the experimental prototype at the time, the scientific outcome (3D spectra of 3C 120) was somewhat limited, but it is interesting to note that about a quarter of a century later the use

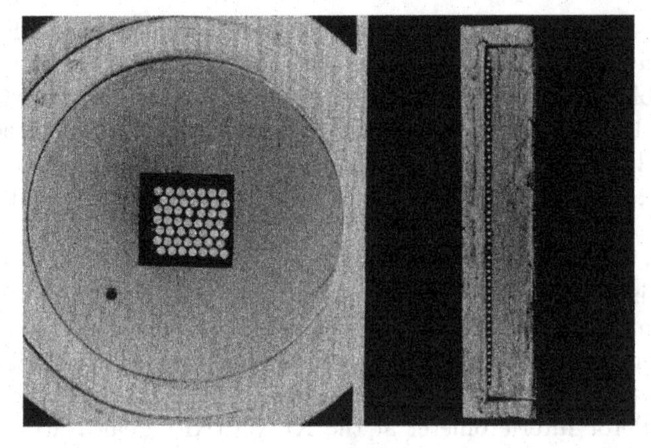

FIGURE 1.23. DensePak-2 at the KPNO 4 m telescope.

of a similar fiber bundle, coupled to a dedicated fiber spectrograph with a modern CCD detector (INTEGRAL at WHT; Arribas *et al.*, 1998a; 1998b) yielded high-quality data of 3C 120, which allowed the application of image-processing techniques to accurately disentangle the radio jet emission from the underlying galaxy (Sánchez *et al.*, 2004b).

Another experiment was the DensePak-1 IFU for the KPNO 4 m telescope (Barden and Scott, 1986), which also had the shortcomings of an early prototype. However, the early experiments stimulated a whole series of new fiber bundle IFU developments at different telescopes; for example, DensePak-2 (Barden and Wade, 1988) was a much improved follow-up at the KPNO 4 m, featuring a rectangular fiber bundle with a total of 49 fibers over a FoV of 16″ × 19″ and a projected fiber diameter of 2.1″ on the sky; SILFID (Spectrographe Integral Linarisation par Fibres de l'Image Directe) for the CFHT (Vanderriest and Lemmonier, 1988); HEXAFLEX/WHT (Arribas *et al.*, 1991), 2D-FIS (2-Dimensional Fiber ISIS System) WHT (García Marín *et al.*, 1994); HEXAFLEX-II/NOT (García Marín *et al.*, 1994); ARGUS (Vanderriest, 1995). Although fiber bundle IFUs suffer from incomplete sampling and sometimes irregular arrangement of the fibers, they have an intrinsically simple structure and can be fitted as add-on modules to existing bench-mounted spectrographs, which probably explains why they represent the most common type of IFS in the early days of 3D spectroscopy. A more recent interesting application is the SparsePak IFU at WIYN (Wisconsin–Indiana–Yale and National Optical Astronomy Observatory; Bershady *et al.*, 2004).

It is worth mentioning that when the topic of IFS had gained considerable momentum in the 1990s, a dedicated International Astronomical Union (IAU) Colloquium was organized in 1994 (Compte and Marcelin, 1995), after which, superseding the initial phase of prototypes, a first generation of common user 3D spectrographs emerged at 4 m-class telescopes.

One of the prominent examples is INTEGRAL at the WHT (Arribas *et al.*, 1998a; 1998b). Unlike several of the early prototypes, INTEGRAL is an optimized fiber bundle IFU, attached to the WYFOS (Wide-Field Optic Spectrograph) fiber spectrograph at WHT (Bingham *et al.*, 1994). INTEGRAL features three rectangular fiber bundles at different magnifications for the purpose of accommodating different observing conditions in terms of seeing, and different user requirements (spatial resolution versus FoV). Each bundle is also furnished with a set of separate fibers for sky background subtraction. The three bundles have parameters as follows (FoV/projected fiber diameter): SB1 7.8″ × 6.4″/0.45″; SB2 16.0″ × 12.3″/0.9″; SB1 33.6″ × 29.4″/2.7″. Besides the different

FIGURE 1.24. INTEGRAL at the Nasmyth focus of the WHT.

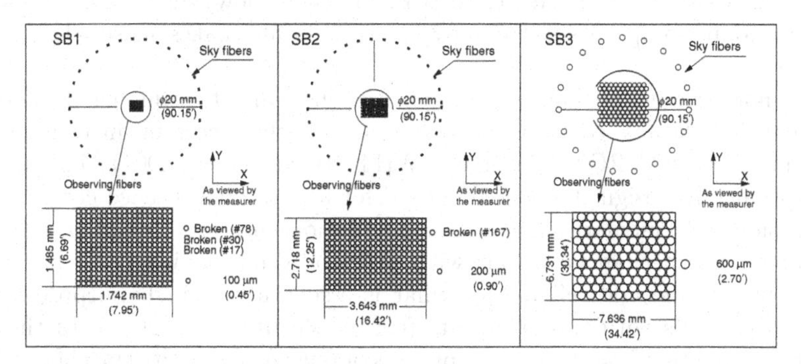

FIGURE 1.25. INTEGRAL fiber bundles.

choice of IFU magnification, INTEGRAL offers a variety of grating options with different dispersion, wavelength coverage and blaze thanks to the many modes available in WYFOS (Wide-Field Optic Spectrograph). This property makes INTEGRAL a very useful common user instrument as it allows observers to a large extent to optimize the instrumental setup and accommodate to their specific scientific problems.

Fiber bundle IFUs have the advantage that they are conceptually simple and that they can be attached a posteriori to existing (preferably fiber) spectrographs. It is therefore probably no surprise that the very first IFUs were indeed based on this principle. However, a few shortcomings soon led to alternate designs, in particular the inevitable drawback of fiber bundles having fill factors of less than unity. The incomplete sampling does not only influence the spatial resolution of seeing-limited observations, but also gives rise to photometric inaccuracies when observing spectrophotometric flux standards: whenever an appreciable fraction of standard star flux falls into the gap space between adjacent fibers, the amount of lost light cannot be straightforwardly accounted for. A way to overcome this shortcoming was to use an optical device that had already been proposed by Courtès (1960), namely a lens array in the optical plane of the telescope. While the principle of operation is explained in more detail in Section 1.4.2, suffice it to say for the purpose of this overview that a contiguous array of microlenses of square or hexagonal shape does not suffer dead space between adjacent spatial elements, unlike a conventional fiber bundle. Such a device was therefore thought to be ideal for the highest spatial resolution of IFS. Using the principle of micropupil imagery and suitable dispersion in combination with a waveband selecting filter, it could also accommodate a large density of spectra on the detector, i.e. a larger number of spatial elements. Despite these advantages, it was not until 1987 when the first lens array IFU 'TIGER' saw first light at the CFHT (Courtès *et al.*, 1988; Bacon *et al.*, 1995). Other implementations were shortly thereafter developed for the Selentchuk 6-m telescope (Afanasiev *et al.*, 1990) and in Kyoto (Ohtani *et al.*, 1994). The TIGER instrument was not only groundbreaking in terms of technological innovation, but also quite successful with scientific achievements, e.g. the kinematic study of the double nucleus in M31 (Bacon *et al.*, 1994), to name just one example. A subsequent follow-up development built by the same group in Lyon, France, was the OASIS (Optically Adaptive System for Imaging Spectroscopy) IFS, which was optimized for very high spatial resolution in combination with the PUEO (Probing the Universe with Enhanced Optics) adaptive optics system at CFHT. Although adaptive optics (AO) operation was difficult to accomplish with natural guide stars only, this development has pointed to what is now generally thought to have become a leading track towards novel spectroscopic instrumentation at large and extremely large telescopes (ELT), i.e. AO-assisted integral field spectroscopy (see below). It has even been claimed that modern adaptive optics spectroscopy in general only makes sense when performed with an IFU.

Although not strictly speaking a common user instrument, a further development of the TIGER principle has led to what became extremely successful in terms of science output, namely the SAURON IFS at the WHT (Bacon *et al.*, 2001). This instrument was very specifically designed and is operated for a particular science case, i.e. absorption or emission line kinematics and stellar population analysis in elliptical galaxies. An overall simple opto-mechanical layout with practically no moving parts has made the instrument very stable and reliable for what is essentially a single-purpose mission; it has accomplished this with superb results (e.g. de Zeeuw *et al.*, 2002). In the trend of an increasing number of 3D spectrographs becoming available at 4 m class telescopes, SAURON may well be counted in the category of post-prototype instrumentation, albeit having the status of a private instrument.

In the process of the development of the baseline designs for IFS as discussed above – i.e. fiber bundle versus lens array – it became clear that the two are largely complementary to each other: large wavelength coverage, relatively coarse spatial sampling and moderate number of spectra, versus moderate wavelength coverage, optimal spatial sampling and large number of spectra. However, it was found that for certain observational problems it would be desirable to combine the superior spatial sampling properties of lens arrays with the large wavelength coverage of fiber bundle IFS. As a result, hybrid 'lens array – fiber bundle' IFUs were developed (see Section 1.4.3). The first implementation of this type of IFU happened in the MPFS (Multi-Pupil Field Spectrograph) instrument at the prime focus of the Selentchuk 6 m telescope in 1997. Other IFSs using this principle are SPIRAL/AAT (Kenworthy *et al.*, 2001), PMAS/Calar Alto (Roth *et al.*, 2005; Kelz *et al.*, 2006) and several other instruments at 8 m-class telescopes as discussed below.

Fiber and lens-array-based designs have governed the layout of the first generation of IFSs. Given the rather limited number of spaxels in comparison with direct imaging detectors, and also considering the inefficient use of detector space, the slicer principle was introduced with the '3D' instrument (Weitzel *et al.*, 1996). This IFS was designed

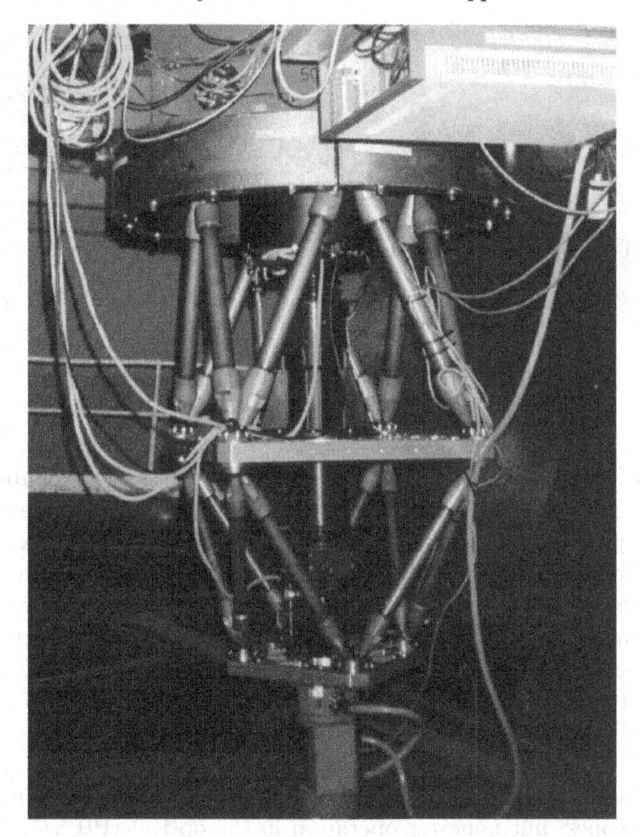

FIGURE 1.26. SAURON at the WHT, La Palma.

FIGURE 1.27. PMAS at the Calar Alto 3.5 m telescope features two IFU options: a fibercoupled lens array IFU, and a bare fiber bundle IFU.

FIGURE 1.28. a) Compact GMOS-IFU. b) VIMOS-IFU, the largest IFU to date, providing a total of 6400 spectra.

for the J, H and K bands in the near infrared (NIR). Since the available detector size was small (256×256 pixels), it was felt that the number of fiber or lens array spectra that could be projected onto the available chip would be too small to warrant the effort of building a conventional IFU. As an alternative, the slicer method generates a stack of long-slit spectra, corresponding to a series of parallel slices in the image plane. Contrary to the fiber and lens array IFUs, the sky is thus imaged directly onto the detector, whereas in the former case the fiber end faces or micropupils are projected onto the focal plane. The detector area is therefore optimally filled with contiguous spectra and no space is lost due to gaps. The optomechanical design of slicers is less critical at long wavelengths, but more difficult in the optical or UV as far as performance in terms of image quality, edge effects and stray light is concerned. A few more NIR slicer instruments were built for 4 m-class telescopes, but none for operation in the optical (PIFS/Hale 5 m, Murphy *et al.*, 1999; UIST/UKIRT, Ramsay Howatt *et al.*, 2004).

The history of IFS as discussed so far included the very first prototypes and a first generation of facility instruments for 4 m class telescopes. The latter already represented a significant step towards establishing 3D spectroscopy as a common user technique. In the late 1990s the light collecting power, superb image quality and demand for new instrumentation of new 8–10 m class telescopes prompted yet another generation of IFSs, some of which were designed as IFUs in the original sense, i.e. add-on modules in front of conventional slit spectrographs (e.g. GMOS-IFU: Gemini Multi-Object Spectrograph – IFU; Allington-Smith *et al.*, 2002; VIMOS-IFU/VLT: Visible Imaging Multi-Object Spectrograph – IFU/VLT; Le Fevre *et al.*, 2003), while others are dedicated 3D spectrographs (SINFONI/VLT, Eisenhauer *et al.*, 2003; OSIRIS/KECK, Larkin *et al.*, 2006). These instruments usually come with their own data reduction pipeline, which is an important prerequisite for opening this non-trivial observing technique to the common user.

Future facilities are presently planned for ground-based and for space telescopes. The most prominent new developments for the ground are MUSE as a second-generation VLT instrument, and VIRUS (Visible Integral-Field Replicable Unit Spectrograph) for HET (Hobby–Eberly Telescope). These new facilities are planned to accommodate a total of as many as 90,000 and 30,000 spectra, respectively. As a consequence, they are quite large and highly complex, requiring novel approaches for manufacture and integration, e.g. replicable devices which can be built in small industrial series, thus reducing risks and saving costs. In space, JWST is planned to host – for the first time – IFUs above the atmosphere, namely as part of the NIRSpec (Near-Infrared Spectrograph) and Mid-Infrared Instrument (MIRI), covering the 0.6–5 μm and 25–27 μm wavelength ranges respectively (Arribas *et al.*, 2005). Of those future instruments, MUSE will come

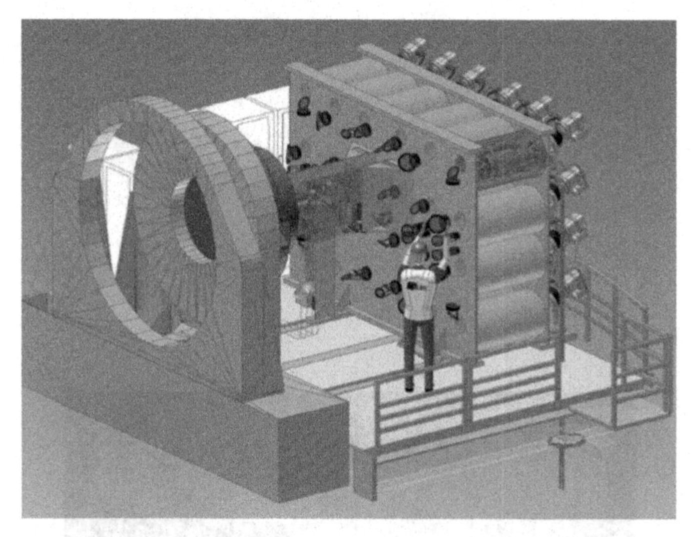

FIGURE 1.29. MUSE at the Nasmyth platform of the VLT (CAD layout).

reasonably close to the ideal of a conventional imager, however creating a full one-octave spectrum in each pixel. Over the wavelength range of 465−930 nm, MUSE will accommodate a square FoV of 1 arcmin with a spatial sampling of $0.2 \times 0.2\,\text{arcsec}^2$ in its wide-field mode. The narrow-field mode will cover $7.5 \times 7.5\,\text{arcsec}^2$ with a sampling of $0.025\,\text{arcsec}^2$. The instrument will be assisted by a ground-layer correcting AO system with 4 laser guider stars and an additional natural guide star (Bacon *et al.*, 2004).

1.4 Technical approaches

In the following, we will discuss in more detail several relevant aspects of technologies that are in use for 3D instrumentation; namely:

- optical fibers;
- lens arrays;
- lens arrays coupled to fiber bundle;
- slicers;
- special techniques;
- non-IFS 3D instruments;
- 3D detectors; and
- figures-of-merit.

1.4.1 *Optical fibers*

Historically, optical fiber bundles represented the first and most simple technology to build up IFUs, often as front ends to existing slit spectrographs. Consequentially, some of the early prototypes were not optimized in terms of efficiency, stability or other parameters. However, a whole family of specifically designed facility instruments following the first generation of prototypes has convincingly demonstrated that optical fibers are quite useful components for 3D spectroscopy. Standard multimode optical fibers are thin waveguides basically consisting of three components:

- the fiber core made of high refractive index glass;
- the cladding as a low refractive index skin around the core; and
- a protective buffer normally made of some sort of plastic material, e.g. polyamide.

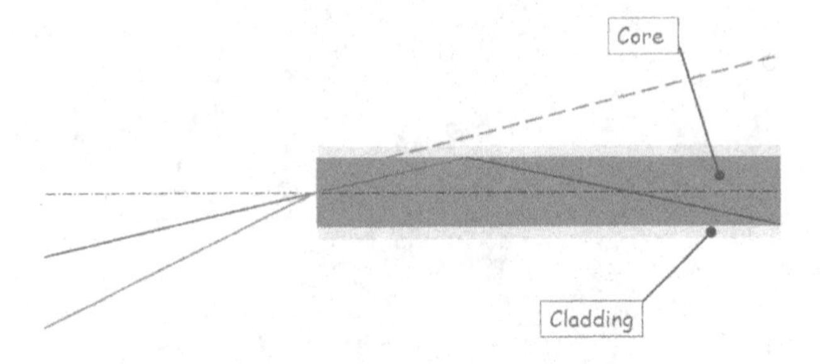

FIGURE 1.30. Fiber 1.

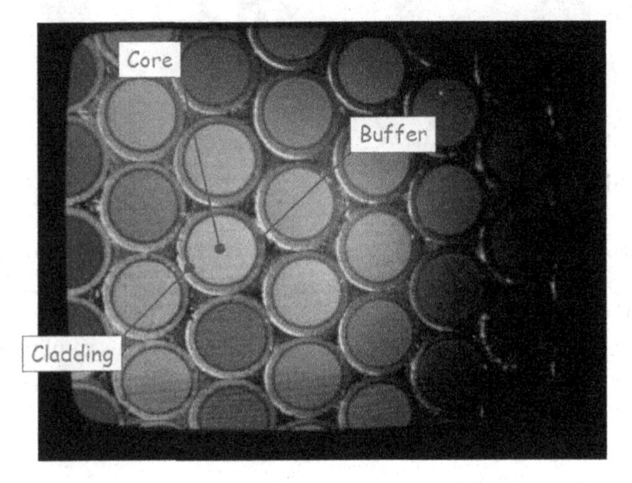

FIGURE 1.31. Fiber bundle.

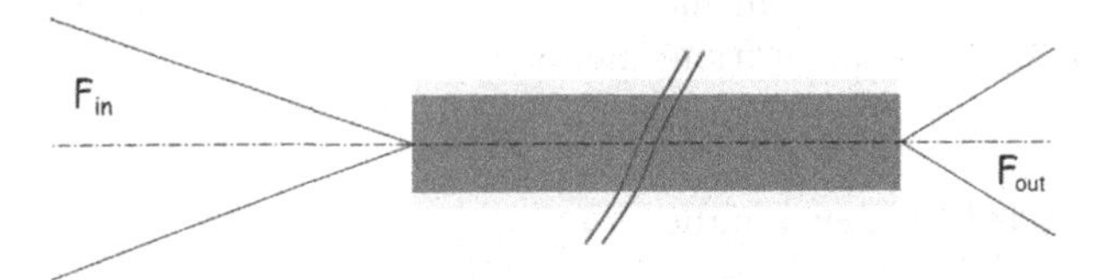

FIGURE 1.32. Focal ratio degradation.

They act as wave guides owing to the effect of total internal reflection, meaning that rays at grazing incidence onto the core–buffer interface will be reflected back into the core. Multiple reflections inside the core lead to a scrambling of the input light pattern. Rays incident onto the fiber input with inclination angles larger than that given by a certain critical angle (numerical aperture, NA) will no longer undergo total internal reflection, but will rather be absorbed in the buffer. Figure 1.31 shows a microscopic view of the polished front face of a bundle formed from several such fibers which were bonded together to make up a hexagonal densest package. In this picture, the bright circular areas are the fiber cores, surrounded by a dark ring (the cladding), and another material (the buffer). Note that this fiber bundle IFU has an incomplete fill factor, as only a fraction of the area is acting as a light-collecting aperture. One of the most important disadvantages of fibers is their property of not normally retaining the input f-number of the incoming beam (Figure 1.32). The emerging light cone has a larger opening angle than the input beam. The effect is called focal ratio degradation (FRD). It is very undesirable as it leads

FIGURE 1.33. Several views of the fiber-optical IFU 'PPak' of PMAS (Kelz *et al.*, 2006).

to overfilling the spectrograph collimator and light losses. Moreover, since the amount of FRD depends on external parameters like mechanical or thermally induced stress, the output *f*-number is not constant with time unless the fiber duct from the IFU to the spectrograph is extremely stable. In certain cases, variable FRD is capable of creating noise, which limits the ultimate S/N obtainable, even under conditions with a lot of light and good photon statistics (Parry and Carrasco, 1990; Baudrand and Walker, 2001; Schmoll *et al.*, 2003).

In summary, fiber-optical IFUs have the following properties: Fibers allow for a very flexible installation of an IFU, and they accommodate bench-mounted spectrographs, in particular, which is a strong argument for stability. There are a few problematic fiber properties, which – with proper care and a suitable design – can be handled fairly well. Fiber-optical IFUs have typically relatively few spaxels, but – depending on the spectrograph used – a large number of spectral bins and a wide wavelength coverage. Fiber bundle IFU can be built with large spaxels to cover a sizeable area on the sky, which is, for example, extremely beneficial for low surface brightness spectroscopy. One of the more serious drawbacks for certain applications, especially when high spatial resolution is intended, is the non-contiguous sampling due to inevitable gaps in the bundle. However, contrary to common belief, fiber-optical IFUs can be built with high efficiency.

1.4.2 *Lens arrays*

The suggestion of using lens arrays for astronomical observations goes back to Courtès (1960); however, it took quite a while until the first implementation appeared at the telescope (Bacon *et al.*, 1995). Figure 1.34 shows an example of a lens array that has a hexagonal geometry (OASIS-IFU). The first 3D spectrograph of the lens-array-type was TIGER, the famous prototype for this class of instrumentation. At first glance, the lens array has one significant advantage over fiber bundles, namely the absence of large gaps between the spaxels. Although even for lens arrays there are small dead zones at the interfaces between adjacent lenslets, the affected area in proportion to the total lens

FIGURE 1.34. Lens Array of the OASIS 3D Spectrograph.

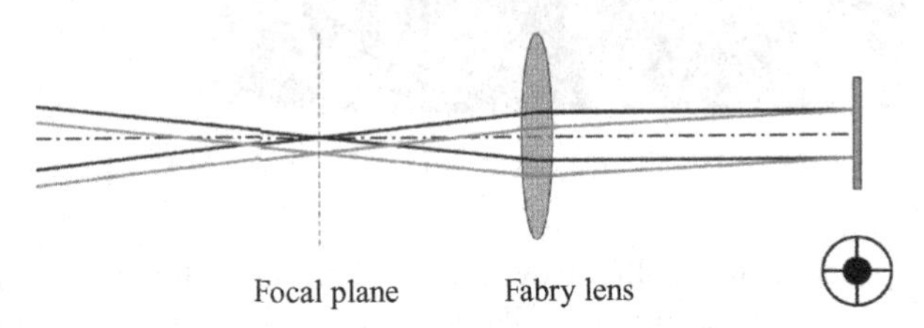

FIGURE 1.35. Principle of a micropupil (not to scale).

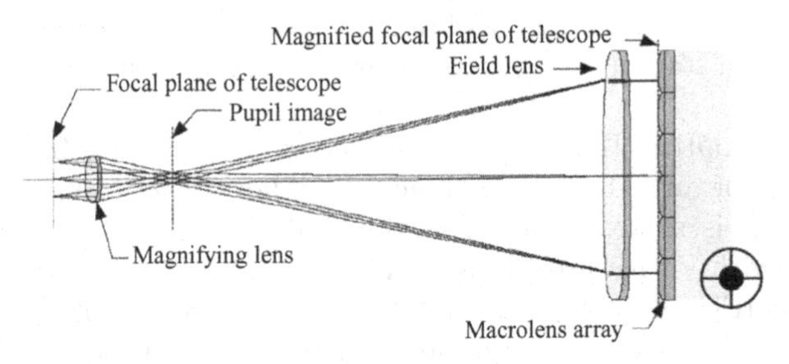

FIGURE 1.36. Optical layout of a lens array with fore-optics (adapted from Dean, 2002).

area is negligible for most lens arrays. Lens arrays are therefore superior devices for high spatial resolution studies of objects on small scales. Another significant advantage of lens array IFUs is often overlooked, namely the principle of micropupil imagery. Instead of imaging the dispersed spaxels of the IFU, it is the microscopic entrance pupil behind each lenslet (= spaxel) of a lens array that is imaged onto the detector (hence the term 'micropupil'). The practical advantages of micropupils for precision spectroscopy are discussed in Section 1.2.1.

The sketch in Figure 1.35 illustrates the optical principle, which is well known from the technology of one-channel photoelectric photometry, where a so-called 'Fabrylens' images the entrance pupil of the telescope onto the photocathode of a multiplier. Instead of having to accept the noise caused by seeing-induced image motion of a stellar image on the photocathode, the static pupil image is unaffected by atmospheric effects or guiding errors. Translating this macroscopic arrangement to the scale of a lens array in

FIGURE 1.37. Example for lens array spectra from SAURON.

combination with some fore-optics having the proper focal length, the result is a set of microscopic pupils behind the lens array, with one micropupil for each lenslet. The size of the micropupils is designed to match the optimal slit width of an imaging spectrograph with a sufficiently large FoV to encompass the physical size of the lens array. The arrangement is then very similar to a multi-object slit spectrograph of the EFOSC (ESO Faint Object Spectrograph and Camera) type (Buzzoni *et al.*, 1984), except that no punch plate is necessary as the slits are defined by the micropupils, and the micropupils are aligned on a totally regular grid. In order to avoid the overlap of spectra coming from neighboring micropupils, the spectral coverage per spectrum must be truncated to approximately the pitch of the lens array, which can be done by using a narrowband filter in the fore-optics train. Moreover, in order to make more efficient use of detector area and to maximize the spectral coverage, the dispersive element (grating, grism) is slightly rotated, resulting in a diagonal arrangement of spectra in the spectrograph focal plane. Figure 1.37 shows a real example from the SAURON instrument (Bacon *et al.*, 2001) as magnified views from:

a) the undispersed micropupils;
b) a continuum flat-field exposure;
c) a spectral line lamp calibration exposure; and
d) an astronomical object exposure.

In summary, lens array IFUs can, in principle, accommodate a large number of spaxels at the expense of wavelength coverage, and some possible problems with truncated spectra near the edge of the detector. The fact that there is contiguous sampling and no gaps makes them very suitable for high spatial resolution observations. Thanks to the micropupil principle, the telescope focal plane is decoupled from the detector, which is a significant advantage for precision spectroscopy. The compact geometry of the spectra allows for compact optomechanical designs and also for highly efficient optical systems. A certain drawback of lens array IFUs is the presence of diffraction causing crosstalk between neighboring spectra, which occurs at different wavelengths (Lee *et al.*, 2001;

FIGURE 1.38. CIRPASS fiber bundle.

Roth *et al.*, 2005). This undesirable effect can, however, largely be corrected for by software in the process of data reduction.

1.4.3 *Lens array – fiber hybrids*

There are applications requiring the sampling properties of a lens array; however, with a wavelength coverage that is more extended than can be accommodated by the truncated spectra of a classical TIGER-type IFU. A solution for this requirement has been realized by combining a lens array with a fiber bundle, where each optical fiber is centered on one micropupil at the exit plane of the lens array. The versatility of the fibers can then be conveniently used to reformat the lens array geometry to a pseudo-slit, thus using the entire width of the detector in the direction of dispersion. Figure 1.38 shows a picture from the fiber bundle attached to the back of the macro-lens-array of the CIRPASS (Cambridge Infrared Panoramic Survey Spectrograph) instrument (Parry, 2004).

As opposed to the pure lens array or pure fiber bundle IFU, the hybrid solution is affected by additional light losses that depend very much on the image quality of the lens array, the precision of the alignment of fibers with regard to the micropupil position, scattered light, diffraction, focal ration degradation, etc. There are instruments that feature an IFU as an add-on module to a classical slit (or MOS) spectrograph, in which case often another set of microlenses is applied to the fiber output ends in order to match the fiber output to the spectrograph beam. In such cases, there are more light losses associated with the additional optical elements. On the positive side, the finite apertures of the fibers in the micropupil plane operate like a spatial filter, blocking diffraction spikes and scattered light coming from the lens array, which, in the case of TIGER type of instruments, would be propagating into the spectrograph. A detailed description of the relevant effects for the example of the PMAS instrument is given by Roth *et al.* (2005).

To summarize, lens-array-fiber hybrids are good for flexible installations; they can accommodate bench-mounted spectrographs, but suffer from problematic fiber properties (which can be handled with proper care). These IFUs typically have relatively few spaxels, but a large number of spectral bins and wide wavelength coverage. They provide contiguous sampling and have no gaps between spaxels, making them good devices for observations that require high spatial resolution.

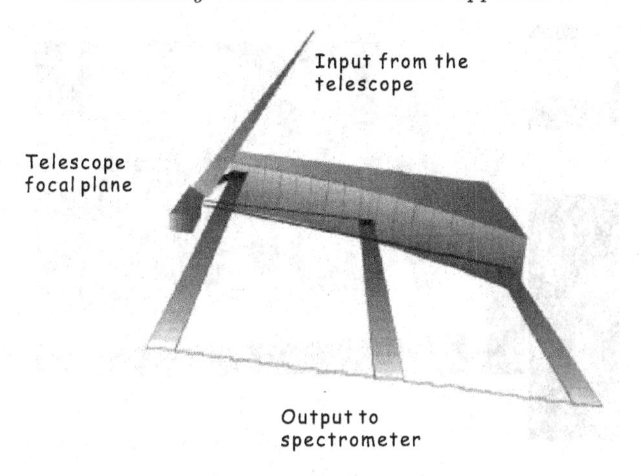

FIGURE 1.39. Slicer principle of operation (from Weitzel *et al.*, 1996).

1.4.4 *Slicers*

The IFU types discussed in Sections 1.4.1 to 1.4.3 have one feature in common: they do not use detector space very efficiently. This obviously has to do with the fact that re-imaging of fiber end faces, or micropupils, respectively, which incurs a finite image quality, and, in fact, a broadened image of those 'slit elements', makes it necessary to provide for a minimal spacing of spectra, unless the crosstalk from the broadened profiles is not regarded as critical. If this is indeed the case, one spectrum would typically cover four rows of spectra in the spatial direction, assuming that Nyquist sampling was given by design. Such a configuration consumes a factor of two more CCD pixels than required in the optimal case, where one spaxel is critically sampled, i.e. it uses just two rows of CCD pixels. If maximization of the number of spaxels is a primary design goal, then the slicer concept is significantly more efficient than the competing IFU types. Slicer IFUs are conceptually based on the optical elements designed to circumvent slit losses in conventional spectrographs; e.g. the Bowen–Walraven image slicer. The image slicer for 3D spectroscopy was first introduced for the MPE (Max Planck Institut für extraterrestrische Physik) instrument '3D' (Weitzel *et al.*, 1996), which was designed as a prototype and a travelling instrument; it was adapted and operated at a variety of 4 m-class telescopes. 3D operates in the NIR, which precludes the use of fibers because of stability issues in a cryogenic environment. Conversely, the slicer as a relatively small optical element is well adapted for operation in a cryostat at low temperatures and allows for an overall compact optomechanical design. The principle of operation of a slicer is shown in Figure 1.39. Light coming from the telescope is deflected by a stack of tilted mirrors (the 'slicer' proper), which are very narrow in one direction and quite extended in another direction. These thin mirrors ultimately act as slits, or, in other words, as spaxels, dissecting the focal plane into narrow and long strips. Another set of mirrors accepts the deflected beams corresponding to each slice and projects the outgoing beams such that an (almost) linear arrangement of the slices is created at the entrance of the spectrograph. The benefit of such a configuration is that the telescope focal plane is directly imaged onto the detector so that, with the choice of a proper scale, one detector pixel can be interpreted as one spaxel on the sky. Of all IFU types, this scheme yields the highest packing density on the detector. However, unlike with micropupil or fiber-based 3D spectrographs, slicer IFUs do not provide image scrambling and therefore do not share the benefits as discussed in Section 1.2.1. The first slicers were used for instruments operating in the NIR, e.g.

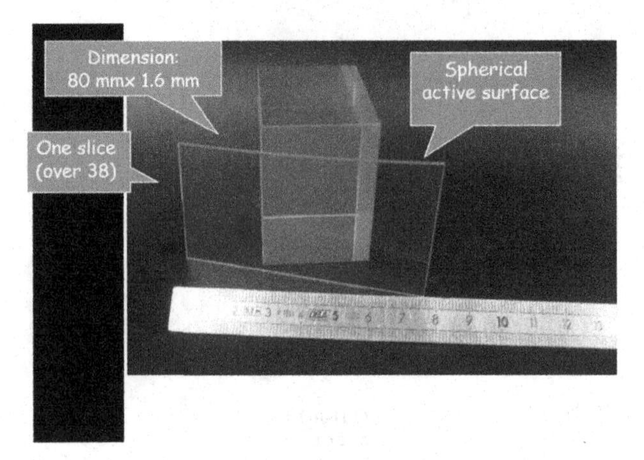

FIGURE 1.40. Discrete Zerodur slicer element (MUSE prototype).

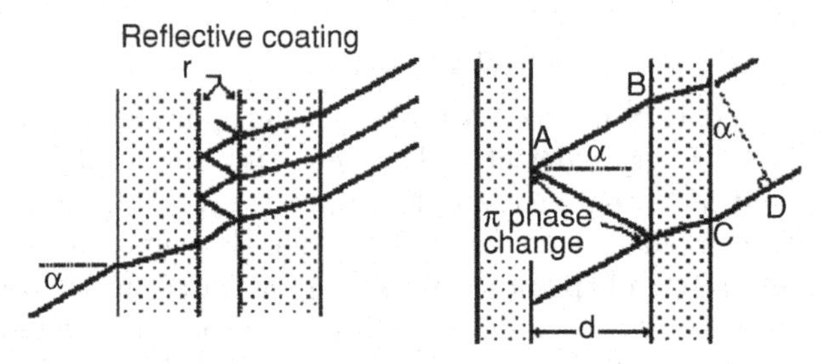

FIGURE 1.41. FP-principle.

3D, SINFONI. With recent advances in diamond-turning technologies, the manufacture of metal optics slicers showing surface roughness values $< 5\,\text{nm}$ has become possible; this means that metal slicers are also useable in the optical (e.g. MUSE, Bacon *et al.*, 2004).

In summary, slicers provide high density of spectra, high efficiency, and a compact optomechanical design for high stability. Slicers have been manufactured from Zerodur, and from metal as monolithic devices. Slicers are routinely operated in the NIR, and there are ongoing research and development efforts to optimize metallic slicers for operation in the visual. In contrast to fiber and lens array IFUs, no scrambling is involved, and the telescope focal plane is projected directly onto the detector.

1.4.5 *Non-IFS 3D instruments*

Without going into much detail, some techniques shall be mentioned that are not exactly 3D in accord with our definition; they do, however, produce data with the same characteristics ('datacubes') as integral field spectroscopy. These non-IFS 3D instruments are generally scanning instruments, i.e. the principal difference is that the measurement process is non-simultaneous.

Fabry–Perot

A very prominent technique, starting from the early days of IFS developments, is the Fabry–Perot interferometer (FP). The basic principle of interference in a FP is shown in the sketch of Figure 1.41: a cavity enclosed by a pair of high-quality plane-parallel glass plates with a reflective coating of reflectance r is acting as a resonator, where rays of

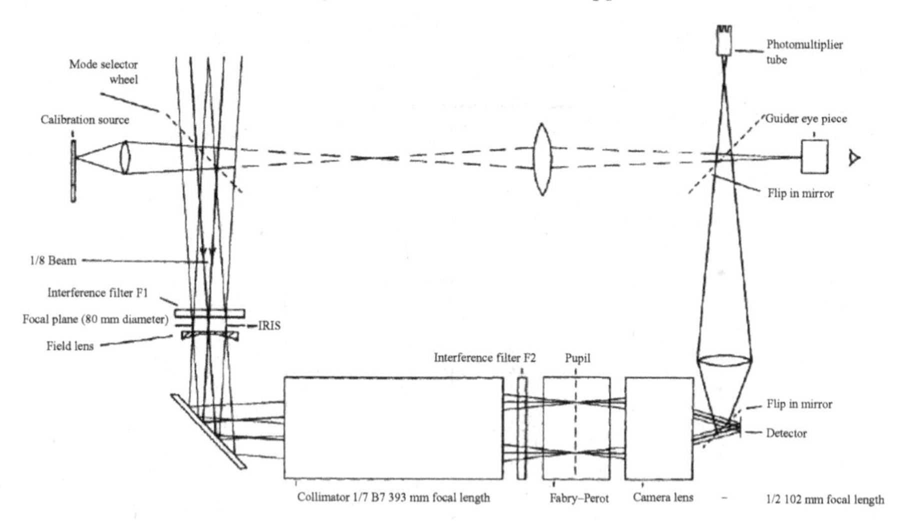

FIGURE 1.42. TAURUS.

equal angles of incidence experience the same phase superposition, and – depending on the optical path length difference – constructive or destructive interference. The optical path length between adjacent rays is given by: $2AB - CD = 2d\cos\alpha$. The condition for constructive interference is then:

$$2d\cos\alpha = m\lambda \tag{1.2}$$

For every wavelength λ satisfying this condition, the interferometer is transmissive, and other wavelengths are extinct. Because of the periodicity of the *cos*-function there are very many orders m with constructive interference when the spacing d is on the order of mm. By varying d, one can tune the interferometer to different wavelengths, and scanning over many values of d is effectively creating a spectrum. Note that in order to avoid ambiguity owing to the many orders m, it is necessary to use a narrowband filter to select the wavelength of interest. The width of the passband depends on the etalon reflectance r, and it can be shown that the spectral resolving power is given by:

$$\frac{\lambda}{\Delta\lambda} = m\pi\frac{\sqrt{r}}{(1-r)} \tag{1.3}$$

A prominent example for a scanning FP imager is the TAURUS instrument (Atherton *et al.*, 1981; 1982), the layout of which is shown in Figure 1.42. Some of the instrumental characteristics of TAURUS were as follows. Mode of operation: seeing-limited direct imaging, FoV = 9 arcmin, spectra resolution: $200 < \mathrm{R} < 100{,}000$, IPCS photon counting detector, wavelength range: 400–700 nm. General features of a scanning FP are: a large field-of-view, high spatial resolution, non-simultaneous measurements, small wavelength coverage (generally one spectral line), high spectral resolution.

Fourier transform spectrometer

A special variant of a scanning 3D instrument is the Fourier Transform Spectrometer (FTS), which is based on the principle of a Michelson interferometer. As opposed to the scanning FP, which produces direct images, the FTS creates an interference pattern that as such is incomprehensible to the human eye. It is only through a Fourier transform performed on a computer that an image is recovered. This instrument type is perhaps

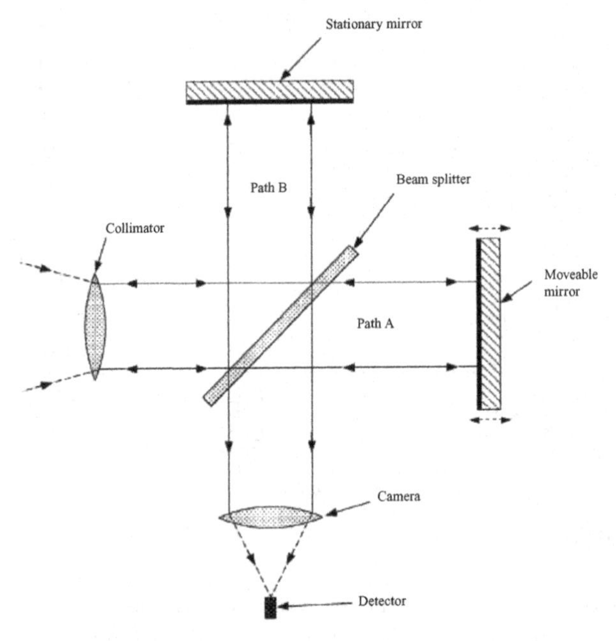

FIGURE 1.43. FTS.

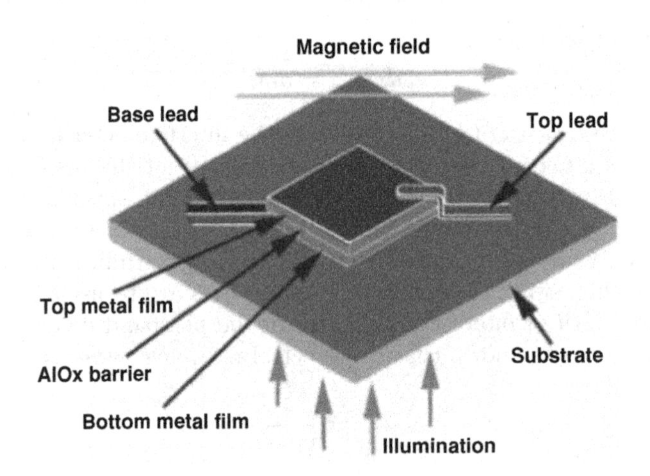

FIGURE 1.44. Superconductive tunnel junction detector.

less straightforward than other 3D spectrographs, but it has a certain number of advantages. These are quoted with several instrumental parameters of the CFHT FTS as an example (Bohlender, 1994): the CFHT FTS is a scanning NIR instrument (1–5 μm), having a high throughput of up to 25%, independent of wavelength. High S/N observations are supported based on a dual channel beam switching mode, which allows for an accurate internal sky subtraction before any signal is detected. The instrumental profile is extremely well controlled (and measurable). The wavelength calibration accuracy is $\approx 100\,\mathrm{m\,s^{-1}}$, and the spectral resolution up to $500\,\mathrm{m\,s^{-1}}$.

1.4.6 *3D detectors*

The dream of observational astronomers would be a photon counting detector that samples all parameters of an incident photon simultaneously; i.e. wavelength, phase and amplitude, polarization and direction. 3D spectrographs as described above are a humble yet very powerful approximation towards this goal. However, a compact detector having these properties would seem to be more appealing intrinsically than the rather complex layout of the various IFU types. Such detectors do exist, and a prominent prototype development is the Superconduction Tunnel Junction (STJ). This device consists of a pair of films that are separated by an insulating layer. The films are made of superconducting materials like niobium, tantalum, hafnium, etc. The detector is operated at a temperature below $T_{crit} \approx 1$K. Owing to the Josephson effect, there is a current through the junction in the presence of a bias voltage and a magnetic field. The current is suppressed by incident light, where an intrinsic energy resolution is provided due to the release of a charge from the incident photon. The typical resolving power that can be achieved, however, is $R = 10-100$, i.e. very low. This limitation is given by the band-gap values attained by selected materials, for which research and development efforts are attempting to provide further improvement. For fundamental reasons, however, spectroscopically interesting resolutions one order of magnitude larger than the present values are not going to become real. Therefore, the STJ will remain a niche application that cannot replace IFUs. Nevertheless, 3D detectors with spatial resolution have been demonstrated as arrays of 10×10 Josephson cells, forming a real 3D detector. First astronomical observation with the STJ were presented by Perryman *et al.* (1999). Their measurements of the Crab pulsar validated that the STJ has advantages for high time resolution multiband imaging, especially in the UV. Other developments include so-called Kinetic Inductance Detectors (KID), which are being developed as large devices with 100×100 elements and $R = 30$, or Transition Edge Detectors (TES) with 8×8, $R = 20$.

In conclusion, intrinsic energy resolution detectors are still at an early stage of development. Their cryogenics to operate at superconduction temperatures around 1 K are complex and expensive. The devices can only be made to achieve low spectral resolution. Although they are sensitive in the blue, they are less suitable for the visual and useless in the NIR. A positive feature is their high time resolution.

1.4.7 *Special techniques*

Multi-slit masks

A rather simple special slit configuration in front of an imaging spectrograph is capable of providing pseudo integral field spectroscopy; pseudo in the sense of very low fill factor. The basic idea is to make a slit mask that has a large number of parallel long slits, each of which is allowed to generate a spectrum over a limited wavelength interval, which is determined by the distance to its next neighbors in order to avoid overlap of adjacent spectra. The wavelength interval can be selected by means of a narrowband filter anywhere in the spectrograph. The example shown in Figure 1.45 was implemented in the FOCAS (Faint Object Camera and Spectrograph) instrument at the SUBARU 8.2 m telescope for the purpose of detecting intracluster planetary nebulae in the Coma Cluster (Gerhard *et al.*, 2005). The chosen slit configuration yielded 3600 spaxels per field diameter at a spatial sampling of 0.1 arcsec pixel^{-1}. The dispersion using the 300B grating was 1.5 Å pixel^{-1}, resulting in 43 spectral bins per spectrum with a pre-filter of 60 Å FWHM, centered at 5121 Å (redshifted [O III] 5007). The total effective survey area was 12,215 arcsec^2, which is 12% of the FOCAS FoV. The remarkable result of a survey in Coma with a total exposure time of no longer than 3 h and a seeing of 0.6–0.8 arcsec FWHM has been that planetary nebula candidates with a limiting flux of 3×10^{-18} erg s^{-1} cm^{-2} were detected. Although the result is impressive in that next to supernovae this discovery is the most distant measurement of individual stars (100 Mpc), a fair comparison with real

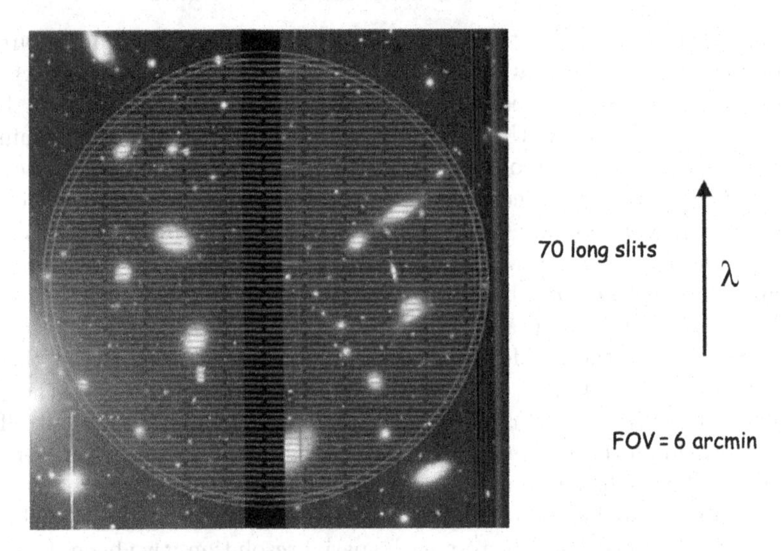

FIGURE 1.45. Multi-Slit-Mask for FOCAS (from Gerhard *et al.*, 2005).

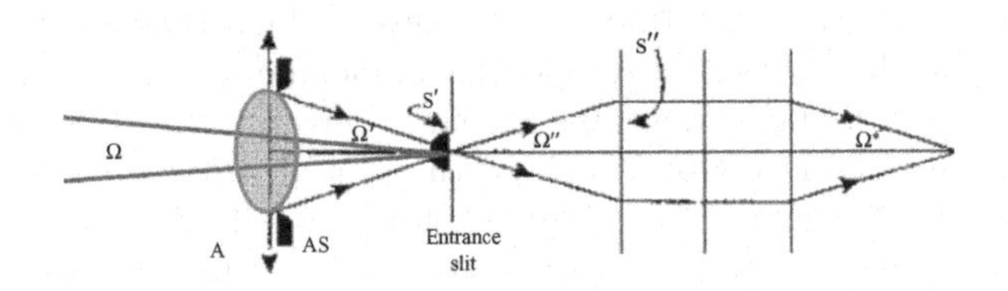

FIGURE 1.46. Definition of Etendue of a Spectrograph.

IFUs shows that the uncertain location of point sources over the slit perpendicular to the slit length obviously introduces large flux uncertainties, making this technique less competitive compared to, for example, the MUSE instrument.

Nod-shuffle 3D spectroscopy

The technique of 'nod-shuffle' spectroscopy is not necessarily restricted to 3D; however, it is a very powerful option for background-limited observations of extended objects, where no patch of empty sky can be found within the FoV of the IFU. As described by Cuillandre *et al.* (1994), consecutive beam switching between object and sky pointings while shuffling charge on the CCD detector between an object and a sky integration zone, and final read-out after the total exposure time has been completed, is capable of eliminating systematic sky subtraction errors in the sense of chopping technique, which are common place in many areas of measuring techniques. A comprehensive discussion is presented by Glazebrook and Bland-Hawthorn (2001), and experience from an implementation in the PMAS instrument was reported by Roth *et al.* (2004a).

1.4.8 *Figure-of-merit*

In order to be in a position to objectively compare competing instrument designs, it is useful to introduce a quantitative measure of performance. The classical measure to this end is 'etendue', which is defined as:

$$\text{Etendue} = \text{area} \times \text{solid angle} \times \text{system throughput} \tag{1.4}$$

Another measure, which takes out the system throughput and thus gauges the performance of an IFU irrespective of the attached spectrograph is 'grasp':

$$grasp = area \times solid\ angle \quad (1.5)$$

Examples for grasp of a number of 3D instruments are presented by Bershady *et al.* (2004). Another figure-of-merit, called 'specific information density' was proposed by Allington-Smith (2007).

While these numbers are useful for planning the optimal design of an IFU for a given science case, observers may also want to consider other parameters, which may be even more relevant than a single figure-of-merit; for example, spatial resolution, spectral resolution, wavelength regime and wavelength coverage.

1.5 Acknowledgments

The Euro3D Research Training Network was funded by the European Commission in Framework Programme 5 under contract number CT-2002-00305. The initiative of the IAC, La Laguna, Tenerife, to organize the XVIIth IAC Winter School on the topic of '3D Spectroscopy' as a task of the Euro3D Work Plan, to provide funding, and to arrange a marvelous school at the beautiful location of Puerto de la Cruz is gratefully acknowledged.

REFERENCES

Afanasiev, V.L., Vlasiuk, V.V., Dodonov, S.N., Silchenko, O.K. (1990), Preprint No.54, Special Astrophysical Observatory, Selentchuk

Allington-Smith, J.R. (2007), in *Science Perspectives for 3D Spectroscopy (ESO Astrophysics Symposia)*, eds. Kissler-Patig, M., Wals, J.R., and Roth, M.M. Springer

Allington-Smith, J., Murray, G., Content, R. et al. (2002), *PASP*, **114**, 892

Arribas, S., Mediavilla, E., Rasilla, J.L. (1991), *ApJ*, **369**, 260

Arribas, S., Mediavilla, E., García-Lorenzo, B. et al. (1995), *A&AS*, **136**, 189

Arribas, S., Mediavilla, E., García-Lorenzo, B., del Burgo, C., Fuensalida, J.J. (1998a), *A&AS*, **136**, 189

Arribas, S., Cavaller, L., García-Lorenzo, B., García-Marín, A., Herreros, J.M., Mediavilla, E., Pi, M., del Burgo, C., Fuentes, J., Rasilla, J.L., Sosa, N. (1998b), in *Fiber Optics in Astronomy III*, eds. S. Arribas, E. Mediavilla, F. Watson, *ASPCS*, **152**, p. 149

Arribas, S. et al. (2005), *Bulletin of the American Astronomical Society*, **37**, 1352

Atherton, P.D., Reay, N.K., Ring, J.R., Hicks, T.R. (1981), *Opt En*, **20 (6)**, 806

Atherton, P.D., Taylor, K., Pike, C.D., Harmer, C.F.W., Parker, R.N. (1982), *MNRAS*, **201**, 661

Bacon, R., Emsellem, E., Monnet, G., Nieto, J.L. (1994), *A&A*, **281**, 691

Bacon, R., Adam, G., Baranne, A., Courtès, G., Dubet, D., Dubois, J.P., Emsellem, E., Ferruit, P., Georgelin, Y., Monnet, G., Pecontal, E., Rousset, A., Say, F. (1995), *A&AS*, **113**, 347

Bacon, R., Copin, Y., Monnet, G. et al. (2001), *MNRAS*, **326**, 23

Bacon, R., Adam, G., Cabrit, S. et al. (2002), in *Scientific Drivers for ESO Future VLT/VLTI Instrumentation*, eds. J. Bergeron, G. Monnet, Springer, Berlin, p. 108

Bacon, R., Bauer, S.-M., Bower, R. et al. (2004), *SPIE*, **5492**, 1145

Balick, B., Preston, H.L., Icke, V. (1987), *AJ*, **94**, 1641

Barden, S., Scott, K. (1986), *BAAS*, **18**, 951

Barden, S.C., Wade, R.A. (1988), *ASPCS 3: Fiber Optics in Astronomy*, 113

Barwig, H., Schoembs, R., Buckenmayer, C. (1987), *A&A*, **175**, 327

Baudrand, J., Walker, G.A.H. (2001), *PASP*, **113**, 851

Becker, T., Fabrika, S., Roth, M.M. (2004), *AN*, **325**, 155

Bershady, M.A., Andersen, D.R., Harker, J., Ramsey, L.W., Verheijen, M.A.W. (2004), *PASP*, **116**, 565

BINGHAM, R.G., GELLATLY, D.W., JENKINS, C. ET AL. (1994), *SPIE*, **2198**, p. 56
BLAND–HAWTHORN, J., TAYLOR, K. (1994), *AAO Newsletter*, **71**, 8
BOHLENDER (1994), *CFHT FTS Manual*
BOWER, R.G., MORRIS, S.L., BACON, R. ET AL. (2004), *MNRAS*, **351**, 63
BUZZONI, B., DELABRE, B., DEKKER, H. ET AL. (1984), *Msngr*, **38**, 9
CAPPELLARI, M., COPIN, Y. (2003), *MNRAS*, **342**, 345
COMPTE, G., MARCELIN, M. (1995), *Tridimensional Optical Methods in Astrophysics*, N°149. IAU Colloquium, 1994
COURTÈS, G. (1960), *AnAp*, **23**, 96
COURTÈS, G., GEORGELIN, Y., BACON, R., MONNET, G., BOULESTEIX, J. (1988), in *Instrumentation for Ground-Based Optical Astronomy*, Proc. 6. Santa Cruz Summer Workshop, ed. L.B. Robinson, Springer, New York, p. 266
CUILLANDRE, J.C., FORT, B., PICAT, J.P. ET AL. (1994), *A&A*, **281**, 603
DEAN, A. (2002), Thesis, IoA University of Cambridge
DE ZEEUW, P.T., BUREAU, M., EMSELLEM, E., BACON, R., CAROLLO, C.M., COPIN, Y., DAVIES, R.L., KUNTSCHNER, H., MILLER, B.W., MONNET, G., PELETIER, R.F., VEROLME, E.K. (2002), *MNRAS*, **329**, 513
EISENHAUER, F., ABUTER, R., BICKERT, K. ET AL. (2003), *SPIE*, **4841**, 1548
EMSELLEM, E., BACON, R., MONNET, G., POULAIN, P. ET AL. (1996), *A&A*, **312**, 777
FABRIKA, S., SHOLUKHOVA, O., BECKER, T., AFANASIEV, V., ROTH, M., SÁNCHEZ, S.F. (2005), *A&A*, **437**, 217
FILIPPENKO, A.V. (1982), *PASP*, **94**, 715
FRUCHTER, A.S., HOOK, R.N. (2002), *PASP*, **114**, 144
GERHARD, O., ARNABOLDI, M., FREEMAN, K. ET AL. (2005), *ApJ*, **621**, L93
GLAZEBROOK, K., BLAND-HAWTHORN, J. (2001), *PASP*, **113**, 197
HILL, J.M. (1988), *ASPCS 3: Fiber Optics in Astronomy*, **3**, 77
HILL, J.M., ANGEL, J.R.P., SCOTT, J.S., LINDLEY, D., HINTZEN, P. (1980), *ApJ*, **242**, L69
HOOK, R.N., LUCY, L.B. (1993), in *Proc. Science with the HST*, p. 245
JACOBY, G.H., KALER, J.B. (1993), *ApJ*, **417**, 209
JERRAM, P., POOL, P.J., BELL, R. ET AL. (2001), *SPIE*, **4306**, 178
KELZ, A., VERHEIJAN, M.A.W., ROTH, M.M. ET AL. (2006), *PASP*, **118**, 129
KENWORTHY, M.A., PARRY, I.R., TAYLOR, K. (2001), *PASP*, **113**, 215
KISSLER-PATIG, M., COPIN, Y., FERRUIT, P., PECONTAL-ROUSSET, A., ROTH, M.M. (2004), *AN*, **325**, 159
LARKIN, J., BARCZYS, M., KRABBE, A. ET AL. (2006), *New Astronomy Review*, **50**, 362
LEE, D., HAYNES, R., REN, D., ALLINGTON-SMITH, J. (2001), *PASP*, **113**, 1406
LE FEVRE, O., SAISSE, M., MANCINI, D. ET AL. (2003), *SPIE*, **4841**, p. 1670
MEDIAVILLA, E, ARRIBAS, S. (1993), *Natur*, **365**, 420
MEISENHEIMER, K., HIPPELEIN, H. (1992), *A&A*, **264**s, 455
MICHLOVIC, J. (1972), *App Opt*, **11**, 490
MURPHY, T.W., JR., MATTHEWS, K., SOIFER, B.T. (1999), *PASP*, **111**, 1176
OHTANI, H., SASAKI, M., AOKI, K., TAKANO, E., KIYOHARA, M. (1994), *SPIE*, **2198**, 229
PARRY, I.R., CARRASCO, E. (1990), *SPIE*, **1235**, 702
PARRY, I.R., BUNKER, A., DEAN, A. ET AL. (2004), *SPIE*, **5492**, 1135
PERRYMAN, M.A.C., FAVATA, F., PEACOCK, A. ET AL. (1999), *A&A*, **346**, L30
RAMSAY HOWATT, S.K., TODD, S., LEGGETT, S. ET AL. (2004), *SPIE*, **5492**, p. 1160
ROTH, M.M., BECKER, T., SCHMOLL, J. (2000), *ASPCS 195: Imaging the Universe in Three Dimensions*, 122
ROTH, M.M., FECHNER, T., WOLTER, D., KELZ, A., BECKER, T. (2004a), in *Scientific Detectors for Astronomy, The Beginning of a New Era*, eds. Amico, P.; Beletic, J.W.; Beletic, J.E., 371
ROTH, M.M., BECKER, T., KELZ, A., SCHMOLL, J. (2004b), *ApJ*, **603**, 531
ROTH, M.M., KELZ, A., FECHNER, T., HAHN, T., BAUER, S., BECKER, T., BÖHM, P., CHRISTENSEN, L., DIONIES, F., PASCHKE, J., POPOW, E., WOLTER, D. (2005), *PASP*, **117**, 620
SÁNCHEZ, S.F. (2004), *AN*, **325 (2)**, 167
SÁNCHEZ, S.F., BECKER, T., KELZ, A. (2004a), *AN*, **325 (2)**, 171
SÁNCHEZ, S.F., GARCÍA-LORENZO, B., MEDIAVILLA, E. ET AL. (2004b), *ApJ*, **615**, 156
SCHMOLL, J., ROTH, M.M., LAUX, U. (2003), *PASP*, **115**, 854
SI'LCHENKO, O., AFANASIEV, V. (2000), *A&A*, **364**, 479

TULLY, R.B. (1974a), *ApJS*, **27**, 415
TULLY, R.B. (1974b), *ApJS*, **27**, 437
TULLY, R.B. (1974c), *ApJS*, **27**, 449
VANDERRIEST, C. (1980), *PASP*, **92**, 858
VANDERRIEST, C. (1995), in *Tridimensional Optical Spectroscopic Methods in Astrophysics*, IAU Coll. 149, *ASPCS* **71**, p. 209
VANDERRIEST, C., LEMONNIER, J.-P. (1988), in *Instrumentation for Ground-Based Optical Astronomy*, Proc. 6. Santa Cruz Summer Workshop, ed. L.B. Robinson, Springer, New York, 304
WALSH, J.R., ROTH, M.M. (2002), *Msngr*, **109**, 54
WEITZEL, L., KRABBE, A., KROKER, H., THATTE, N., TACCONI-GARMAN, L.E., CAMERON, M., GENZEL, R. (1996), *A&AS*, **119**, 531
WELLS, D.C., GREISEN, E.W., HARTEN, R.H. (1981), *A&AS*, **44**, 363
WILKINSON, A. SHARPLESS, R.M., FOSBURY, A.E.M., WALLACE, P.T. (1986), *MNRAS*, **218**, 297
WILLIAMS, R.E., BLACKER, B., DICKINSON, M. ET AL. (1996), *AJ*, **112**, 1335

2. Observational procedures and data reduction

JAMES E.H. TURNER

2.1 Introduction

In this chapter, I give an introduction to observing with integral field units and performing basic reduction of the resulting data, prior to scientific analysis. After briefly considering the context of the lectures, I begin by discussing strategies for observing. This is followed by a short tutorial on sampling theory and its application to integral field unit (IFU) data, before continuing with an overview of the requirements for each stage of data reduction. I finish by considering the data reduction process as a whole, along with associated issues such as error propagation and file formats.

2.2 Background

Techniques for integral field spectroscopy (IFS) have been in development for at least two decades (Vanderriest, 1980).[1] During the 1980s–1990s, numerous prototype IFUs and even a few public instruments were deployed at observatories and used for scientific work. Nevertheless, IFS has only become widely available at major telescopes during the past five years or so, following two centuries of slit spectroscopy. Experience in observing with IFUs and processing the data is just starting to become commonplace within the community, but will be spread more widely by the current generation of postdoctoral and student astronomers.

In terms of data reduction and analysis, IFS poses some non-trivial new requirements. The most obvious factor is the introduction of 3D datasets to mainstream optical and near-infrared (NIR) (as opposed to radio) astronomy. Although older scanning methods such as Fabry–Perot interferometry produce higher-dimensional datasets, these techniques are relatively specialized by comparison. IFS is an extension of traditional spectroscopy, covering a range of spatial and spectral sampling regimes, increasingly so as larger-format units become available. This creates a need for non-specialists to apply spectroscopic and image processing to 3D datasets. Perhaps a more important issue than dimensionality, however, is that IFUs scramble spatial information in the process of mapping three dimensions onto a 2D detector. To add to this, the spatial pixels of an IFU rarely form a square grid to begin with. Special software is therefore needed to recover spatial or 3D images. Depending on the processing strategy, this may affect multiple data reduction stages. Finally, a third consideration in dealing with IFU data is homogeneity, given variations in transmission, spectral resolution or background counts across the field. This leads to additional flat-fielding requirements, for example, compared with a simple aperture.

Although instruments with IFUs have become more common in the past few years, the availability of matching general-purpose software has lagged behind somewhat. A key task of the Euro3D Research Training Network (2002–05) has been to create data visualization and analysis tools to help address this issue, as well as a standard file format for working with reduced IFU data. Other recent progress includes adaptation of image-viewing programs to work with 3D data. The proliferation of high-level environments such

[1] Ignoring the even earlier use of image slicers to avoid slit losses (Bowen, 1938).

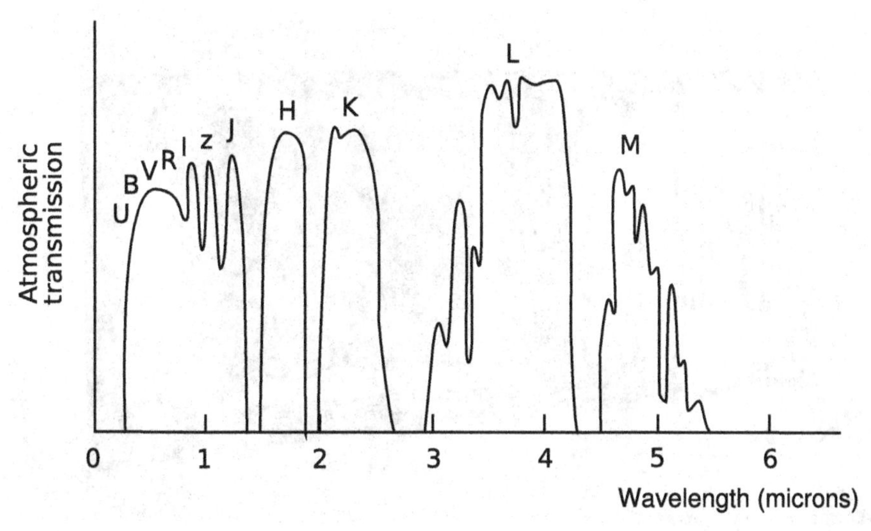

FIGURE 2.1. Sketch of the relative atmospheric transmission as a function of wavelength in the optical and near infrared. Some common photometric bandpasses are labelled with the appropriate letters. (Based on a figure from a US Naval publication, reproduced at http://wikipedia.org.)

as Python and interactive data language (IDL), with multi-dimensional array handling capabilities, also provides some flexibility for manipulating 3D datasets. There is still progress to be made, however, before the same range of tools exist for working with IFU data as for slit spectroscopy.

2.3 Observing strategies

2.3.1 *IFUs versus other instruments*

Current IFUs work at optical wavelengths (~0.4–1.0 μm, with CCD, charge-coupled device, detectors) or in the near-infrared (~1.0–2.5 μm with HgCdTe or ~1–5 μm with InSb arrays). Future mid- and far-infrared IFUs are also planned, respectively, for the James Webb Space Telescope (*JWST*) (Wright *et al.*, 2004) and Sofia (Raab *et al.*, 2004). Figure 2.1 is a rough sketch of the atmospheric transmission over the range of wavelengths covered by existing IFUs, showing the relation to common photometric bandpasses.

The details of how an IFU is used depend on its integration with the host instrument, as well as on wavelength and the type of IFU. Some integral field spectrographs are implemented as stand-alone instruments (e.g. Bacon *et al.*, 2001), whereas others are general-purpose spectrographs with an IFU module that can be inserted in place of an ordinary slit mask (e.g. Allington-Smith *et al.*, 2007; Figure 2.2 herein). Sometimes an IFU can be retro-fitted to feed an existing long slit, typically using optical fibres. A hybrid of multi-object and integral-field spectrograph designs involves individually deploying multiple small IFU fields within the telescope field; several such instruments have been proposed and one is already in routine operation (Pasquini *et al.*, 2002). In addition, some 'single-object' optical spectrographs include a second IFU field or separate sky fibres for sampling the background at the same time as a target object (e.g. Allington-Smith *et al.*, 2002).

Different types of IFUs have a few common characteristics compared with other instruments. In particular, there is a trade-off between the numbers of spatial and spectral pixels that will fit on the detector (these may also be limited by optical size constraints).

FIGURE 2.2. The image slicing IFU of the GNIRS spectrograph at Gemini South (right), mounted alongside various slit masks in a motorized sliding mechanism that inserts the appropriate mask into the beam.

More often than not, this translates into small field sizes (arcseconds) compared with typical long slits or imagers (arcminutes or even degrees). Some IFUs, on the other hand, are designed with coarse spatial pixels and/or short spectra, giving a larger field (~tens of arcseconds) and potentially much better sensitivity for low surface brightness objects, instead of resolving fine spatial detail (e.g. Bershady *et al.*, 2004). Sometimes it is possible to choose between multiple spatial scales.

A second common characteristic of IFUs is that they introduce non-uniformities over the field, as mentioned in Section 2.2, especially with optical fibres. Together with a small field size or coarse sampling this makes dithering and mosaicking important strategies for observing with IFUs, both to average over variations and to cover a larger area or improve the sampling (see Section 2.3.2 *Dithering and mosaicking*, and Section 2.4). Moreover, the IFU field response has to be 'flat fielded', in addition to that of the detector, unlike the step response of a slit.

Finally, it is worth noting that high-spatial-resolution IFUs are a natural companion for adaptive optics systems, both scientifically (many spectroscopic projects using adaptive optics benefit from 2D spatial coverage) and technically (the difficulty of using a slit at such small spatial tolerances). This is another source of special requirements for observing and data reduction, albeit outside the scope of the current chapter.

2.3.2 *The observing process*

Here and in Section 2.5, I shall concentrate on the case of a single target field, observed at optical or near-infrared wavelengths, since this applies to nearly all current integral field spectrographs. Many of the points discussed are generic issues in spectroscopy, considered here in the context of IFS for completeness. Precise details vary significantly between instruments, with factors such as wavelength, flexure, repeatability and the available modes or configurations; thus much of the following text is general discussion relating to each observing or data reduction stage, but Section 2.3.2 (*Overview*) and Section 2.6 tie the steps together with examples of specific sequences.

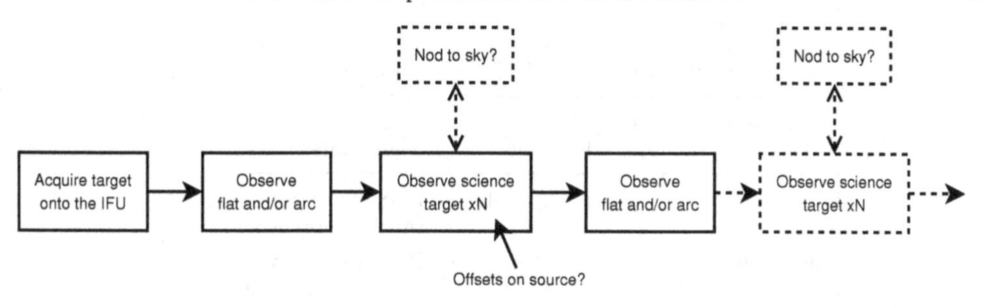

FIGURE 2.3. Sketch of a typical observing sequence at optical wavelengths.

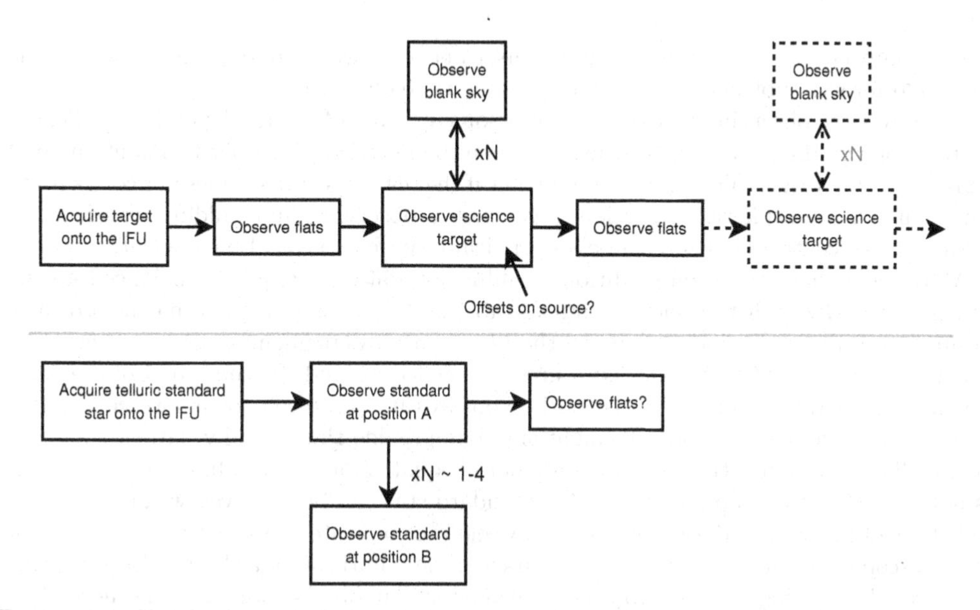

FIGURE 2.4. Sketch of a typical observing sequence at near-infrared wavelengths.

Overview

Figure 2.3 is a sketch of a typical on-sky observing procedure at optical wavelengths. After centring the target on the IFU, some number of exposures of its spectrum are taken, often interspersed with flat and/or arc lamp calibration spectra (where flexure is important). If the target is extended enough to cover the whole field, it may be necessary to observe it alternately with blank sky; however, this is usually avoidable with separate fibres or lenses at some distance from the field centre, sampling the background contemporaneously with the object. Smaller spatial offsets between target exposures are sometimes used for mosaicking or averaging over flat-field variations.

Figure 2.4 is a sketch of a typical on-sky observing procedure in the near-infrared. Unlike in the optical case, it is usual to alternate exposures on the science target with blank sky spectra (perhaps in an ABBA sequence, to reduce offsetting overheads) in order to sample the object and background with the same optical path and detector pixels. Given the abundance of telluric emission lines in the near infrared, there is typically less need to take arc lamp spectra for wavelength calibration, but frequent flat fields may still be required in the presence of instrumental flexure. Usually a separate standard star observation is taken – close to the science target in both airmass and time – to calibrate the telluric absorption features found at longer (red to infrared) wavelengths.

TABLE 2.1. Typical 'off-line' calibration data

Day time or twilight (with a dark dome)	Twilight	Night time
Lamp flats Arcs	Twilight flats	Flux standard (mainly optical)
Biases (optical) Darks (mainly IR)		Radial velocity or line index standards
Special calibrations		

Being compact, the star can normally be offset between two or more positions within the IFU field for sky subtraction, without spending time off source.

Whatever the observing wavelength, after some number of hours (depending on flexure or telescope guiding), a re-acquisition step may be necessary in order to maintain good centring on the target. This is more important if the telescope guider does not compensate for changing atmospheric dispersion between the observing and guiding wavelengths. Nevertheless, centring is likely to be less critical than for a typical narrow slit.

More often than not, some additional calibration data are required that need not be taken on the sky with the science target (Table 2.1). This may include flat-field or arc-lamp frames if flexure does not make the illumination dependent on time or telescope pointing. For optical CCDs, bias exposures are needed, to find the zero point of the data. For infrared arrays or older CCDs, dark exposures may be used to remove counts due to thermally excited electrons. Twilight sky flats provide the spatially flattest source of strong illumination if artificial sources are not adequate alone. If absolute flux calibration is not needed, a spectrophotometric flux standard star can be observed within a reasonable time of the science data (days or even weeks) to calibrate the sensitivity spectrum of the telescope+instrument. (In the near infrared, the telluric standard may also serve this purpose.) Depending on the project, additional stellar observations may be needed as spectral templates for velocity or line index measurements. Either kind of standards may be taken in poor observing conditions if absolute fluxes are not required. Finally, special calibrations are required for some instruments, for example to calibrate the alignment of the IFU with the detector.

Acquisition

Blind telescope pointing is rarely accurate enough to centre a target perfectly within a small IFU field, so acquisition often involves iterative centring at the required position – usually the middle of the IFU – as for a slit. At each step, an exposure is taken, the target position is measured and the telescope (or in some cases the IFU) is offset slightly to adjust the pointing.

Measurement of the target position can be done either with a normal imaging camera that is fixed with respect to the IFU (perhaps a different mode of the same instrument) or by reconstructing a 2D image of what the IFU itself is looking at (Figure 2.5). In the former case, the target is offset until it reaches pixel coordinates known to correspond to the IFU centre, based on prior measurements. This requires that the position of the IFU on the camera is repeatable, without too much flexure or other sources of significant error. In the latter case, the best sensitivity is achieved by taking images undispersed through the IFU (or, if applicable, using the wavelength of an emission line). The 1D slit or micropupil spots are then rearranged to form a 2D image. If saturation is a concern

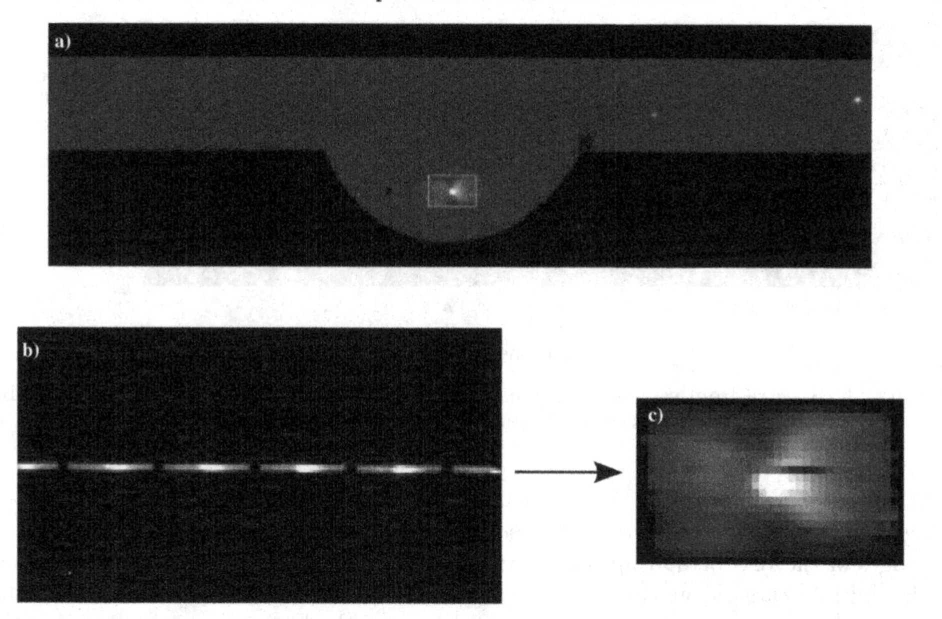

FIGURE 2.5. a) Acquiring a target object into a box marking the IFU position in the GNIRS direct
imaging field; b) section of an image of the same target, taken undispersed through the IFU; c) a reconstruction of the IFU field, made by stacking the image slices from the exposure in the previous panel.

or the spectrograph does not allow taking undispersed images, spectra can be summed over wavelength before reconstructing a 2D image, or a 3D datacube can be constructed first and then collapsed in wavelength (if not too slow). Finally, if flexure is of concern and initial pointing is poorer than a few arcseconds – or the target is faint or difficult to identify – the two approaches can be combined by using a direct camera image to place the target within the IFU field first and then fine-tuning the pointing by taking exposures through the IFU.

When combining exposures from an instrument with large spatial pixels it may help to have repeat acquisitions (e.g. from different nights) accurately aligned; if spatial information is missing due to the coarse pixel size, co-alignment of data during later processing is likely to be problematic (Section 2.4). Differences in atmospheric dispersion between exposures, however, tend to limit how well alignment can be maintained over a range of wavelengths, so some error is often unavoidable. For spatially well-sampled IFU data, on the other hand, one can correct for both pointing offsets and relative changes in atmospheric dispersion afterwards; thus, in many cases periodic re-acquisition is less important than for a slit, where miscentring on the aperture is difficult to correct afterwards and can lead to loss of signal.

Object and sky spectra

Once the target is at the right place in the IFU field, the observer can keep taking spectra until the required signal-to-noise level is reached (or some practicality interrupts the observation). As well as integrating on the target itself, one normally has to observe blank sky as a reference for subtracting out telluric emission lines and other sources of background counts. Given that IFU dimensions are typically smaller than slit lengths, the probability of having to nod completely off the target to obtain blank sky is greater.

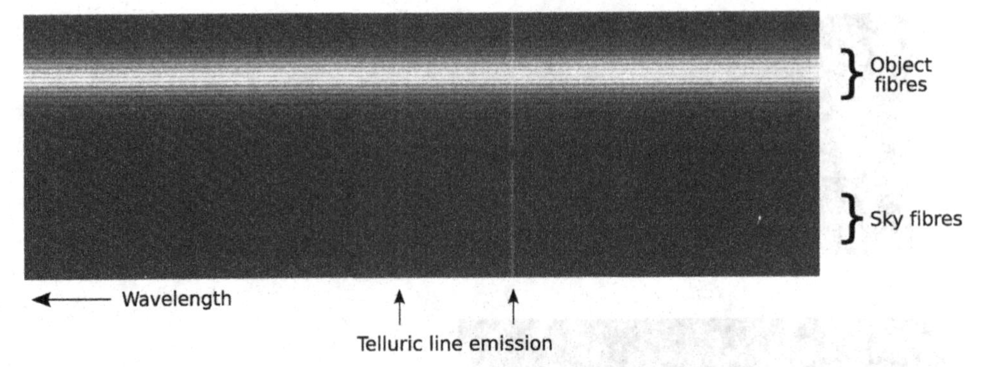

FIGURE 2.6. Spectra from separate, interleaved blocks of object and sky fibres in the GMOS IFU (this detector subsection overlaps 2 of a total of 30 blocks of fibres).

At optical wavelengths, the sky can usually be sampled at the same time as the object, using one of several methods. If the target is compact enough, one can use blank sky from the edges of the IFU or nod up and down the field in order to subtract pairs of frames. With multiple targets one might dither around the field and later co-add exposures with rejection to remove the objects. Often, however, the instrument is designed with a separate background field or sky fibres to take care of the problem. These are placed far enough away from the main science field (e.g. a few arcminutes away) to fall on blank sky in most cases, or at least to contain negligible signal from the outer regions of an extended target, compared with the science field. If the target is particularly extended, however (or maybe the sky field is contaminated by an object), it is necessary to nod off the science target completely from time to time and expose on blank sky elsewhere. Figure 2.6 shows an example of separate object and sky spectra from a fibre IFU.

In the near infrared, standard practice is to nod to sky, usually every other exposure, to allow removal of telluric and thermal emission and detector dark current by subtracting pairs of raw exposures. Since sky lines are stronger than in the optical and vary on timescales as short as a few minutes, it is usual to nod to sky more frequently for optimal subtraction. For compact targets such as standard stars, we can again nod within the IFU field to subtract pairs of exposures without spending time off source, but more often than not half of the observing time is spent on blank sky.

At non-thermal wavelengths (~1 μm) and high enough spectral resolution, it may be sufficient for some projects to spend all of the observing time on source and later mask out or interpolate over sky lines, after subtracting separate dark frames instead of sky.

Integration times

IFS exposure times are mainly determined by the same factors as for other spectroscopic modes. For faint targets, the minimum exposure is, as always, limited by the need to integrate long enough for the background noise to overcome detector read-out noise. With IFUs it often takes longer to reach the same number of counts per pixel as for a slit. One reason is that high spatial or spectral resolution IFUs have smaller apertures than a typical slit, being able to sample finer detail without losing light overall (but, of course, spreading the light over more apertures ultimately leads to higher read noise). IFUs also place extra optics in the beam, with a corresponding loss of efficiency (although throughputs of up to ~90% have been achieved, compared with a slit in the same spectrograph). A third factor is that IFU optics sometimes introduce extra magnification, which again means spreading the light over more pixels. On the other hand, IFUs with large spatial elements – designed for high sensitivity rather than fine sampling – can reach the read-noise limit very much more quickly. For very bright objects, the minimum exposure time is sometimes set by instrumental factors such as detector read-out speed.

Maximum exposure times are affected by several factors. In the infrared we have to start a new exposure often enough to allow for variations in sky lines (on multiple characteristic timescales) for accurate subtraction. For bright targets, one can take short exposures of a few seconds to a few minutes, to sample the fast telluric variations. For fainter targets, longer exposures of 10–30 minutes are used, to average over short-timescale variations without going so long as to be affected by slower changes. Exposures much longer than these usually lead to poor sky subtraction, although the stability varies from night to night. When working between sky lines at higher resolutions, the subtraction accuracy may be less important. Another common time limitation for bright objects (or perhaps bright sky lines) is the need to avoid reaching non-linear count levels or saturating the detector capacity altogether with too many photo-electrons. Finally, there are various other practical reasons for dividing up observations, such as allowing for changes in flexure between the slit and detector or the telescope image and IFU, the use of repeated samples to help remove cosmic rays, spatial dithering on the source or simply avoiding excessive time loss in the event that something goes wrong with an integration.

Given the above considerations, typical spectroscopic exposure lengths for scientific targets range from a few minutes up to 60 minutes (in the optical) or ~ 20 minutes (in the near infrared).

Dithering and mosaicking

There are, broadly, three scenarios for offsetting the position of an IFU with respect to the target between exposures; I shall refer to them here as dithering, mosaicking and sub-pixel dithering.[2] These are usually carried out by moving the telescope (or tilting its optics) slightly, but occasionally it is possible to move the IFU itself within the telescope field. In any of the three scenarios, an accurate way of registering the relative shifts is needed in order to combine data correctly afterwards. If the system is known to apply (or report) offsets precisely enough, the nominal shifts can be applied blindly during data reduction, possibly in addition to wavelength-dependent atmospheric dispersion corrections. Otherwise one or more image features must be used for reference, to measure the alignment afterwards.

Dithering involves making small spatial offsets (e.g. 1–2 IFU elements), to average over instrumental variations. This is useful because IFU optics introduce artificial structure into the focal plane compared with a clear slit aperture (although even slits can have small non-uniformities in width). It also helps with detector defects, as for other observing modes. Whilst flat-fielding corrects throughput differences across the field, dithering evens out the resulting noise variations and 'fills in' any dead elements, especially for fibre IFUs. It may also smooth over any differences in spectral-line profiles from one IFU element to another. For short exposures, restricting offsets to an integer number of spatial pixels sometimes allows multiple spectra to be co-added without interpolation; this only applies if atmospheric dispersion is similar between the exposures, the IFU grid is regular and offsetting errors are small, but can help avoid problems due to undersampling (see Section 2.4).

Mosaicking uses offsets comparable to the size of the IFU field (especially for fields of only a few arcseconds) to increase spatial coverage. For mosaics of up to ~ 4 times the native field of view, the most reliable approach is often to keep the centre of the target (e.g. galaxy nucleus) within the field of view at every pointing, as a reference for

[2] Sub-pixel dithering is referred to as 'subsampling' in the original lectures, but that was perhaps not the best choice of terminology, given prior use of 'supersampling' for the same idea.

aligning exposures. If larger mosaics are needed, one either has to rely on the hardware for accurate movements or use multiple reference peaks with well-known relative positions to calculate the offsets.

For IFUs with coarse spatial pixels, sub-pixel dithering may help avoid loss of spatial information and the resulting processing artifacts. This idea is best known in connection with the Drizzle technique for HST Wide-Field Planetary Camera (WFPC) data (Fruchter and Hook, 2002). A likely example would be stepping by one half or one third of a fibre, lenslet or image slice between exposures (much smaller increments do not necessarily gain extra detail, given their size relative to the smoothing effect of the larger pixel area, but may improve stability when combining data with offset errors; see e.g. Lauer, 1999). It is important to know the offsets accurately in order to interleave exposures correctly. Unlike the Hubble Space Telescope (HST), ground-based observatories suffer from variable seeing and cloud, which are likely to limit how accurately a higher-resolution image can be reconstructed from multiple poorly sampled datasets. To the author's knowledge, sub-pixel dithering has not yet been well tested with IFUs, although a 3D-capable version of the Drizzle algorithm has recently been written by National Optical Astronomy Observatory (NOAO) and is being tested for use with the Near Infrared Integral Field Spectrograph Integral Field Unit (NIFS IFU) on Gemini North.

At near-infrared wavelengths, dithering and mosaicking could potentially allow spending a larger fraction of observing time on source than for a single target pointing plus sky. The most common observing sequence for nodding to sky is probably sky–object–object–sky, repeated as many times as necessary. This optimizes the signal-to-noise ratio for faint targets, since there is equal background noise from sky and object exposures. When offsetting, however, data are combined with relative shifts, so a single sky exposure can be subtracted from multiple object exposures and the results co-added without using the same measurement repeatedly at any given position. Thus a sequence such as sky–object–object (repeated) can be used. The signal-to-noise ratio is the same as when using equal numbers of object and sky exposures, but only if data do not need binning or fitting spatially. This does not, of course, remove the need to observe sky frequently enough to sample temporal variations. In practice, the most conservative schemes (i.e. sky–object– or sky–object–object–sky– repeated) tend to give the best results.

Flat-fielding

As for other observing modes, flat-field frames are needed as a reference for correcting instrumental efficiency variations (with x, y or λ). Since an IFU introduces structure in the focal plane, there are two components to the overall flat-field response: the usual detector flat, including pixel-to-pixel variations and other multiplicative features, and an IFU flat, accounting for differences in transmission between the optical fibres, lenslets or image slices that make up the IFU, or along the length of slices. Variations over the IFU can be due to several factors, such as fibre stresses (focal ratio degradation), alignment variations between optical elements, optical bonding, differences in surface roughness and diffraction losses. The IFU and detector are often flat-field corrected as a single system, but can be treated separately if necessary due to flexure. In order to have the correct spatial and spectral characteristics simultaneously, a final flat can be constructed from more than one source (see below). The spatial structure of the flat corresponds largely to the IFU, whilst spectral variations are mainly due to the detector and spectrograph.

Spectrally, (detector) flats require a dispersed source of illumination with a smooth continuum: usually a bright calibration lamp. Dispersing the light mimics science data,

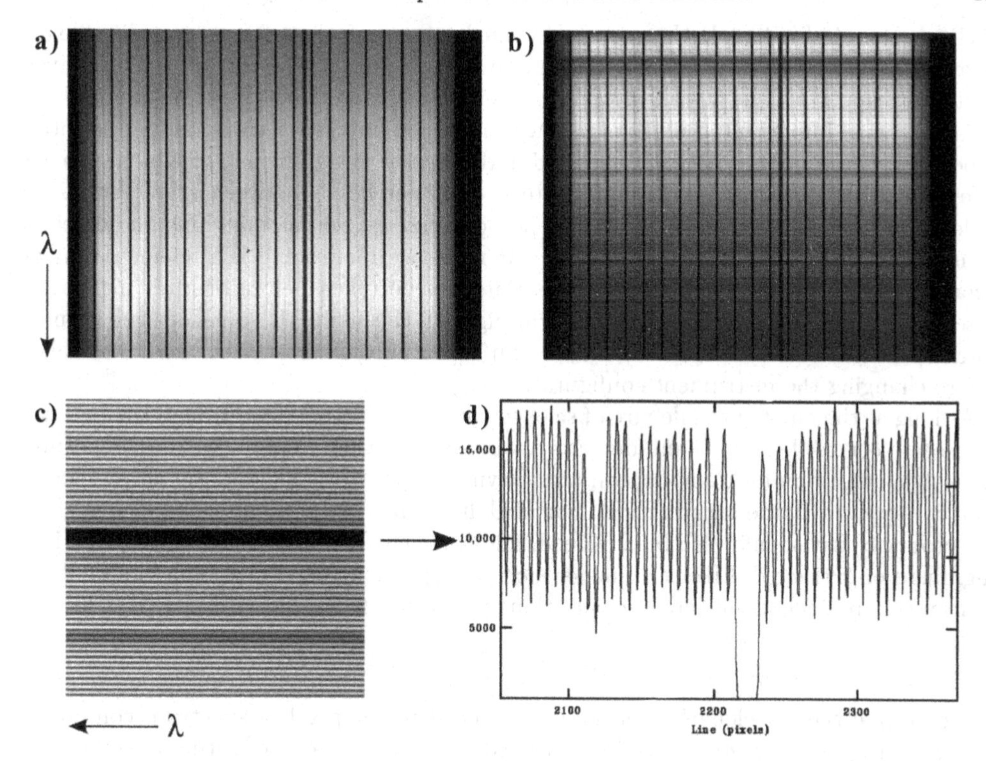

FIGURE 2.7. Some example IFU flats: a) a calibration lamp spectrum from the GNIRS image slicer; b) a twilight sky spectrum, showing telluric structure; c) small subsection of a lamp spectrum from the GMOS fibre IFU, covering two blocks of fibres with a gap between them; and d) a vertical cross-section of the previous panel, showing the throughput differences and overlaps between optical fibres.

allowing for any wavelength dependence of the flat-field pattern (e.g. due to quantum efficiency or fringing). An intrinsically smooth spectrum is needed to distinguish small-scale flat-field structure from the shape of the spectrum itself, which is typically fitted and removed during processing. The detector flat may incorporate a fringe pattern due to internal optical interference, especially for CCDs at longer wavelengths. For the IFU flat component, calibration lamp data can also be used to measure differences in spectral response (rather than just overall throughput) between IFU elements, if such differences happen to be significant. If separate detector and IFU flats are needed, on the other hand, one could take lamp exposures through a clear long-slit aperture, matching the spectral setup of the science data without the IFU in the focal plane.

Spatially, a flat illumination source is needed for correcting IFU structure and any vignetting by the spectrograph. The most uniform reference is the twilight sky (although the sky has a usable brightness level for an interval of only a few minutes, twice per day; for high spatial or spectral resolution IFUs, the brightness may even be optimal before sunset or after sunrise, making for long observing nights). Lighting a blank, neutral area inside the telescope dome with appropriate lamps is another common technique for achieving relatively flat illumination. Over the small area of a typical IFU field, however, even an inbuilt calibration lamp may be sufficiently uniform for the spatial content of the flat, if designed appropriately. Figure 2.7 shows some examples of different types of flat-field spectra.

Given that individual IFU spectra are much narrower than for a long-slit, flexure perpendicular to the dispersion axis is more of an issue; a shift of only a few pixels could correspond to the width of a fibre/lenslet or a large fraction of an image slice. It is therefore important that flats are well matched to their corresponding scientific exposures. Often flexure between the IFU and detector is significant, in which case the safest strategy is to take frequent flats (e.g. once per hour) between science data, or at least to take them at the same telescope pointing. As well as matching the detector illumination accurately, this provides a clear reference for tracing IFU elements in the science spectra. In the absence of flexure, flats can be taken before or after night-time observations and applied to data from throughout the night. If some optical element (such as the tilt of the disperser) moves non repeatably, flats or arcs should be taken before changing the instrument configuration.

As long as they match the flexure of science data, lamp flats taken through the IFU can be used to correct both IFU and detector variations together. Where the lamp exposures are not spatially uniform enough on their own, a separate twilight observation can be used to apply a flatness correction. For IFU flats that do not match science data, a detector-only flat should also be obtained, allowing them to be reduced and shifted into alignment or extracted. Likewise, image slicer twilights may need matching lamp flats to remove flexure-dependent detector variations that differ from science data.

Wavelength calibration

To determine the wavelength accurately at each detector pixel, a spectrum containing well-known features at zero redshift is needed, covering the range of the detector. The features must both be recognizable (by their relative or absolute positions and strengths) and have accurately catalogued wavelengths. We can then measure their pixel coordinates and interpolate to find wavelengths at intermediate positions. For IFUs, each separate 1D fibre/lenslet or point in a 2D slitlet spectrum needs calibrating.

Often the best reference source is an arc lamp. Some common examples are argon, copper-argon, thorium-argon, xenon and krypton lamps, which have many strong lines with precise laboratory measurements (e.g. Figure 2.8). In the red or near infrared, an alternative is to use telluric emission lines in the raw science spectra, with the advantage that flexure is then not an issue. Where this is not an option (such as for high resolutions or short exposures at thermal wavelengths), well-defined telluric absorption bands can sometimes be used instead.

Typically, at least one arc exposure is taken in order to derive a detailed wavelength solution, including non-linear terms. Sometimes this is observed during the day time or twilight. Where there is flexure, any small zero-point shifts in wavelength can often be corrected using even a single strong telluric line in the raw science spectra. Sometimes, however, there are no suitable sky lines at all, especially for high dispersions in the blue. If wavelength calibration is important in those cases, any flexure must be tracked using frequent arc spectra between science exposures (ensuring that arcs are taken before changing the disperser tilt, if it is non-repeatable). Occasionally, multiple arc lamps may be needed to obtain enough lines over certain short wavebands.

Telluric calibration

At far red to infrared wavelengths, scientific data tend to be affected by telluric absorption lines. The common photometric bandpasses (e.g. I, z, J, H, K, L, M) are defined in spectral regions with relatively good atmospheric transmission, but still contain many minor absorption features due to water vapour and other molecular species. Moreover,

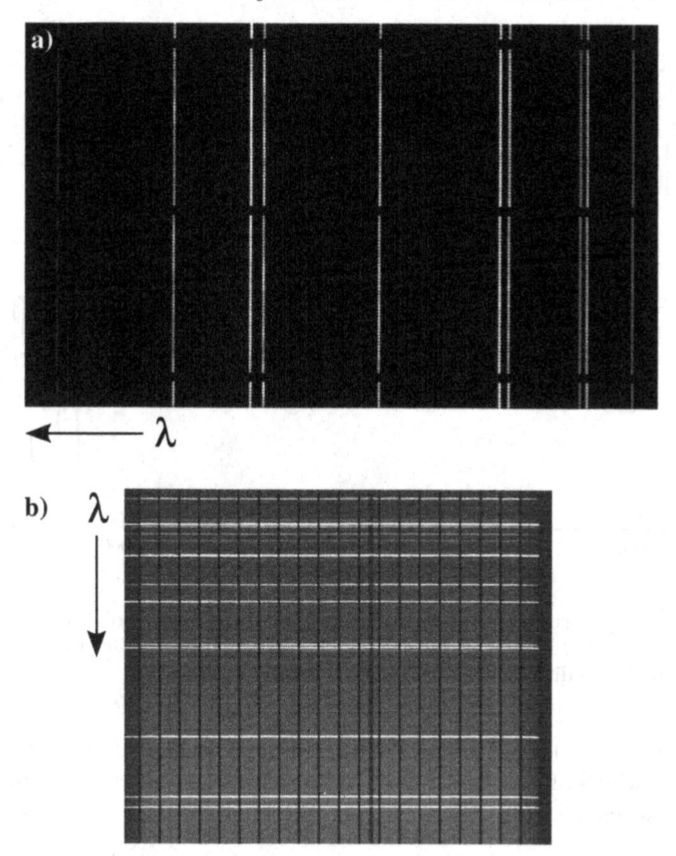

FIGURE 2.8. a) A subsection of a typical copper-argon arc spectrum, taken with the GMOS optical fibre IFU; b) an argon arc from the GNIRS near-infrared image slicer.

there is sometimes occasion to observe in between these clean regions, perhaps to measure a strong emission line that is redshifted from the visible.

For many purposes, telluric absorption left in scientific spectra would be problematic. Telluric lines can be confused with real stellar features, for example, when using an automatic algorithm to measure velocities. More importantly, telluric lines can overlap real spectral features, changing their profiles so that spurious line widths, centres or strengths would be measured from the raw data.

In order to correct for telluric absorption, a star of known spectral type is observed with little or no intrinsic absorption at wavelengths of scientific interest; typically, this is an A-type star (Figure 2.9), which has an intrinsic spectrum that is almost featureless apart from hydrogen lines. The amount of absorption in telluric bands scales with airmass and varies in time, particularly with changes in water vapour. Telluric standards are therefore observed immediately before or after the corresponding science object, with coordinates chosen to match the airmass, or zenith distance, at which it is observed. Where the range of airmass variation during a science observation is small (e.g. < 0.3 airmasses), a single star can be observed, matching the average airmass. For longer observations, the range of airmass can be bracketed with telluric standard spectra taken before and after.

If the spectral profile of the instrument varies substantially across the IFU field, one might take telluric spectra with the standard star at multiple positions on the IFU to attempt to calibrate different regions separately. For image slicers, point-like and diffuse

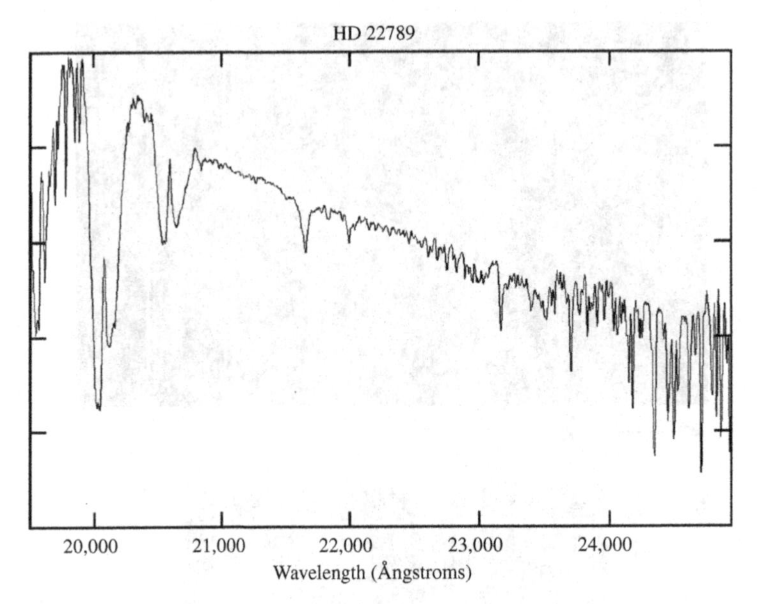

FIGURE 2.9. K-band spectrum of an A-type star with telluric absorption lines. With the notable exception of the Brackett-γ line at 2.17 μm, most of the structure in the spectrum is introduced by the atmosphere. Normalization is arbitrary.

image features will produce slightly different spectral profiles anyway, having different shapes across the width of an image slice; this is likely to be a relatively minor source of error, however, especially after summing over the telluric observation spatially.

Flux calibration

To allow meaningful comparison of fluxes between different wavelengths – or with previous datasets – a stellar calibration spectrum is normally needed. Often this is just a relative calibration of the instrumental efficiency variation with wavelength, but where necessary it may also serve as an absolute throughput or sensitivity measurement for the instrument+telescope+sky.

Relative flux calibration is the spectral equivalent of flat-fielding on large spatial scales (whilst flat fielding itself takes care of smaller spectral structure from pixel-to-pixel differences and fringing). It is needed for taking the ratios of lines at different wavelengths, or measuring the true continuum slope of a target, for example. Given photometric observing conditions, flux calibration additionally allows the brightness of a source or spectral feature to be measured in absolute units (such as $\mathrm{W\,m^{-2}\,Å^{-1}\,arcsec^{-2}}$). Alternatively, line strengths can be measured relative to the continuum, as equivalent widths, avoiding dependence on accurate flux calibration.

To derive the sensitivity spectrum for a set of scientific observations, the reference star needs to be observed with the same instrument configuration and have a known intrinsic brightness as a function of wavelength. At visible wavelengths, numerous spectrophotometric standard stars are available, with wavelength-tabulated fluxes in the literature (e.g. Oke, 1990). In the near infrared, one often just uses a convenient star of well-known spectral type and with a reliable broad-band magnitude measurement; the intrinsic continuum can then be modelled with a black-body curve for the appropriate temperature and magnitude. Figure 2.10 is an example of a near-infrared absolute throughput spectrum, derived from a G-type star.

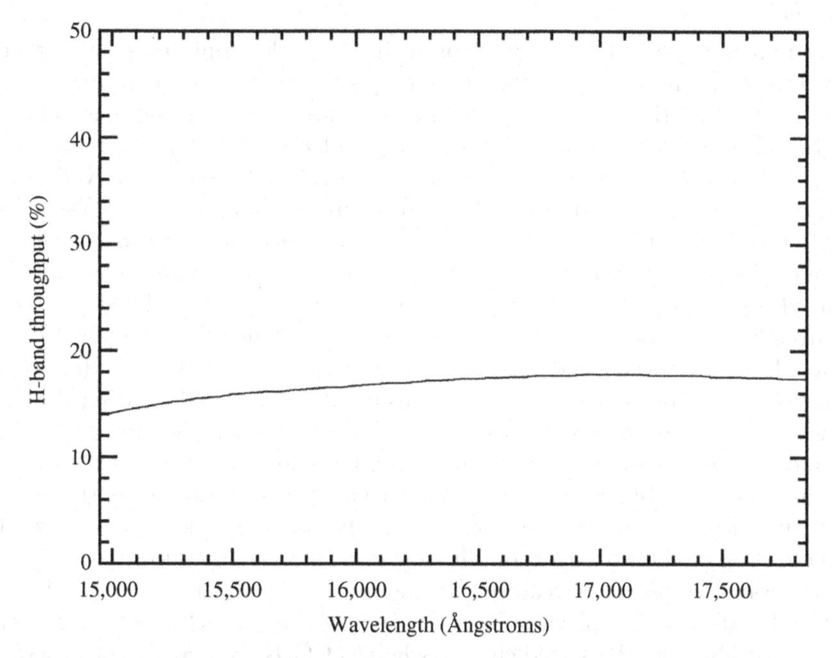

FIGURE 2.10. Absolute throughput spectrum for the NIFS integral-field spectrograph at an H-band setting; the curve was derived from a G-star observation by comparing the observed counts with a black-body model of the star's intrinsic flux.

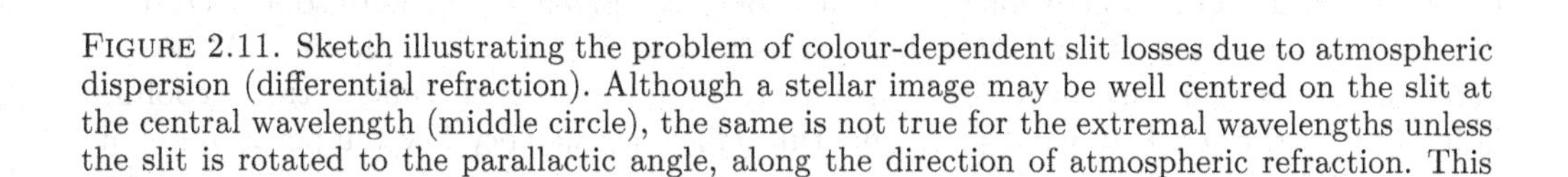

FIGURE 2.11. Sketch illustrating the problem of colour-dependent slit losses due to atmospheric dispersion (differential refraction). Although a stellar image may be well centred on the slit at the central wavelength (middle circle), the same is not true for the extremal wavelengths unless the slit is rotated to the parallactic angle, along the direction of atmospheric refraction. This effect scales with airmass and is strongest at blue wavelengths, in the optical.

Flux standards can be taken one or more times during a given observing run (a short enough period that the instrument and telescope characteristics will not change significantly). They need not be observed at the same time as scientific spectra, unless absolute spectrophotometry is critical or the same data are to be used for both telluric and flux correction. Where absolute flux measurements are not required, standards can be observed through cloud, which to first order is neutral in colour.

Because an IFU normally captures all the light from an incident stellar image, without slit losses, observing standards is relatively straightforward. Absolute fluxes can be measured without any special setup and, in principle, a single observation may be used for both telluric and flux calibration. For long-slit spectroscopy, on the other hand, absolute flux calibration requires using a special wide slit to capture all the light, resulting in a lower spectral resolution than for scientific data; since a telluric spectrum must match the scientific resolution, this has to be taken separately. Moreover, atmospheric dispersion causes colour-dependent throughput losses for narrow slits, unless observing at the parallactic angle (Figure 2.11); thus even relative measurement of a sensitivity spectrum is easier and more accurate with an IFU.

Detector bias

Raw measurements from a CCD detector include an electronic bias level. Since this varies slightly over the array, a separate zero point needs to be determined for each pixel, to allow calculation of the true counts accumulated above the bias. This simply involves taking a few very short exposures (ideally of zero duration, i.e. immediate read out), in the dark. The resulting images directly provide the zero-point value, with no photoelectrons stored, for each pixel. These exposures include read-out noise, of course, so several are taken and later averaged together, to reduce the noise to a level that will not degrade scientific data when subtracted. A typical bias frame looks flat, with random variations of just a few counts superimposed; depending on how well the electronics are isolated from electrical noise sources, the image may also include some faint structure.

The bias level produced by detector electronics can drift with time, but is usually stable enough for reference exposures to be taken only occasionally, during the daytime. Sometimes the detector controller also creates an overscan region for each exposure, by continuing to read out the detector after shuffling and retrieving all the accumulated charge. This adds extra blank image columns, for monitoring variations in the overall bias and read-out noise. If necessary, the overscan can be used to make an overall zero-point correction, on top of the pixel-to-pixel differences from bias exposures.

Infrared arrays are normally read out by measuring the difference in counts between the start and end of each exposure. This is possible because each pixel can be read out rapidly without affecting the stored charge, whereas CCDs are read out destructively, by shuffling the charge from one column to the next and finally off the detector. Thus, the bias level is removed automatically for infrared observations and does not need measuring separately.

Dark current

Over the duration of a long enough exposure, detector pixels accumulate some electrons due to thermal excitation and array defects, as well as from incident photons. Although detectors are cooled with nitrogen or helium to minimize this thermal current, cooling is limited by the need to avoid poor quantum efficiency at very low temperatures. For modern CCDs, the effect may nevertheless be negligibly low (e.g. $1\,e^-\,\mathrm{hour}^{-1}\,\mathrm{pixel}^{-1}$). For infrared arrays, on the other hand, dark current is more important and tends to affect some pixels much more than others. This creates variations in the raw data that obscure real features (Figure 2.12). A reference is therefore needed for distinguishing dark counts from those due to photons from a target object.

For any given exposure time that is used in observing, the dark current component of images can be estimated by exposing the detector in complete darkness for the same length of time. Multiple exposures are usually taken, as for bias frames, to account for read noise and to allow removal of cosmic rays. Since, however, the exposures may be much longer than biases (and may therefore be best taken at night or in twilight if light leaks are an issue), it is not always practical to observe more than a few frames.

For scientific observations with separate, equal sky exposures, the same dark current is present in both object and sky data. Thus in the infrared, subtracting object and sky pairs removes dark current automatically from those spectra, without the need for special calibrations. This usually does a good job of cleaning up most hot pixels, leaving just a statistical increase in noise; any pixels that remain affected after subtraction are masked out during reduction. Separate dark calibrations are required, however, in cases where background subtraction is attempted without nodding to sky, or where sky frames will be scaled to match on-target data more accurately.

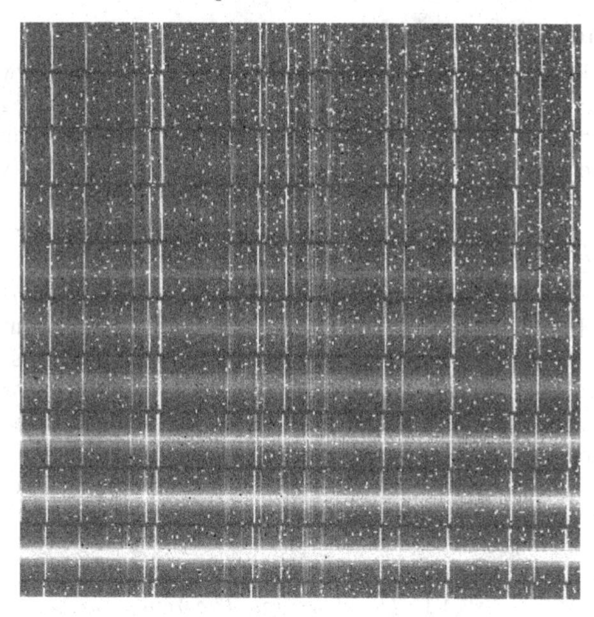

FIGURE 2.12. Section of a raw scientific exposure from an HgCdTe array. Most of the numerous bright specks superimposed on the target and sky spectra are 'hot' pixels, with high dark current.

For infrared calibration data, matching darks are usually required unless the exposures are short enough not to have significant dark counts. For flat fields of more than a few seconds, it is common to take pseudo-darks in the same configuration as the flats themselves, but with the lamp turned off. For arc spectra, darks are needed if uncorrected hot pixels could otherwise be confused with emission lines in a 1D section.

In summary, darks are often needed at infrared wavelengths for calibrations, but are optional (and time consuming) for scientific data, depending on the reduction method. With modern CCDs, darks are often not needed at all (but since separate sky frames are usually not available as a reference, they would be more critical if required).

Summary of observing strategies

On the whole, observing with a single IFU is not radically different from standard slit spectroscopy. As usual, different techniques apply at optical and near-infrared wavelengths. A few issues stand out, however, compared with other spectroscopic modes.

Since IFUs perform imaging as well as spectroscopy, their use reflects both aspects. Target acquisition allows 2D image reconstruction through the IFU, for example. Spatial dithering and mosaicking are also important, involving offsets of varying sizes in both dimensions. Some other practical differences relate to flat-fielding requirements, sky subtraction and data inspection. The IFU itself has to be flat fielded, as well as the detector, carefully accounting for any flexure. At optical wavelengths, sky subtraction may involve separate sky fibres or a background field, instead of interpolating between the ends of a slit. In the near infrared, one typically has to nod off to sky instead of up and down a slit. When viewing data, one has to learn how to interpret scrambled images and/or use special image or datacube reconstruction software. Finally, sensitivity is often an issue for high-resolution IFUs, even on 8 m telescopes (although large-pixel designs can do well compared with a slit).

2.4 Sampling images

Sampling theory helps us understand the effects of dividing an image or waveform into discrete points, such as detector pixels. This is fundamental in digital imaging, but appears rarely to be taught or discussed outside the engineering and information-theory communities. In astronomy, one sometimes encounters the assertion that 'Nyquist sampling' requires two detector pixels per full-width-at-half-maximum (FWHM) of a point source. Here I show where this criterion comes from and consider how the results of sampling theory relate to IFU data reduction.

Sampling theory is particularly relevant to IFS data because of the way in which images are dissected both by the IFU and the detector. This not only introduces an extra sampling stage, in effect, but often results in a spatial grid which is non-rectangular, or even completely irregular. Thus resampling in up to three dimensions is generally necessary if we want to get data onto a cubic or square grid that is easy to work with. This is complicated by the fact that IFUs sometimes undersample telescope images with large spatial pixels, for sensitivity reasons or to maximize the field of view. It is important to understand the consequences of this last case in order to avoid introducing artifacts during processing.

2.4.1 *The Sampling Theorem*

The Sampling Theorem deals specifically with sampling on a regular grid. It is often named after one or more of the following researchers: Kotel'nikov (1933) of the Russian Academy of Sciences; Nyquist (1928) and Shannon (1949) of Bell Labs, USA; and Whittaker (1915) of Edinburgh University, UK. Various naming combinations are used in the literature, such as 'Shannon's Theorem' or the 'Shannon–Whittaker–Kotel'nikov Theorem'. Roughly speaking, the idea is that an analogue signal containing a finite amount of information can be recovered completely from a large enough set of discrete samples.

For simplicity, consider a one-dimensional function, $f(x)$. Sampling at a particular point, x', can be represented using a Dirac delta (impulse) function:

$$f(x') = \int_{-\infty}^{\infty} \delta(x - x')\, f(x)\, dx. \tag{2.1}$$

In order to have this property, $\delta(x)$ can be any function of unit integral that is 'infinitely narrow', i.e. zero everywhere except at $x = 0$. Multiplying a function by delta therefore produces an impulse whose area is normalized to the value of the function at the relevant point (in this case x'):

$$f(x)\, \delta(x - x') = f(x')\, \delta(x - x'). \tag{2.2}$$

Graphically, we represent this delta function as an arrow whose height, $f(x')$, indicates its normalization (Figure 2.13).

A comb (or shah) function is an infinite series of equally spaced delta functions:

$$\mathrm{III}\left(\frac{x}{a}\right) = |a| \sum_{n=-\infty}^{\infty} \delta(x - an), \tag{2.3}$$

(where the multiplicative $|a|$ just preserves the unit area of δ when scaling the coordinate argument by the same factor). Thus, sampling $f(x)$ at regular intervals of a is equivalent to multiplication with a comb function:

$$\mathrm{III}\left(\frac{x}{a}\right) f(x) = |a| \sum_{n=-\infty}^{\infty} f(an)\, \delta(x - an). \tag{2.4}$$

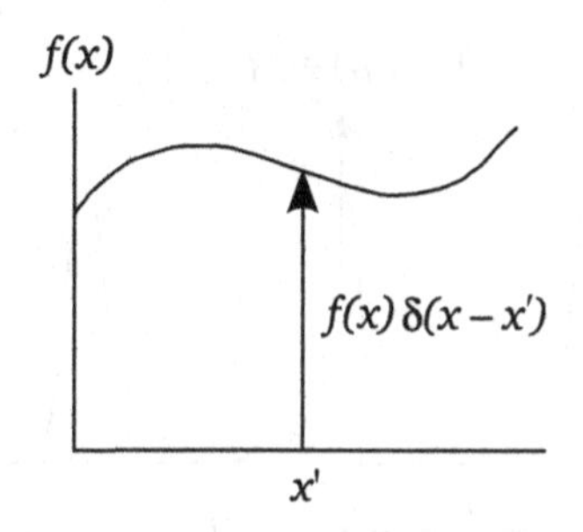

FIGURE 2.13. Pictorial representation of sampling with $\delta(x - x')$.

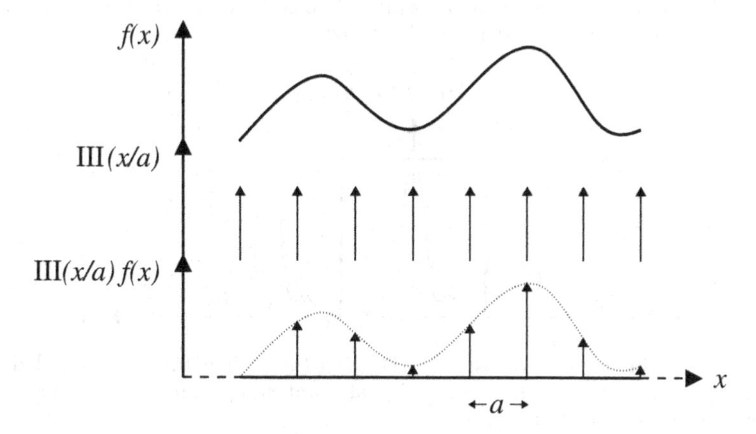

FIGURE 2.14. Sampling a continuous function, $f(x)$, at regular intervals of a, by multiplication with a comb function.

Clearly, III f is a series of impulses, normalized to the sampled values: it contains the same information from $f(x)$ as would point-like measurements (Figure 2.14).

The comb function is a useful tool for representing sampling because its Fourier transform is also a comb. Working with frequencies in reciprocal units of x (rather than $2\pi/x$) gives:

$$\mathrm{III}\left(\frac{x}{a}\right) \overset{\mathcal{F}}{\rightleftharpoons} a\,\mathrm{III}(au). \tag{2.5}$$

According to the Convolution Theorem, multiplication of functions in the spatial domain is equivalent to convolution of their Fourier transforms (and vice versa). Thus the representation of sampled data in frequency space is:

$$\mathcal{F}\left(\mathrm{III}\left(\frac{x}{a}\right) f(x)\right) = a\,\mathrm{III}(au) * F(u). \tag{2.6}$$

This is the frequency spectrum of the original, continuous $f(x)$, replicated infinitely at intervals of $1/a$ (Figure 2.15); convolving $F(u)$ with each $\delta(u - u')$ simply shifts the origin to u'.

A function is said to be 'band limited' if its Fourier transform is zero above some cut-off frequency, u_c. In Figure 2.15 and Equation 2.6, the replicas of $F(u)$ do not overlap if $f(x)$ contains a maximum frequency component of $u_c \leq 1/2a$; thus we can separate out $F(u)$ from the transform of the sampling pattern and recover the original $f(x)$ without any loss of information. If the replicas do overlap, on the other hand, the function cannot be determined uniquely.

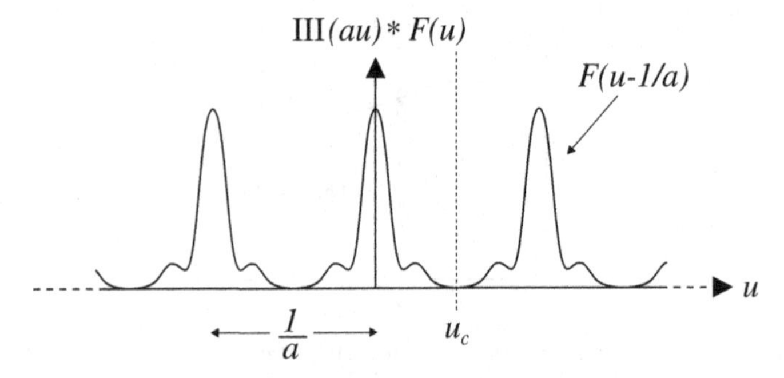

FIGURE 2.15. The Fourier spectrum of a uniformly sampled function, $\text{III}(x/a)f(x)$, is the Fourier spectrum of $f(x)$, repeated at regular intervals of $1/a$.

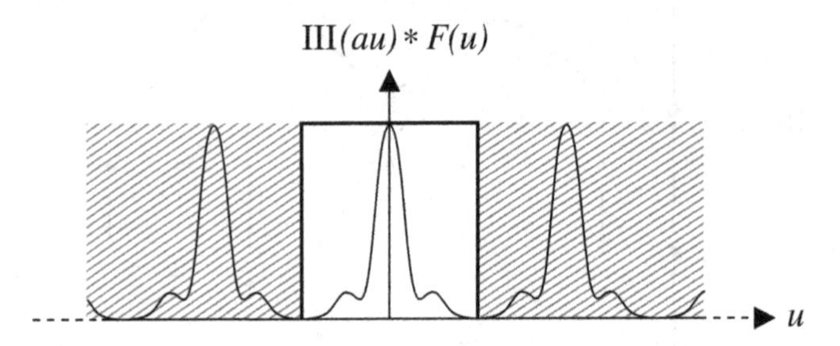

FIGURE 2.16. Given the Fourier transform, $\text{III}(au) * F(u)$, of a function sampled above the Nyquist rate, the transform, $F(u)$, of the original, continuous function can be recovered by multiplication with a box function, $\Pi(au)$.

The Sampling Theorem can therefore be stated as follows: 'A band-limited function can be recovered completely from a set of regular samples, as long as the sampling rate is at least twice the highest component frequency.' This is clearly also equivalent to saying that the separation between samples must be no greater than half the shortest component wavelength.

For a given sampling interval, a, the maximum distinguishable frequency of $1/2a$ is known as the Nyquist frequency. Likewise, the critical sampling rate of $2u_c$ for a fixed bandwidth of $f(x)$ is often called the Nyquist rate. Sometimes these terms are interchanged, however, or used to refer to u_c.

To recover $f(x)$, we can isolate $F(u)$ by multiplication with a box function, $\Pi(au)$, that has unit value between $\pm 1/2a$ and is zero elsewhere (Figure 2.16).[3] This has the Fourier transform $\text{sinc}(x/a)/a$, where $\text{sinc}(x) = \sin(\pi x)/\pi x$ is an infinite function with zeros at integer x (Figure 2.17). Thus, by the Convolution Theorem, $f(x)$ is reconstructed in the spatial domain by a summation of regularly spaced sinc functions (Figure 2.18):

$$f(x) = \left(\text{III}\left(\frac{x}{a}\right) f(x)\right) * \frac{\sin(\pi x/a)}{\pi x} \tag{2.7}$$

$$= |a| \sum_{n=-\infty}^{\infty} f(an) \, \frac{\sin(\frac{\pi}{a}(x - an))}{\pi(x - an)}. \tag{2.8}$$

Having reconstructed the original analogue signal, we can then resample it arbitrarily onto a convenient grid (such as a regular datacube, for IFU data).

[3] The box cut-off can be anywhere between u_c and $1/2a$ (if different), but the latter is convenient because it is known and fixed for a particular sampling.

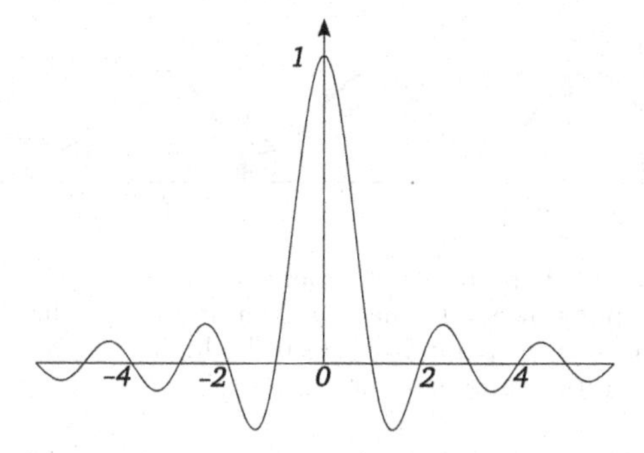

FIGURE 2.17. The sinc function, $\sin(\pi x)/\pi x$, which peaks at unit value at $x = 0$ and oscillates towards small values at infinity, with zeros at integer x.

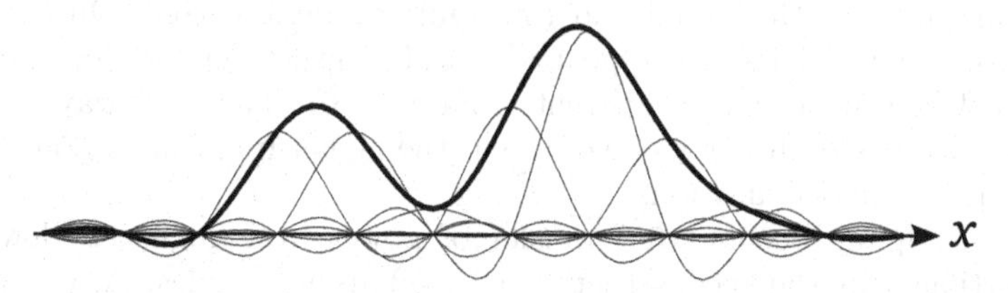

FIGURE 2.18. Reconstruction of a continuous function from regularly spaced samples, by convolution with a sinc function. Each grey curve is a sinc function centred at the appropriate sample point and normalized to the corresponding sample value (cf. Figure 2.14). The thicker black curve shows the overall sum as a function of x.

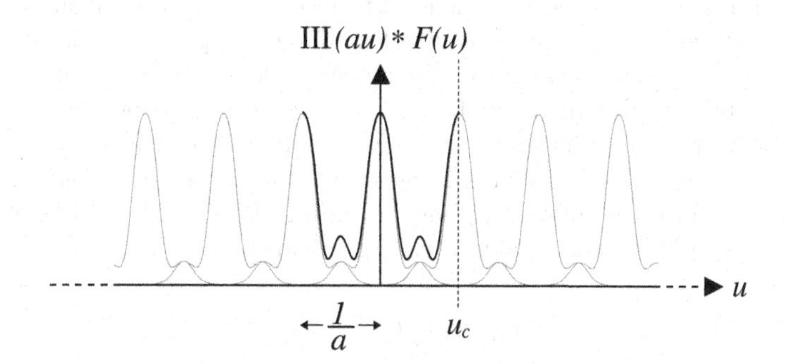

FIGURE 2.19. When $f(x)$ is undersampled, such that it contains frequency components $> 1/2a$ for a sampling interval of a, replicas of $F(u)$ overlap in the Fourier spectrum of the sampled data.

Equation 2.8 specifies $f(x)$ at all intermediate points between the samples. This is possible because the number of frequency components was finite to begin with, so $f(x)$ can be represented by a corresponding number of basis functions. The basis could instead be a Fourier series, but finding the coefficients would then be more complex than above: sinc functions are a natural basis, because the required coefficients are simply the sample values.

2.4.2 *Undersampling*

For a given sample separation, a, if $f(x)$ contains frequencies greater than the Nyquist frequency, $1/2a$, it is said to be undersampled. In the Fourier spectrum of the sampled signal, $F(u)$ then overlaps its neighbouring replicas (Figure 2.19); the spectrum is effectively

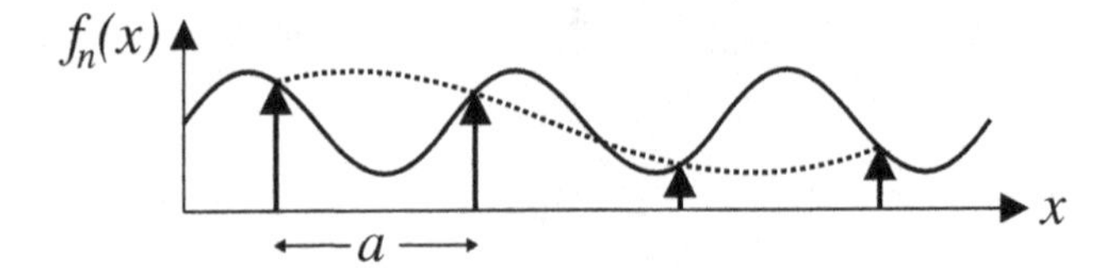

FIGURE 2.20. An undersampled frequency component and its lower frequency alias have the same values at regular sampling intervals, making them indistinguishable. The Nyquist, or critical sampling, criterion requires the sampling rate to be high enough with respect to the signal bandwidth that such ambiguities cannot occur.

wrapped around or reflected at $1/2a$, such that power at frequencies above this limit is aliased to lower frequencies. The aliasing can be thought of as beating between high-frequency components of the signal and the sampling period itself. On a regular grid, high- and low-frequency pairs can appear identical at every sample point, even though their values in between are quite different (Figure 2.20). The only way to distinguish between these aliases is to disallow one of them (the higher frequency in our case); hence the requirement for band limitation.

Where undersampling occurs, the transfer of power from higher to lower frequencies creates artifacts in the reconstruction of $f(x)$ from samples. Any super-Nyquist frequencies are not only unmeasurable, but contaminate otherwise good data. We can end up with less useful information than if $f(x)$ had been smoothed to a lower resolution/bandwidth before sampling.

With irregular sampling patterns, the similarity between alias frequencies is broken – as can be seen from Figure 2.20 – so there is no direct transfer of power from one frequency to another. Nevertheless, information is still lost if the sampling rate is too low, which unavoidably leads to reconstruction error. Moreover, if the number of available samples is too small, the problem of finding a set of basis coefficients may become ill conditioned. In general, the theory of reconstruction from irregular samples is considerably more involved than that discussed here (see references in Section 2.4.4, *Irregular sampling*).

Further information on regular sampling, including proofs of the theorems relied on here, can be found in Bracewell (2000, or earlier editions).

2.4.3 *Sampling images*

The Sampling Theorem deals with infinite and conveniently band-limited signals, so the next step is to relate this to real-world images.

As will be familiar to the reader, when a plane wave passes through an aperture, it is diffracted, so that the light ray leaving the aperture is spread out in angle (Figure 2.21). The image produced on a screen or detector by the resulting optical interference pattern depends on distance from the aperture. In the far field, where the aperture is effectively a pinhole, the amplitude of the diffraction pattern as a function of angle is given by the Fourier transform of the function describing the aperture transmission (i.e. shape). What we actually measure is intensity ($I \propto A^2$), so in practice we are interested in the square of the Fourier transform. For a rectangular aperture (described by a 2D version of the box function discussed in Section 2.4.1), this is a product of sinc functions. For a circular telescope mirror, it is the closely related Airy pattern, shown in Figure 2.22:

$$I(\theta) \propto \left[\frac{J_1(\pi D\theta/\lambda)}{\pi D\theta/\lambda}\right]^2, \tag{2.9}$$

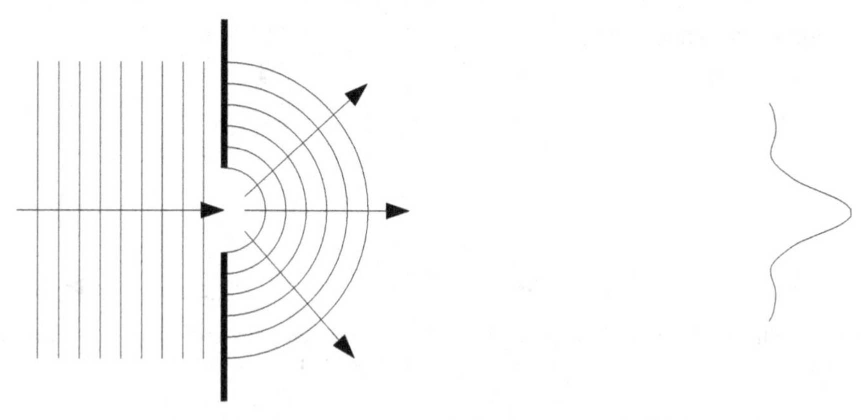

FIGURE 2.21. Diffraction of a plane wavefront by a finite aperture. The resulting interference produces a far-field intensity distribution similar in cross-sectional form to the sketch on the right-hand side (where intensity increases to the right).

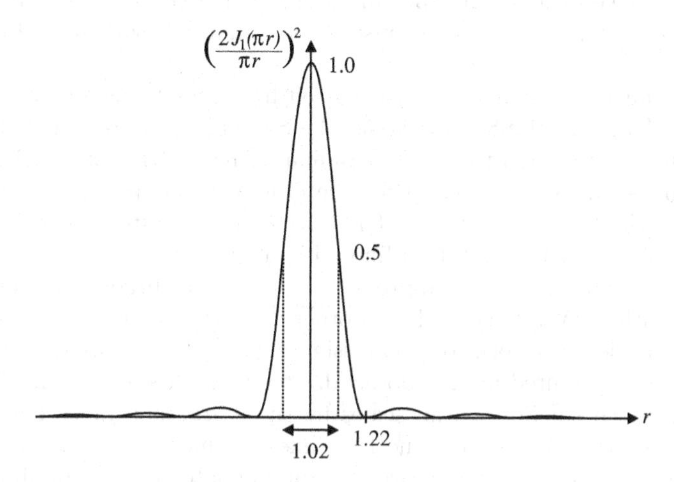

FIGURE 2.22. Cross-section of the Airy intensity pattern produced by diffraction of a plane wave by a circular aperture.

where D is the mirror diameter, λ is the wavelength of light, θ is the angular displacement in radians from the central direction of propagation and J_1 is a first-order Bessel function of the first kind. A derivation can be found in Goodman (1968) or Hecht (1987).

The finite image of a point source produced by a given optical system is known as its point-spread function (PSF). For a diffraction-limited telescope, the PSF is therefore the Airy pattern described above (ignoring a smaller correction for the central obscuration). For a seeing-limited telescope, the PSF is a superposition of atmospheric speckle images that tends towards a Gaussian-like form over long enough ($\sim 60\,\mathrm{s}$) exposures.[4] Since the PSF describes the distribution of light from any point-like element, the 2D image formed by a telescope is equivalent to an infinitely detailed projection of the sky, convolved with the PSF.

The Fourier power spectrum of the PSF is known as the modulation transfer function (MTF). This describes the sensitivity of the system to different spatial frequencies,

[4] In practice, this form is often better described by a Moffat profile (Moffat, 1969), which is centrally similar to a Gaussian curve, but allows for broader wings.

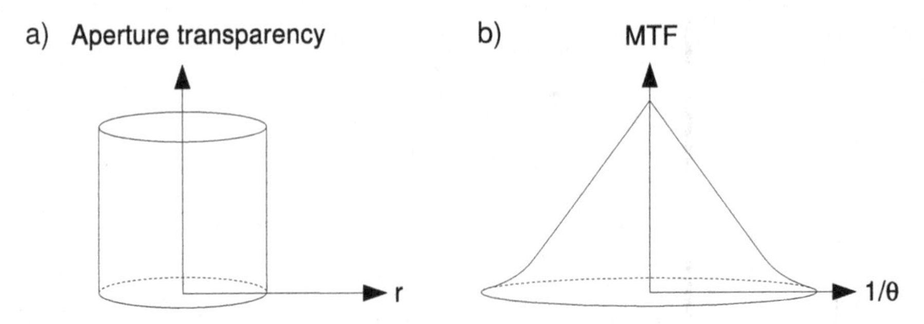

FIGURE 2.23. The MTF of a telescope is described by the self-convolution of the transmission over its aperture (so there is always some high-spatial-frequency cut-off).

convolution with the PSF being equivalent to multiplication by the MTF in the frequency domain. Since a telescope's diffraction PSF is the squared Fourier transform of its aperture transmission, the Convolution Theorem also tells us that the MTF is given by the self-convolution of the aperture transmission (Figure 2.23). It follows that, for a finite aperture, telescope images must be band-limited, with some high-frequency cut-off.

For a circular mirror of diameter D, the cut-off frequency turns out to be D/λ cycles per radian (see Goodman, 1968). To satisfy the Sampling Theorem, the detector pixel spacing must therefore be less than $\lambda/2D$ radians. The FWHM of the Airy pattern is $1.02\,\lambda/D$ radians, so in terms of the PSF the critical pixel spacing is $0.49 \times$ FWHM. Thus we confirm the two-pixel (really 2.04 pixel) sampling criterion. Strictly speaking, however, this applies only to diffraction-limited imaging.

For seeing-limited images, we can approximate at least the bright core of the PSF by a Gaussian curve with a FWHM typically an order of magnitude greater than the diffraction limit. The Fourier transform of a Gaussian curve is also Gaussian, so is infinitely extended. Thus seeing-limited images do not have a strict frequency cut-off, other than that imposed at much smaller spatial scales by the diffraction limit. Theoretically, we would therefore need a FWHM of something like 20 times the pixel size to reach the Nyquist rate. Fortunately, in practice, we can approximate critical sampling with a less stringent criterion.

The Gaussian distribution and its Fourier transform can be written

$$e^{-\pi(ax)^2} \overset{\mathcal{F}}{\rightleftharpoons} \frac{1}{a} e^{-\pi(u/a)^2}. \tag{2.10}$$

For unit a, these have a standard deviation of $1/\sqrt{2\pi}$ and a FWHM of $2\sqrt{(\log_e 2)/\pi} \approx 0.939$. Although both curves are infinitely extended, their values die away quickly after a few standard deviations. If a Gaussian PSF is imaged with a FWHM of n pixels, then, rewriting its transform in terms of the FWHM using $na = 0.939$, we have

$$F(u) = 1.06ne^{-3.56n^2u^2}, \tag{2.11}$$

with u in cycles per pixel and a standard deviation of $\sigma = 0.375/n$. For a FWHM of $n = 2$ pixels, the Nyquist frequency of $u = 0.5$ cycles/pixel therefore corresponds to 2.7σ, where the Gaussian MTF is at only 2.8% of its peak value. Super-Nyquist frequencies are therefore heavily suppressed in amplitude, to $< 3\%$. The total integrated power that is aliased back into the main spectrum is only 0.7%. Thus, even for seeing-limited images, the '2 pixel per FWHM' criterion is good enough in most cases. A pixel density slightly greater than this will reduce aliasing to very low levels.

2.4.4 *Practical issues*

Finite detector

The above discussion outlines how the Sampling Theorem relates to imaging, but several idealizations still need justification. Perhaps most obvious is the fact that sampling is done with $\sim 10^3$ detector pixels or $\sim 10^1$ spatial IFU pixels per dimension, rather than an infinite comb function.

A finite detector image is equivalent to a comb function multiplied by the telescope image (as previously) and truncated with a box function of width na, where n is the number of pixels and a their separation. The Fourier transform is therefore an infinite comb convolved with $F(u)$ (compare Figure 2.15), convolved again with a narrow sinc function whose first zeros are at $\pm 1/(na)$. This zero location corresponds to twice the Nyquist frequency divided by n (i.e. two pixels of the discrete Fourier transform). The extra convolution reflects the fact that a finite band-limited image is described by a finite number of frequency components; otherwise, the transform of the sampled data would be infinitely detailed. Thus, to first order the finite detector size makes no difference; the transform of sampled data is still, approximately, $F(u)$ repeated at intervals of $1/a$. Convolving $F(u)$ with something much narrower than itself does not extend the frequency band appreciably, so the same sampling criterion applies.

To second order, things are a bit more complicated. The sinc function is infinite and slowly decaying at lower power levels, so the transform of the sampled data actually has no hard cut-off frequency. In fact, a finite function, such as a sampled image, can never be strictly band-limited. This frequency extension is manifested as reconstruction error towards the edges of the data, similar to the oscillation seen at the ends of a Fourier series when approximating a sharp edge. The behavior can be improved somewhat by suitable filtering.

Finite pixel size

Whereas delta functions are infinitely narrow, pixels are as wide as their separation. At each sample point, the telescope image is averaged over the area of a pixel, which to first order can be a 2D box function, a hexagonal lenslet or a circular fibre. The measured sample value is therefore the convolution of the telescope image at the relevant point with the pixel shape. This just contributes an extra term to the overall measured PSF, along with the natural seeing, telescope diffraction PSF and instrumental aberrations. Since the effect is to convolve with the pixel shape prior to sampling, the idea of point-like samples remains intact; the comb representation and pixel size are mathematically separable.

Noise

Our discussion of the Sampling Theorem assumed noiseless data, so here we must consider the effect of random errors. Poisson noise is incoherent and spread over all frequencies, rather than contributing power at any particular frequency. Noise aliasing is not a concern, because it would just fold noise back into the main spectrum, affecting the statistical realization of the errors without systematic effect. The question is therefore how the noise level propagates through the resampling process. When interpolating images, the value at a particular x is a weighted sum of sinc functions, one per sample (see Equation 2.8 and Figure 2.18). Since noise terms add in quadrature, each of the original sample values contributes to the total variance at x according to the square of its weight. Since the values of $\mathrm{sinc}^2(x)$ at unit intervals add up to one (Bracewell, 2000), the noise level in the interpolated image is the same as the original. Noise therefore gets carried through

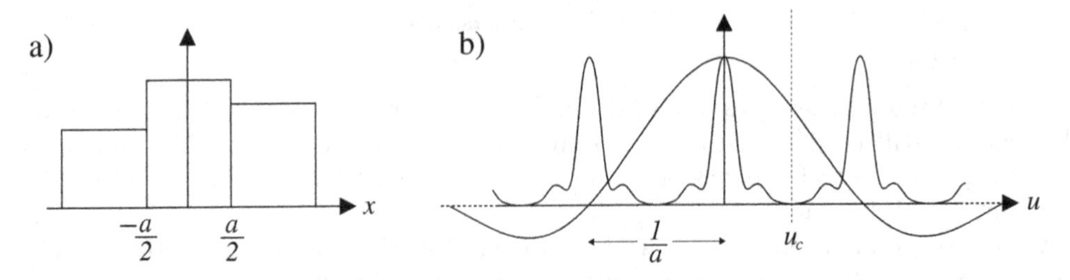

FIGURE 2.24. a) Nearest-neighbour interpolation uses the closest sample value to each x and is therefore equivalent to convolving a series of point samples with $\Pi(x/a)$. b) The corresponding envelope in the spectral domain is $\propto \mathrm{sinc}(au)$, whose zero values coincide with the sampling replicas of the original image spectrum, $F(u)$.

the system without systematic effects, to first order. In practice, of course, the noise level will vary over the image and the sinc function may allow bright peaks to dominate further out.

Practical interpolants

Although a sinc function is the 'ideal interpolant', it is infinite and strongly oscillating, so is not used in practice without modification. Discontinuities such as edges and bad pixels can cause severe diffraction-like 'ringing' effects (Gibbs' Phenomenon). Tapering the sinc function with an envelope such as a Gaussian or cosine multiplier provides a more robust and computationally feasible solution. This type of interpolant is good for well-sampled data, albeit relatively slow to compute unless look-up tables are used. It still tends to be more sensitive to spikes and discontinuities than other interpolants, but the envelope can be tuned to make a trade-off between suppression of ringing and avoiding smoothing. An equivalent technique is to take the fast Fourier transform (FFT) of the sampled data, apply a bandpass filter in the Fourier domain, shift in phase and convert back again.

Nearest-neighbour (zero-degree spline) interpolation is the simplest and fastest to compute method. This is equivalent to convolving the sampled data with a box function, $\Pi(x/a)$, instead of $\mathrm{sinc}(x/a)$. Since the Fourier transform is a sinc function, it does not actually cut off $F(u)$ at the Nyquist frequency, although it attenuates higher frequencies (Figure 2.24). The first zeros are at $\pm 1/a$, or twice the Nyquist frequency. Beyond that, $\mathrm{sinc}(au)$ goes negative and therefore reverses the phase of some frequencies. The frequencies passed above $1/2a$ correspond to the 'blockiness' of the nearest-neighbour interpolant, which can be considered a reconstruction error.

In terms of reproducing the analogue telescope image, the nearest-neighbour algorithm is about the least accurate method. Nevertheless, the values are guaranteed to be reasonable, since they are the same as the original samples: blockiness is an artifact, but a well-behaved and understood one. It should also be noted that this interpolant does not smooth the data, as each value comes from a single sample. Resampling with the same interval at new points effectively just introduces a small coordinate shift. Since both the values and gradient are discontinuous, the reconstruction cannot be used directly for fitting smooth contours or velocity fields (although it might be possible to resample first at the same rate and then use a smooth interpolant, if the sampling grid is regular).

Linear (first-degree spline) interpolation is equivalent to convolving the sampled data with a triangle, or tent, function, $\Lambda(x/a)$ (Figure 2.25). This is the self-convolution of the box function, $\Pi(x/a)$, and has width $2a$. Its Fourier transform is therefore $\propto \mathrm{sinc}^2(au)$, which dies away much quicker than $\mathrm{sinc}(au)$ in the wings and is positive everywhere, but still does not reach zero until $\pm 1/a$. The frequencies passed above the Nyquist limit

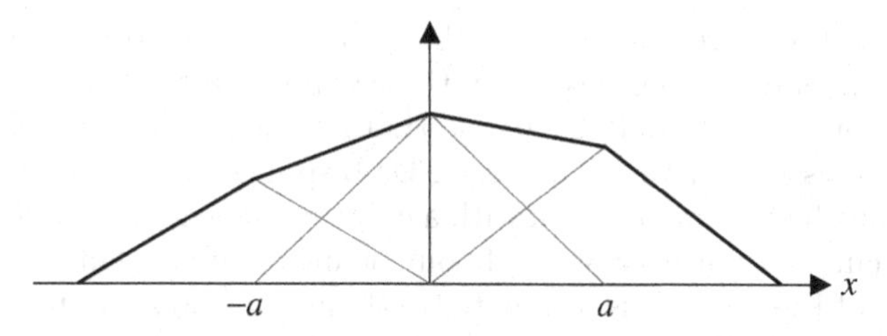

FIGURE 2.25. Linear interpolation is equivalent to convolving a series of point samples with $\Lambda(x/a)$. In this example, the dark line shows the summation of three individual triangle functions, one per sample.

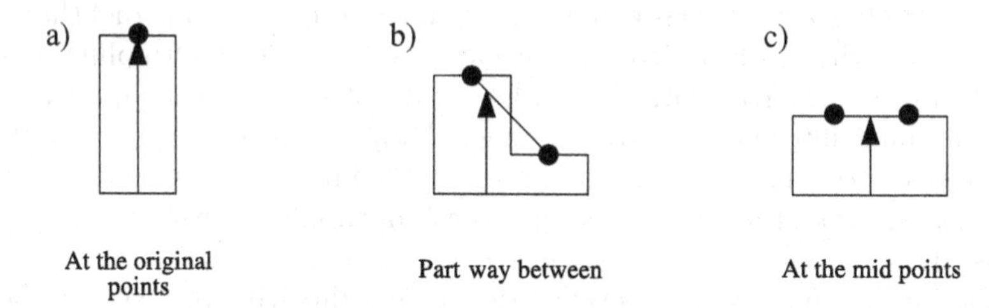

FIGURE 2.26. Resampling a linear interpolant at the original rate is equivalent to sampling in the first place with pixels of a size and sensitivity distribution that depend on the offset. In each of the three cases above, the height at each point represents the relative contribution to the interpolated value at the position of the arrow.

correspond to the 'jaggedness' or discontinuous gradient of the reconstruction. The interpolated values themselves are continuous, however, and follow the original curve more closely than in the nearest-neighbour case. Linear interpolation is still relatively inaccurate compared with higher-order interpolants, but once again the result is robust and guaranteed to be 'reasonable', since the interpolation is bounded by the original sample values.

It is instructive to consider what happens when resampling a continuous linear interpolant (which is what people often think of as 'interpolating'). Compared with nearest-neighbour, convolving the data with an additional $\Pi(x/a)$ term (or multiplying the transform by an extra $\mathrm{sinc}(au)$) suppresses high frequencies more. Depending where we resample with respect to the original points, the Fourier components effectively combine with different phases, producing different degrees of smoothing. Resampling at the original points clearly just reproduces the original values, with no smoothing at all. Resampling at the mid-points, on the other hand, is the same as averaging pairs of samples and gives maximal smoothing; this is exactly equivalent to sampling with pixels that are twice as large to begin with[5], so the values are guaranteed to be sensible even when undersampled. Finally, resampling at some arbitrary position is equivalent to initial sampling with pixels that have a spatial step in sensitivity from one half to the other (Figure 2.26).

Cubic (third-degree) splines are a special case of piece-wise cubic polynomials, where the coefficients of each piece are constrained to give continuous first and second derivatives at the knots, hence a smooth curve. Following on from the nearest-neighbour and linear cases, the cubic b-spline kernel is $\Pi(x/a)$ convolved with itself three times. Its

[5] Since the sampling rate is unchanged, the larger pixels would have to overlap, as for multiple dither positions.

Fourier transform is therefore $\propto \text{sinc}^4(au)$, which quickly drops to ~ 0 beyond the central peak, but also suppresses frequencies below the Nyquist limit substantially. In fact, the Central Limit Theorem (Bracewell, 2000) asserts that the result of repeated convolution tends towards a Gaussian curve, so higher-order b-splines and their transforms are all Gaussian-like. Convolving sampled data with a cubic b-spline gives a well-behaved result, which is more accurate than nearest-neighbour or linear interpolants but which causes strong smoothing of the data. It is also not strictly an interpolation because it does not pass through the original samples.

Better results can be achieved using a cubic spline curve that *does* interpolate the data. Instead of simply convolving samples with the cubic b-spline kernel, a set of coefficients can be derived such that the interpolant behaves more like the ideal sinc reconstructor (e.g. Lehmann *et al.*, 1999). This is equivalent to convolution with a kernel that is similar to a sinc function, with both positive and negative values but less 'rippling' beyond the central peak. Its Fourier transform is box-like, but with more of a tapered cut-off. In practice, cubic spline interpolation provides a good approximation to sinc reconstruction, but with better cut-off characteristics and a reasonable number of values to compute at each point. It is therefore the best compromise for many applications involving reasonably well-sampled data.

Polynomial interpolants can also be constructed that directly approximate the form of a sinc function, for use as convolution kernels. These can again be simple and relatively compact, producing sinc-like results with less computation and less oscillation than a summation of sinc functions. Some other possible basis functions include wavelets, Fourier series, Taylor series and Gaussian summations.

Irregular sampling

It is possible to formulate more general theories for irregular sampling, but the relatively advanced mathematics is beyond the scope of this lecture (for an overview, see Marvasti, 2001). In principle, a band-limited signal can be specified by as few samples as it contains unknown frequency coefficients – at least the average interval between samples must meet the regular Nyquist criterion (e.g. Beutler, 1966). However, the vectors describing sample weights in the space of the basis functions are no longer orthogonal for non-uniform spacing, so reconstructing the signal can become an ill-conditioned problem. For guaranteed stability, some methods require that the maximum separation between samples is no greater than the Nyquist interval.

Reconstruction from irregular samples can be approached in more than one way. For specific sampling patterns, an analytical solution can be derived; a simple case is that of interlaced grids, dealt with by Bracewell (2000). Alternatively, one can approximate the signal by fitting regular basis functions, perhaps iteratively (e.g. Strohmer, 1993). Either way, any clustering of samples can lead to strong amplification of noise in the reconstruction, especially when working close to the Nyquist limit (e.g. Lauer, 1999).

Irregular sampling is most likely to be of concern where the spacing varies between adjacent elements at the IFU input, particularly with optical fibres (assuming the sample density is high enough in the first place to allow interpolation). It may also occur indirectly in the rare case where fibre profiles are undersampled at the detector (Turner, 2001). The spatial pixels of an image slicer form a grid that appears irregular in two or more dimensions, but which can be separated into a series of regular grids in 1D. For datasets that are only slightly non-uniform on large scales due to geometric distortions, it may be acceptable just to treat the sampling as regular, as long as the Nyquist criterion is still met, and then resample the interpolant at variable intervals corresponding to a regular grid in the appropriate system.

2.4.5 *Summary and application to astronomical data*

For well-sampled data, it is possible to resample images with very little degradation, as if we had sampled with a convenient regular grid to begin with. A diffraction-limited telescope image can be recovered in its original analogue form as long as we sample with ≥ 2 pixels across the width of the Airy PSF. For seeing-limited images, at least 2 pixels/FWHM are needed to reduce aliasing to a level comparable with other errors, such as flat-fielding systematics or noise. For accurate reconstruction with minimal smoothing, a tapered sinc function or smooth piecewise interpolant such as a cubic spline is likely to be the best option.

Close to the Nyquist limit or in the presence of image defects, it is best to avoid convolution kernels with too sharp a cut-off in the frequency domain, especially truncated sinc functions; these accentuate the ringing close to sharp features that occurs when idealizations such as diffraction-limited resolution are not met. Smooth spline-like interpolants with compact kernels are better behaved, but still prone to some oscillation. Convolution kernels that are positive everywhere are insensitive to such issues, but can cause strong smoothing, especially near the Nyquist limit.

For undersampled data, one only knows the image intensity at the sample points, with insufficient information to say what happens in between. The ideal strategy is therefore to avoid interpolating altogether. This is not always practical, however, when data are on a non-rectangular grid with distortions and perhaps mutual offsets. It may be necessary to interpolate in order to combine datasets or work with a square grid (otherwise analysis would require special end-to-end treatment, incorporating basic processing). Unfortunately, sinc-like interpolation of undersampled data would introduce serious aliasing artifacts, whilst smooth, positive convolution kernels could cause excessive loss of detail. The best compromise may therefore be to use simple nearest-neighbour or linear methods, or the related Drizzle algorithm (Fruchter and Hook, 2002). These are relatively inaccurate in terms of reconstructing a smooth image, but ensure that values remain meaningful. They can, however, introduce their own peculiarities in the presence of image distortions, such as skipping nearest neighbours or variable smoothing.

Whilst the above discussion considers sampling telescope images, the same principles apply to sampling a dispersed spectrum. The spectral PSF of a slit spectrograph or image slicer is a rectangular slit (for extended sources) or a narrow section of the telescope PSF, truncated by the slit (for stars). Although the edges are smoothed slightly by the finite resolution of the optics, these spectral profiles pass high frequencies more strongly than a Gaussian or Airy profile of similar width, so 2 pixels/FWHM does not work quite as well. One may therefore need to be careful to use an interpolant that damps high frequencies without a sudden cut-off. This is less of an issue for spectra with intrinsically broad features, where the sampling may be good even though the instrumental PSF for narrow lines would be undersampled. Fibre or lenslet spectra, on the other hand, often produce profiles closer to a Gaussian curve, so can be treated more like seeing-limited images.

2.5 Overview of data reduction issues

The aims of basic IFU data reduction are fairly straightforward. Ideally, we would like to generate an end product similar to that which a '3D' colour-sensitive detector would provide. The actual data storage format may (or may not) be different in detail, but functionally equivalent. At the same time, we would like to preserve the integrity of raw data, which, amongst other things, means avoiding degradation through excessive

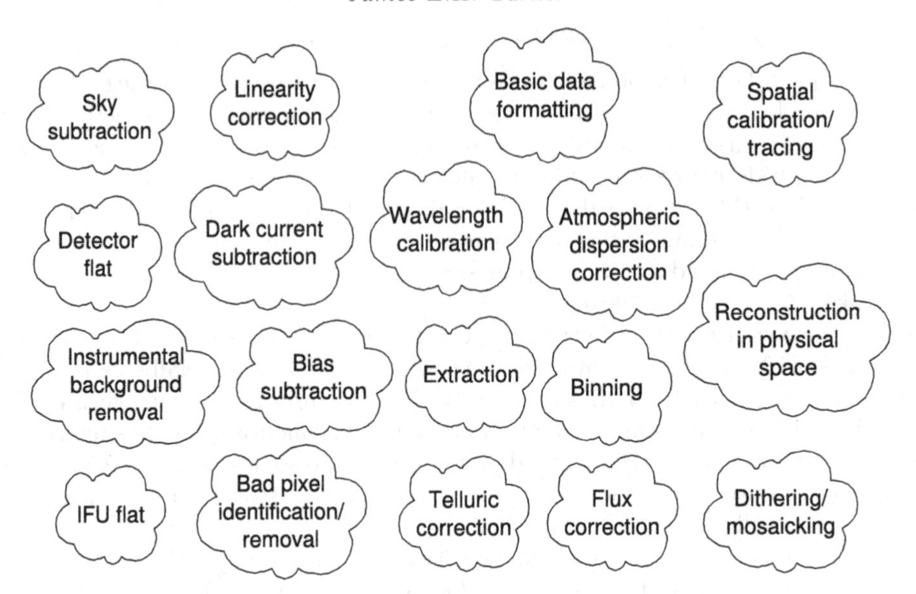

FIGURE 2.27. Key IFU data reduction steps, in no particular order.

or inappropriate resampling. The whole process should be optimized, depending on the project, to meet suitable signal-to-noise or resolution criteria. Finally, as well as rectifying individual frames, we often need to combine exposures taken at different telescope pointings, airmasses or instrument configurations.

Given that the first of these goals usually demands resampling, whereas the second suggests avoiding it, one often has to prioritize one of these over the other, especially where images are undersampled by the IFU. This is a key reason for the co-existence of 'datacube' and 'Euro3D' formats for processed data (see Section 2.6).

2.5.1 *Common steps*

The sequence in which data are reduced varies significantly depending on wavelength and instrument, but there are a number of typical steps (Figure 2.27); many are common to other spectroscopic modes, but have different details for IFS. In the remainder of this lecture, I consider each operation in turn, roughly in the order of application. Although details vary between instruments, the concepts are fairly generic. In the next lecture, I shall give some examples of how the steps can be combined in a complete reduction process.

2.5.2 *Detector linearity and saturation*

Optical CCD detector counts often scale almost linearly with flux until close to saturation, which typically occurs when the counts reach a fixed maximum level (such as $2^{16} - 1 = 65{,}535$ counts), corresponding to the limit of the analogue-to-digital converter. Any pixels read-out at the saturation limit can be flagged as bad before further reduction takes place. Clearly it is preferable to avoid saturating in the first place, however, especially where bleeding into neighbouring pixels is likely to be an issue.

Infrared detector arrays are significantly non-linear in their response to the number of photons accumulated. At the start of the reduction process, the measured count level must therefore be corrected to scale linearly with flux, referring to a pre-determined calibration curve. When IR arrays approach saturation, at their full well capacity, the counts first become unusably non-linear and often then 'wrap around', dropping back

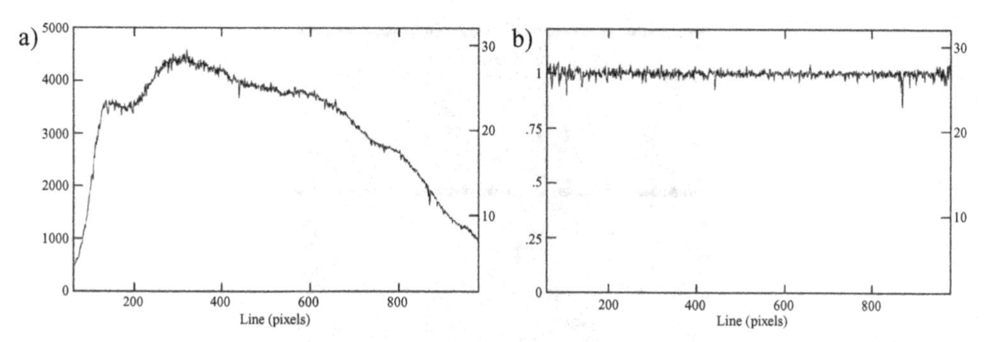

FIGURE 2.28. Fitting and dividing out the continuum shape in a calibration lamp spectrum a), to produce a detector pixel flat field b).

towards zero again; it may therefore be more difficult to identify highly saturated pixels automatically than in the optical.

2.5.3 *Detector bias subtraction*

For CCDs, the average of several bias frames can be subtracted directly from raw data, to correct the zero point for each pixel. If the mean bias drifts measurably with time, one can also average or fit the bias level in the overscan region of science exposures or flats (where applicable) and use the value to correct their overall zero point.

For IR arrays, there is usually no need to do anything regarding bias, unless the detector controller saves 'before' and 'after' reads separately, in which case those pairs need subtracting here.

2.5.4 *Consolidating and formatting the data*

By an early stage in the process, data should be packaged in an optimal format for further reduction. If the detector is read out in separate quadrants, those might need pasting together. Any overscan columns can be trimmed after subtraction, since they do not contain real data. If a standard bad-pixel mask is available for the detector, one may want to store that information with each exposure as part of a data quality array (see Section 2.6) and/or blank out or interpolate over the corresponding pixels. A variance array can be created from the almost-raw pixel values if one intends to propagate error information (Section 2.6). Finally, a description of the IFU sky-to-detector mapping needs to be stored somewhere; it is often convenient to include this as part of each image file, as a flexible image transport system (FITS) table or in header keywords. Such information can also be attached at this stage, if not included automatically when exposures are read out.

2.5.5 *Flat-fielding*

As discussed in Section 2.3, it is necessary to correct for differences in sensitivity both between detector pixels and across the IFU field. Depending on the instrument, these aspects can be treated separately or together.

To create a detector flat, a set of several dispersed calibration lamp flats are usually combined, giving a signal-to-noise ratio high enough that the flat will not contribute random noise when applied to science data. The bias should, of course, be removed as well. The spectral continuum shape of the co-added flat may be fitted and divided out at each spatial position, leaving only small-scale pixel-to-pixel differences (Figure 2.28); any larger-scale variation in detector sensitivity with wavelength is usually dealt with during flux calibration, rather than here. The resulting flat should be normalized to an average

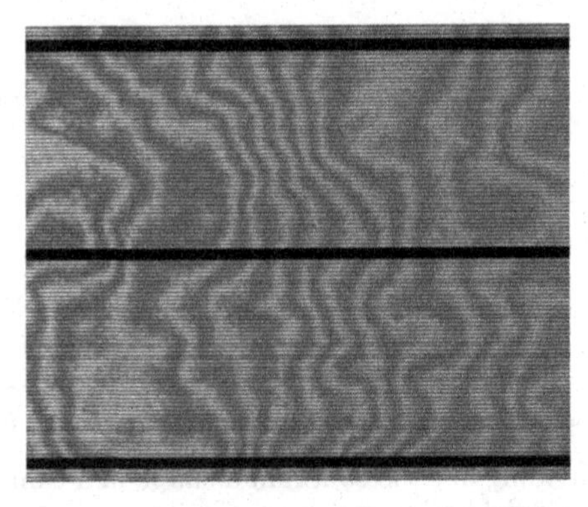

FIGURE 2.29. Detector fringing in optical fibre IFU spectra: wavelength increases to the left; the narrow horizontal stripes are individual fibres; the thicker dark bands are gaps between blocks of 50 fibres; a small subsection of the detector is shown.

of one, of course, to preserve flux. Raw science spectra are then divided pixel-for-pixel by the flat, either separately or as part of a combined pixel+IFU correction (see below).

Particularly for CCDs at redder wavelengths, it may be necessary to deal with fringing due to optical interference (e.g. Figure 2.29). This occurs when light is reflected back and forth within the detector substrate (or perhaps between filter surfaces) before being absorbed. For spectroscopy, this is a multiplicative effect; each point on the detector receives quasi-monochromatic light, so fringing is a function of the total amplitude at the relevant wavelength. In principle, fringing can therefore be removed by flat-fielding the detector. However, the process then becomes sensitive to any small drifts in flexure or grating tilt, since differences in illumination between science exposures and flats alter the pattern, leaving systematic residuals.

For spectrographs that use a mosaic of separate detectors, any chip-to-chip differences in flat-fielding must be treated carefully. Colour sensitivity curves can vary between CCDs, for example, leaving discontinuities that cannot be corrected by scaling each chip or fitting a smooth lamp continuum across the chips. Instead, the quantum efficiency of each CCD may need to be factored in as a function of wavelength. Fringing properties can also vary significantly between chips.

An IFU flat corrects the relative transmissions of different IFU elements, or variations along image slices, to the same system (Figure 2.30). In the absence of relative flexure between flats and science exposures, pixel-to-pixel detector variations can also be left in, giving a combined pixel+IFU flat. This requires a calibration lamp spectrum, since twilight flats have many sky lines and are thus unusable for calibrating spectral variations. It is usually possible to match the flexure of science data by taking flats often enough. This also allows IFU elements to be traced easily by using the flats as a reference. Where there is flexure between IFU flats and science data, however, both should be divided by a normalized detector pixel flat, removing super-Nyquist variations, before shifting them to match. To obtain sufficient signal in between the centres of IFU elements, the detector flat may need to be taken without the IFU in the beam, perhaps using a long slit matching the spectral illumination of the detector by the IFU.

Where a calibration lamp does not accurately mimic the illumination of the IFU by the sky, it can be combined with a twilight exposure to correct the spatial content. The lamp flat provides detector pixel variations and any differences in wavelength response between IFU elements, whilst the twilight contributes spatial variations between IFU elements, or just a large-scale flatness correction to the spatial information from the lamp flat. The

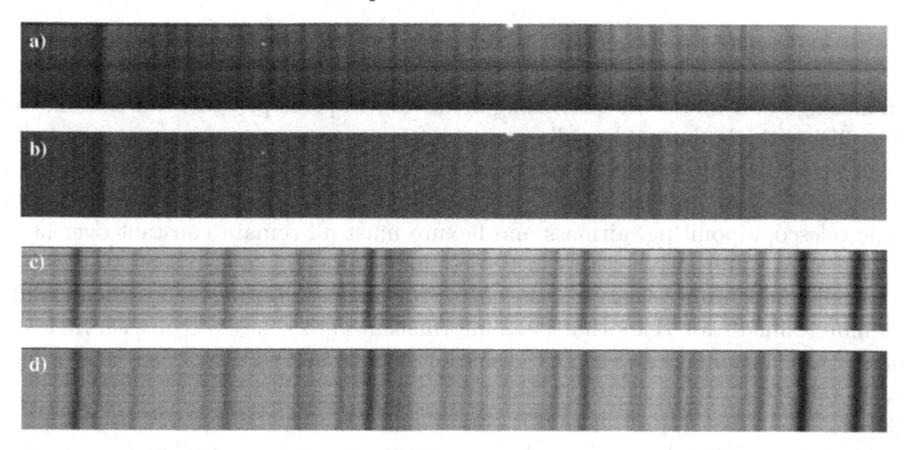

FIGURE 2.30. Example IFU spectra before and after flat-fielding out detector and IFU structure. From top to bottom, the panels are a) a single image slice, with structure from the slicing optics, b) the same exposure after a lamp flat is applied, c) a stack of extracted fibre spectra, one per row and d) the same data after a combined lamp and twilight flat is applied. Twilight spectra have been used for illustration, since they are bright and highlight structure clearly.

spectral content can be removed from twilight spectra by summing in wavelength, which also helps with signal-to-noise requirements. For consistent results, it is important to sum over the same range of wavelength for each element, rather than a fixed number of pixels. One can then compare the ratios of spatial variations in twilight and lamp flats, adjusting the latter to match the former.

If IFU elements have differing transmission spectra, the lamp flat should be normalized to account for this, dividing each spectrum by the *average* over the elements of the continuum spectrum for the lamp. This preserves spectral corrections to the overall flux calibration (which is derived from a standard star). It may be necessary to shift and scale the continuum to match each element. Otherwise, if spectral differences are negligible, the continuum can be divided out separately from each IFU element, leaving only pixel variations. One would not normally expect to encounter large wavelength-dependent differences across an IFU, but the possibility is worth considering, to avoid extra systematic errors.

When reducing data from optical fibres and lenslet arrays, one can normally extract each fibre or lens spectrum to 1D before flat-fielding. The flat can still incorporate pixel variations, as long as science and flat-field spectra are extracted with the same weights. If neighbouring lenses or fibres are to be deconvolved in science spectra, the same procedure should be applied to the flat. Each spectrum is then divided through by the final flat as normal. An exception is the unusual case of unresolved fibres, which can be treated more like image slicer data (Turner, 2001). In principle, a possible complication with fibres is that their relative throughputs could vary with flexure, depending on exactly how each one is stressed (e.g. Oliveira *et al.*, 2005). In practice, this has not been noticeable in the author's experience, but it gives yet another motive for taking flats to match the flexure of science exposures.

In image slicer data, each slice can be treated much like a slit spectrum, but accounting for differences in transmission between slices as well as along their lengths. A 2D flat can be derived for each slice by dividing out the spectral continuum of the lamp from each image row. To preserve spatial variations, a consistent normalization should be maintained both along and between slices when doing this.

2.5.6 *Cosmic ray and bad pixel correction*

Before interpolating or co-adding data, any bad pixels should be masked out to avoid contaminating good values (or generating artifacts). In principle, this can be done either in 2D or 3D, with the latter providing more information at some expense in simplicity. Any exposures that are genuinely identical (apart from noise) can simply be co-added with statistical rejection in 2D. This often applies only for short integrations, however, since the telescope pointing, airmass and flexure must all remain constant over at least three frames. An alternative is to use PSF (band limitation) constraints to detect features in individual frames that are too narrow or steep to be real (e.g. Rhoads, 2000). This is more difficult for IFUs that pack information onto the detector with little or no redundancy, such as image slicers, but easier for smooth and well-sampled fibre or lenslet PSFs.

Following a similar approach in 3D, one could identify bad pixels in the process of constructing a datacube. This makes a greater number of neighbours available for constraining (and interpolating) values, but does not allow for interpolation prior to 3D reconstruction. When co-adding data in 3D, statistical rejection can be applied to the interpolated values and used alongside data quality arrays from an earlier stage, to identify bad samples.

Given the small number of spatial elements and high density of information in a typical IFU dataset, defective pixels may have a relatively high impact. At the same time, they can be less straightforward to identify than when co-adding direct images, extracting a 2D spectrum of a point source or working with more extended slit spectra. This reduction stage may therefore require some care.

2.5.7 *Instrumental background*

To ensure that measurements reflect the true object+sky intensity, any instrumental background counts must be removed. Especially with fibres, nearby elements at the IFU input can be widely separated at the detector, so it is possible for varying background levels to introduce structure on small spatial scales. Where there are gaps between IFU elements (or blocks of elements) at the output, the background level in the gaps can be used to model and remove any diffuse scattered light. In the infrared, however, dark current and any other target-independent background sources are normally taken care of by sky subtraction, without special treatment. Note that background light can have different colour characteristics from the dispersed object signal and change between mosaiced detectors.

2.5.8 *Spatial calibration*

Before data can be transformed into a physical coordinate system, it is necessary to identify IFU elements (fibres, lenses or slices) on the detector and calibrate their spatial coordinates. This usually involves locating the spatial elements in a flat-field spectrum matching the science data, with reference to a known IFU mapping.

For individually resolved optical fibres, one can identify fibre peaks in a cross-section perpendicular to the dispersion axis (see Figure 2.7) and trace them along the length of the detector. If individual spectra are heavily blended together, one can use any gaps between blocks of fibres or perhaps cross-correlation of the fibre throughput pattern to calculate where individual elements are located. If there is no flexure (e.g. at a Nasmyth focus), the positions can be pre-determined by masking out groups of fibres in a flat and tracing one group at a time.

For lenslet IFUs, we are again looking for peaks in cross-section, but with the spectra distributed around the detector in 2D rather than forming a slit. An optical model is typically used to determine the relative positions and lengths of spectra for a given sky-to-detector mapping, reducing the problem to one of measuring zero-point offsets from the peaks.

For image slicers, any gaps between the slices at the output can be used to find their edges. A special Ronchi screen (consisting of regular parallel lines) may also be placed in front of the IFU input, to calibrate the alignment and scale of each slice along its length. If slices are staggered in wavelength, steps in the wavelength solution could provide an alternative indication as to where the edges are. In the absence of flexure, pre-determined pixel positions can again be used. Having found the edges of the slices (or Ronchi bands), the tilt and curvature of each one can be traced along the dispersion axis of the detector. The usable length of each slice can be delimited according to some data threshold or a fixed number of pixels from the centre.

2.5.9 *Wavelength calibration*

Depending on the type of IFU, a set of 1D or 2D spectra have to be calibrated in wavelength. Fibre spectra can usually be calibrated individually after extraction to 1D. Lenslet spectra are sometimes calibrated to first order as part of the extraction process; the results can be refined later in 1D if needed. For image slicers, we can fit coordinates separately for each 2D spectrum. Whatever the format, the wavelength reference is most often an arc lamp spectrum (Figure 2.8), processed in the same way as science data. This is used to fit a relatively low-order curve in 1D or 2D for each spatial element, describing the wavelength as a function of pixel index. The procedure is similar to that for multi-object spectroscopy with fibres or slits, except that manual interaction may not always be practical for IFUs with thousands of spectra to calibrate, in which case reliable automatic line identification is more critical.

For science data containing identifiable sky lines, these can be used to account for any flexure with respect to the arc exposures, applying a small zero-point correction to the full wavelength solution from the arc (otherwise, arcs must be taken often enough to match science data). At infrared wavelengths, there are often enough sky lines over the length of a spectrum to fit a full wavelength solution without having to use an arc at all. Once a wavelength has been determined for each raw or extracted sample point, there are two options for rectification (see also Section 2.5.10). Typically, individual 1D or 2D spectra are transformed to a common, linear grid before continuing with further processing. It is also possible, however, to map the points directly to 3D without an intermediate resampling stage.

Given the presence of optical distortions across the spectrograph and/or step offsets between image slices in some designs, it is quite common for different IFU elements to have slightly different spectral ranges. This leads to a dataset covering only parts of the IFU field at the ends of the spectra.

2.5.10 *Extraction*

This stage involves extracting IFU elements from the raw data format to something convenient for later processing. For fibres or lenses, we usually want to extract each spatial element to a 1D spectrum, integrating in cross-section with or without weights (Figure 2.31). This is complicated by the fact that adjacent spectra often overlap substantially in cross-section (e.g. Figure 2.7). For fibres aligned in wavelength, this amounts to an extra convolution in one spatial dimension: the simplest option is to accept a small loss of spatial resolution, summing over each fibre along with some signal from its neighbours. Where spatial resolution is exceptionally important, the alternative is to try deconvolving neighbouring spectra, at a significant cost in extra noise. Deconvolution is sensitive

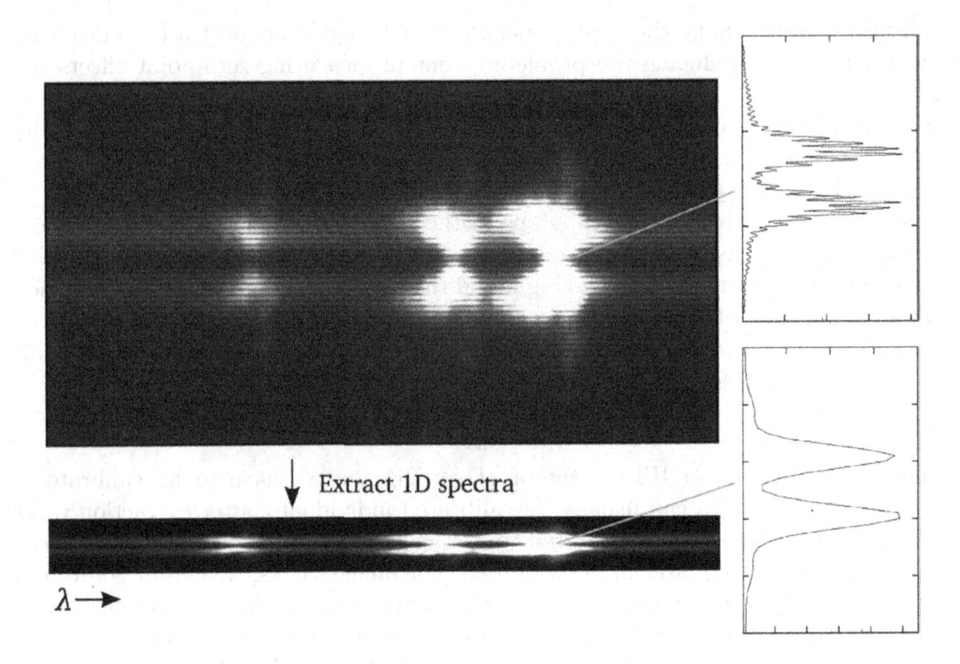

FIGURE 2.31. Example of GMOS fibre spectra before and after extraction to 1D by summing over each fibre in cross-section.

to knowing the PSF accurately, which is complicated by variations between fibres or lenslets, but has been used successfully for IFUs such as CIRPASS and OASIS. The unusual case of very closely spaced (~ 1 detector pixel) fibres can be treated more like an image slicer, without extracting to 1D (Turner, 2001).

Since adjacent fibres can have different throughputs, spectra extracted without deconvolution receive slightly different contributions from the fibres on either side. This perturbs the effective spatial coordinate of each fibre slightly along one axis, producing a sampling grid that is not quite regular on the sky. In practice, this effect may be very small, except close to dead fibres, and is sometimes safe to ignore.

For lenslet IFUs (or misaligned fibres), any overlaps between spectra are offset in the dispersion direction. Rather than just causing slight spatial smoothing, this contaminates spectra with light from the wrong wavelengths. It is therefore important to deconvolve any signal from neighbouring lenslets in this scenario, to avoid spurious measurements.

For image slicers, extraction generally does not involve summing over pixels, but sooner or later one wants to 'cut out' the useful regions of raw images corresponding to the slices. This is done after tracing the slice positions, before or after wavelength calibration. One approach is to straighten each slice, producing a traditional 2D extracted spectrum, as for a slit. This may be convenient for further processing with existing software. All the slices should be resampled onto the same grid in physical (e.g. x, λ) coordinates, so they can simply be stacked later in 3D along the other spatial axis (y). Even so, a further resampling stage may still be necessary, to correct atmospheric dispersion or combine datasets from different pointings. The other possibility for 'extraction' is to map pixel values directly from the original raw format to a 3D representation, after calibrating in wavelength, without any intermediate resampling in 2D. The latter method is appropriate for minimizing the interpolation of undersampled data.

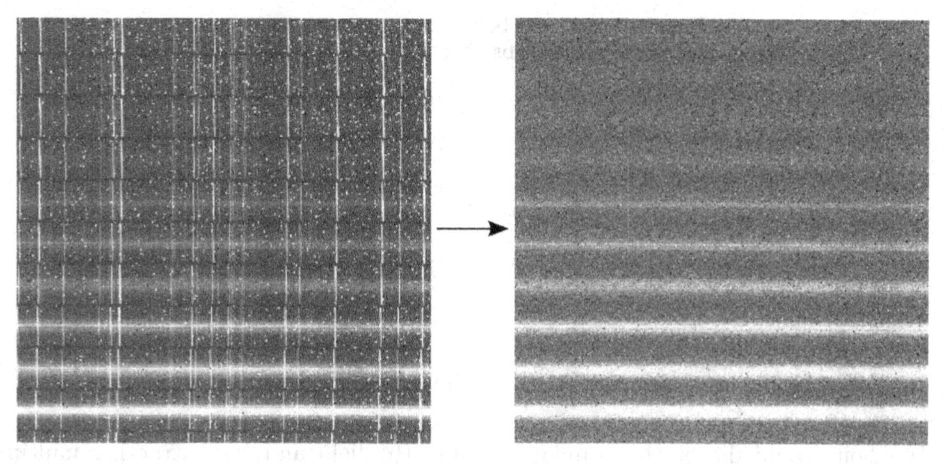

FIGURE 2.32. Section of a near-infrared image slicer spectrum, before and after sky subtraction. Many bright pixels with high dark current also subtract out well.

2.5.11 *Sky subtraction*

To measure the signal from a science target, sky emission lines and continuum must first be removed. For infrared (or optical 'nod-and-shuffle') data, sky and any other background are normally removed by subtracting pairs of consecutive raw exposures early on (e.g. Figure 2.32). This means that corresponding object and sky measurements are taken with the same IFU elements and detector pixels, avoiding any subtraction residuals due to flat-fielding errors or slight mismatches in spectral profiles between different IFU elements. If the sky varies significantly between exposures, one can attempt to interpolate between sky frames taken before and after the target object, or if darks are available, it may be feasible to scale the sky to the correct level. Telluric line species can vary differently, however, so may not always scale well together.

For most optical data, sky is subtracted from 1D/2D calibrated spectra or a 3D datacube. The sky spectrum can be averaged over a number of spatial pixels with no object signal, to reduce the noise level, before subtracting it from all the spatial pixels. Thus the process adds little extra noise to the result. Depending on the instrument and target, the sky spectrum can be taken from separate sky fibres, the edges of the science field or a blank sky integration. Since different optical fibres or lenslets can have slightly different PSFs, subtraction residuals may be an issue (Allington-Smith and Content, 1998). Convolving object and sky spectra to the same resolution could help avoid this, at a cost. One could also attempt to match up IFU elements with similar PSFs, either empirically or nearby on the detector, where optical variations within the spectrograph are an issue. This has not yet been tested, to the author's knowledge.

2.5.12 *Telluric correction*

In red or infrared spectra, any telluric absorption lines that obscure the true target spectrum need removing. To obtain a 1D telluric reference spectrum, a standard star observation is reduced in the same way as scientific data and its flux is summed up spatially. If necessary, one can try to match the airmass of a science exposure by interpolating between telluric spectra taken at different airmasses. The spectrum can be normalized by fitting and dividing out the continuum shape (interpolating over any weak stellar absorption lines). The result is then used to divide out telluric features from 2D calibrated science spectra or a 3D datacube (Figure 2.33). If the spectral PSF varies over

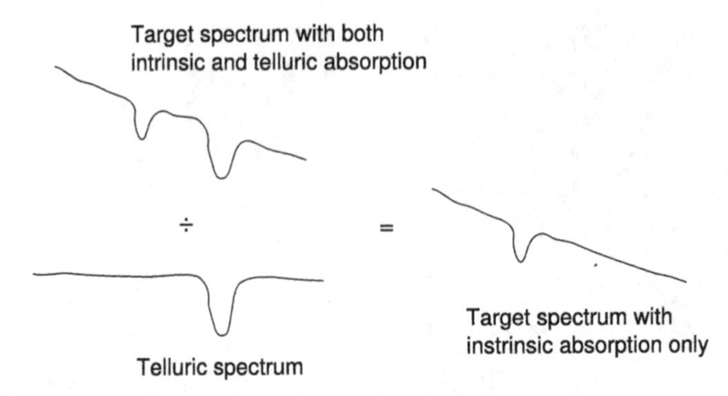

FIGURE 2.33. Sketch illustrating correction for telluric absorption.

the IFU, one could dither the standard around the field and try extracting multiple spectra, for calibrating different groups of elements.

2.5.13 *Flux calibration*

Here, a standard star is used to correct measured counts for the combined transmission of the instrument, telescope and atmosphere as a function of wavelength. As for telluric calibration, we reduce a standard star observation with the same instrument configuration as the corresponding scientific data and sum up flux spatially, to get a 1D spectrum. The same dataset might, in fact, be used for both purposes. To derive a sensitivity or throughput spectrum (e.g. Figure 2.10), one must somehow determine the intrinsic flux of the standard star as a function of wavelength and compare this with the measured counts. In the optical, one can refer to tabulated narrow-band fluxes that are available for numerous stars in the literature (and in standard reduction software). In the infrared, one can use a published broad-band magnitude and spectral type (i.e. temperature) to generate a model black-body curve matching the continuum of the standard. Reduced scientific data in 1D–3D are then divided through at each spatial point by the derived sensitivity function, converting measured counts into fluxes on a consistent system such as $\mathrm{W\,m^{-2}\,\AA^{-1}\,arcsec^{-2}}$. Where either standard star or scientific observations have been taken through cloud, normalizing to some arbitrary scale might be more appropriate. Although much variation of the sensitivity spectrum over the IFU field is unlikely, appropriate flat fielding can take care of any differences (Section 2.5.5).

2.5.14 *Reconstruction in 3D*

Reduced data in 1D or 2D arrays can be transformed back to a physical coordinate grid either before or after scientific analysis, depending on the application. In general, this requires interpolation. One approach is to resample once onto a 3D datacube in (x, y, λ), for subsequent analysis and visualization. Special tools (such as the Euro3D software) can also be used to interpolate 'on the fly', as needed during visualization or analysis. Where applicable, a third possibility is to analyse individual 1D spectra and interpolate the results onto a 2D map. Finally, one may prefer to perform analysis directly in 3D at the original, calibrated pixel positions, perhaps by fitting a physical model of the target. This last method avoids resampling and any associated uncertainties, but makes it difficult to use existing, general-purpose software.

The best process for transforming data to a uniform grid depends on both the initial sampling and IFU geometry. For well-sampled datasets, one can interpolate the data with smooth surfaces that reconstruct the dispersed telescope image accurately, allowing

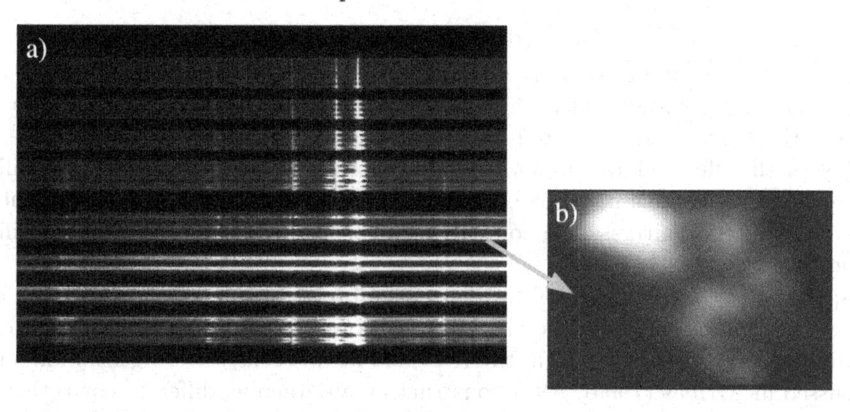

FIGURE 2.34. a) Example of extracted fibre IFU data from a Seyfert galaxy observation and b) one image plane of the datacube produced by resampling spatially with smooth, piecewise polynomials.

resampling onto datacubes or 2D spatial maps with minimal degradation (Figure 2.34). The result is therefore similar to what a 3D detector might provide more directly. For undersampled data, extra care must be taken to avoid introducing aliasing artifacts or excessive smoothing; it is not possible to reconstruct exactly what a 3D detector would see. One should resample as little as possible – maybe once, to get to 3D – and stick to positive, compact interpolants, such as nearest-neighbour, linear or Drizzle weightings.

Depending on the geometry of the IFU, one has a choice as to how many dimensions to interpolate in at once. Resolved fibre or lenslet spectra can be extracted to 1D and resampled onto a common grid in wavelength before interpolating spatially. Where the IFU input consists of square lenslets, the spectra can then simply be arranged to form a 2D grid spatially. For lenslets or fibres packed in a triangular (hexagonal) configuration, a square spatial grid is obtained by interpolating between the spectra at each wavelength element, in either 1D or 2D. To deal with one dimension at a time, it is possible to interpolate first along each linear row of hexagons or circles and then across the rows.

With image slices, the (x, λ) coordinates of each 2D spectrum are not trivially separable, due to image distortions. It therefore makes more sense to interpolate spatially and in wavelength at the same time. This is the same problem as straightening and wavelength calibrating a long-slit spectrum in 2D, except that all the slices should be transformed to the same (x, λ) grid, so cannot be treated completely independently. Once the slices are rectified, they need just to be stacked in y to get to 3D. Image slicers often have rectangular spatial pixels, however, with the slice width being different from the detector pixel size. In those cases, one can interpolate between slices in y, to get to a square spatial grid.

For any type of IFU data, the alternative to resampling each element individually is to interpolate processed detector pixel values directly in 3D, after calibrating x, y and λ for each pixel. This is typically more complicated (except for the simplest interpolants) than working in lower dimensions, especially where samples are distributed irregularly in physical coordinates. For interpolating smooth curves, it is significantly easier to find pre-existing software that works in fewer than three dimensions, on regular grids and with homogeneous coordinates. For undersampled data, however, transforming straight to 3D may be the best approach, using something like a nearest-neighbour or Drizzle scheme to map pixels directly onto a cubic grid. As of 2006, the Gemini IRAF (image reduction and analysis facility) package includes a task, 'gemcube', that can do this using a version of the Drizzle algorithm in 3D.

2.5.15 *Atmospheric dispersion*

Atmospheric dispersion, or differential refraction, causes the position of a target within the IFU field to vary with wavelength. This is typically less problematic than for a single slit, where the target can be shifted entirely out of the aperture at some wavelengths. Nevertheless, the effect can distort spectra and cause different-sized variations at different airmasses. Where significant, a correction must therefore be applied before combining datasets from different airmasses or comparing the positions of spatial features at different wavelengths.

The spatial offset as a function of wavelength can be determined either using an atmospheric refraction model (given the airmass, position angle and other parameters) or by centroiding on a point source within the IFU field at different wavelengths. Some details are discussed in Arribas (1999). After constructing a datacube, different corrective shifts can then be applied to each wavelength plane. Alternatively, the sky coordinates of each pixel (as calculated during spatial calibration) can be recalculated before transforming to 3D or comparing with a model. At low airmass or infrared wavelengths without adaptive optics, it may be safe to assume dispersion is negligible. Also, when integrating flux over a spatial aperture centred on a point source, dispersion is taken care of without special treatment.

2.5.16 *Dithering and mosaicking*

Dithering and mosaicking are discussed in the accompanying lectures by Pierre Ferruit, so are mentioned here only briefly. This stage often involves shifting and adding already-resampled data in 3D, but can also be integrated with 3D reconstruction, resampling multiple exposures onto the same grid. The latter approach would be better with spatial undersampling and/or for combining sub-pixel dithers. Occasionally, where exposures are taken with integer-spatial-pixel offsets and negligible differences in flexure or airmass, it may be possible to co-add extracted spectra instead of 3D data.

2.5.17 *Spatial binning*

For IFUs with small spatial pixels, it is often difficult to obtain a good signal-to-noise ratio across the full field. Targets such as galaxies can vary spatially over a large range of brightness within a few arcseconds. To make optimal use of the available information, one may want to bin data adaptively, in variably sized spatial regions. The Voronoi tessellation method of Cappellari and Copin (2003) divides IFU data into irregular bins of approximately constant minimum signal-to-noise ratio, allowing reliable spectral mapping over the whole field (Figure 2.35). This sacrifices critical sampling of the seeing, of course, so it is important to interpret spatial content accordingly, perhaps by comparison with a similarly binned model of the target. If the object is known to be intrinsically smooth, this may be less of a concern.

2.5.18 *Summary of data reduction issues*

A number of reduction stages are the same as, or similar to, those for other spectroscopic modes. IFUs have a few rather special requirements, however, notably for flat-fielding and mapping data from 2D to 3D. The details of how sky subtraction is done in the optical and how spectra are located and extracted are also somewhat different. Dithering and mosaicking steps are more akin to imaging reduction than slit spectroscopy. Lastly, unlike conventional imaging or spectroscopy, IFS allows (and sometimes requires) after-the-fact correction of atmospheric dispersion.

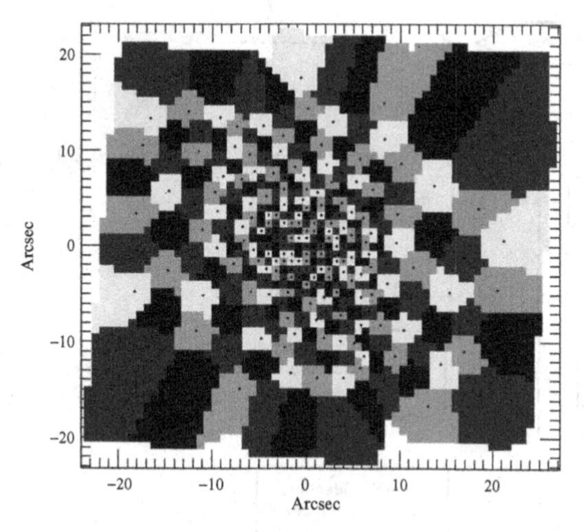

FIGURE 2.35. A spatial Voronoi tessellation of IFU data, using bins of approximately constant signal-to-noise ratio (Cappellari and Copin, 2003).

As a final note, it is worth emphasizing that, with the correct treatment, one can often use IFS data for high-fidelity imaging, as well as spectroscopy.

2.6 Data reduction process

The last lecture gave an overview of the different stages of basic IFU data reduction. Here, I discuss the data reduction process as a whole, showing some brief example sequences for optical and infrared data and considering some associated practicalities, namely error and data quality propagation and file formats.

2.6.1 *Reduction sequence*

Figure 2.36 and Figure 2.37 show typical optical and near-infrared data reduction sequences, respectively, for two of the Gemini IFUs. Data from other well-sampled optical fibre or infrared image slicer IFUs could be reduced similarly. Separate sequences for calibration data such as flats, biases and standard stars are not shown here, but these usually follow the same process part way through, before or after combining exposures. For example, optical flats are reduced up to the stage of bias subtraction and co-added. Standard stars are usually treated like scientific exposures, except for flux or telluric correction, before extracting a spectrum from the 2D or 3D dataset.

2.6.2 *Error propagation*

At the end of a data reduction sequence, it is often important to have a good estimate of the errors in data values. This allows us, for example, to estimate the statistical significance of a faint detection, the reliability of stellar ages and metallicities derived from line strength indices (Cardiel *et al.*, 1998) or the intrinsic random errors in velocity measurements.

Raw data have relatively well-defined errors, due to photon statistics and detector read noise – but after numerous processing stages it is at best difficult and at worst impossible to estimate errors directly from the data values. The solution is to track the errors throughout processing. For each detector pixel, an error value is stored in a separate image, alongside the main data array. During each processing step, this array

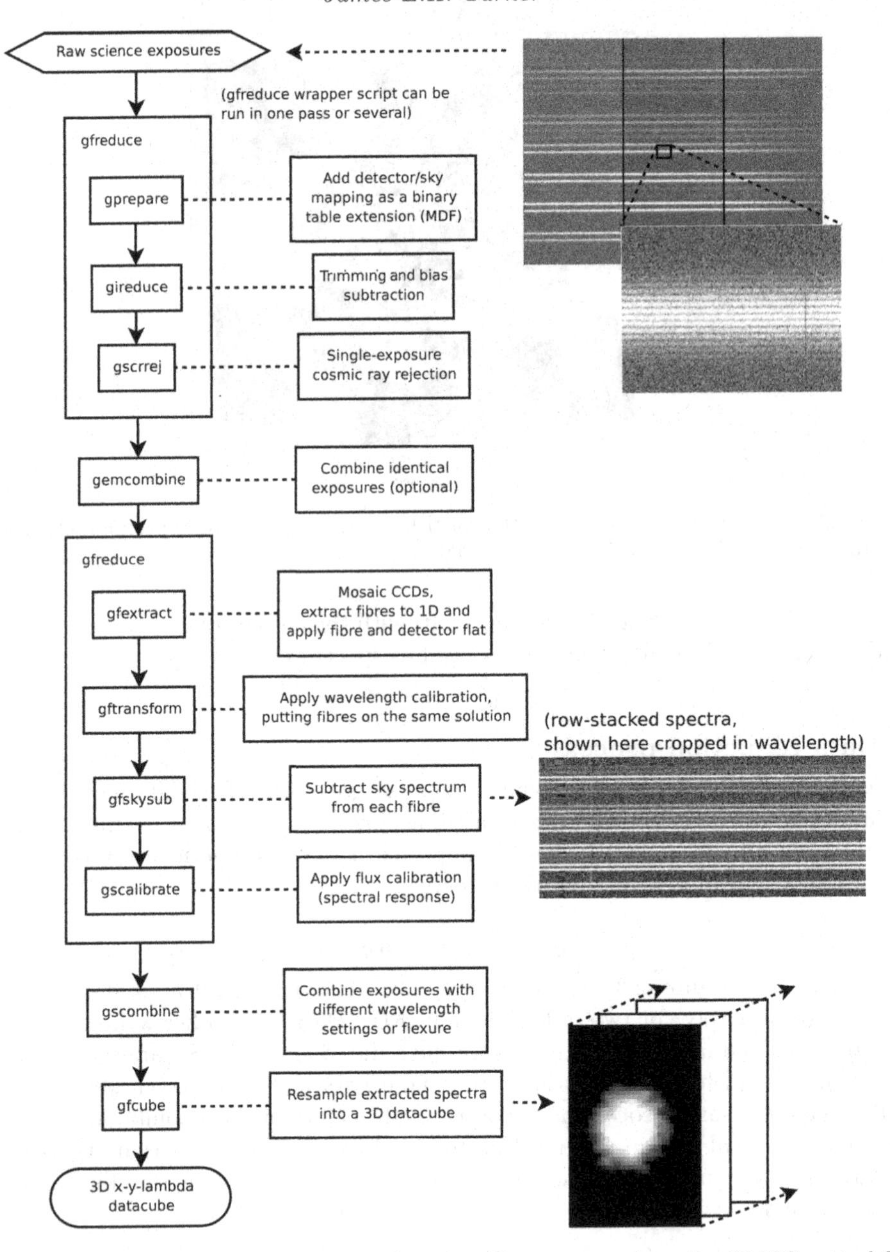

FIGURE 2.36. Example reduction sequence for scientific exposures from the GMOS optical fibre IFU, using Gemini IRAF package tasks.

is processed in parallel with the scientific image, to reflect how the errors change. When adding two images, for example, the corresponding errors are added in quadrature.

Sources of error

A process where discrete values vary statistically around a well-defined mean, such as counting photons, is described by a Poisson distribution:

$$P_{\mu}(n) = e^{-\mu}\frac{\mu^n}{n!}, \tag{2.12}$$

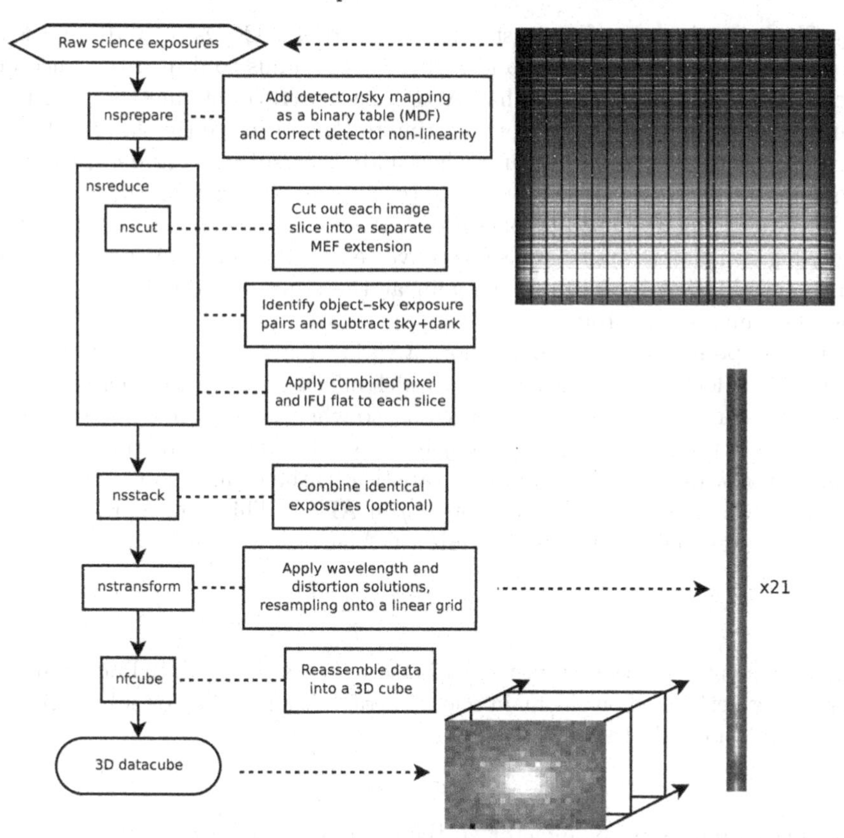

FIGURE 2.37. Example reduction sequence for scientific exposures from the GNIRS infrared image slicer, using Gemini IRAF package tasks. Telluric and flux correction are not shown, but in this case would be performed between nstransform and nfcube.

with a mean (expected number of photons) of $n = \mu$. The standard deviation from the mean is simply $\sigma \equiv \sqrt{\mu}$, so when counting photons (really photo-electrons), the statistical error is the square root of the expected number of photons (electrons). In practice, the error is estimated using the square root of the *measured* number of electrons, since that is what we know. For large μ, the Poisson distribution is Gaussian, with $\sigma = \sqrt{\mu}$.

The main random sources of measurement error in data are read noise from the detector electronics, Poisson noise from the science target and sky (or other background) and Poisson noise from detector dark current. The last two of these can be estimated as discussed above, directly from the pixel values, whilst read noise is a fixed property of the instrument. We also have systematic errors – notably inaccuracies in flat-fielding – introduced during processing or by instrumental variations. These are usually present at the level of a few percent and are difficult to reduce to zero. Such effects can be more difficult to account for than noise, but are typically dominated by statistical errors.

Creating and propagating error arrays

Detector electronics usually do not report one count per stored electron. The number of electrons corresponding to a single count, or 'analogue-to-digital unit' (ADU), is known as the detector gain (really an inverse gain). This is set to some value low enough for read noise to dominate digitization errors, but high enough to avoid reaching the maximum value of the analogue-to-digital converter at too small a fraction of the detector full-well capacity. Typical gains vary from just a few e^-/ADU for CCDs to > 10 for IR arrays, the latter having greater read noise and higher backgrounds to deal with. Since Poisson statistics apply to electrons, rather than ADUs, measured counts must be multiplied by the gain before taking the square root to estimate Poisson noise (in electrons).

Given that errors add in quadrature when adding or subtracting data, the numbers stored in an error array are actually variance (σ^2) values, rather than the noise (σ) itself, to avoid repeated square and root operations. If science data are stored in ADUs, the corresponding variance arrays must also be converted to those units. Alternatively, science data can be multiplied through by the gain at the start of reduction and everything propagated in units of electrons.

As a starting point for error propagation, a variance array can be created containing the square of the detector read noise, $\sigma^2_{\rm read}$, which affects every pixel independently of its count level. Read noise is measured in electrons, so when working in detector counts, the initial variance is then $(\sigma_{\rm read}/g)^2$, where g is the gain. After subtracting the electronic bias level from raw data, one can add the statistical variance in the measured counts, n, for each pixel, which is ng in electrons or n/g in ADUs. This sum represents the total variance for the raw data, which is therefore (in ADUs)

$$\sigma^2 = \frac{\sigma^2_{\rm read}}{g^2} + \frac{n_{\rm total} - n_{\rm bias}}{g}. \tag{2.13}$$

At each subsequent reduction step, the variance array is manipulated according to the operation being performed on the main science array. When adding or subtracting pixels, their errors add in quadrature:

$$\sigma^2 = \sigma_1^2 + \sigma_2^2. \tag{2.14}$$

When scaling an image by some factor, a, the error is scaled accordingly:

$$\sigma^2 = a^2 \sigma_1^2. \tag{2.15}$$

When one image is multiplied or divided by another, the fractional errors add in quadrature. Each variance array must therefore be divided by the square of the corresponding science image, adding the results and multiplying the sum by the square of the final science image:

$$\sigma^2 = n^2 \left(\left(\frac{\sigma_1}{n_1}\right)^2 + \left(\frac{\sigma_2}{n_2}\right)^2 \right). \tag{2.16}$$

For more complex operations on science data, represented by some arbitrary function, $n = f(n_1)$, the first derivative of f can be used to estimate how output values vary with small changes in the input for a given n:

$$\sigma^2 = f'(n_1)^2 \, \sigma_1^2. \tag{2.17}$$

At the end of the complete reduction process, error propagation allows the final noise values to be determined relatively accurately just by taking the square root of the final variance array.

Resampling

In raw data, each pixel has an independent statistical error. If resampling causes smoothing, however, the errors of different pixels become correlated. As an example, linear interpolation at the mid-point between samples is an averaging of pairs. The error of each interpolated value is therefore reduced by $\sqrt{2}$, without halving the number of pixels.

Clearly the average signal-to-noise ratio over all pixels cannot have increased, however, nor can summing two adjacent resampled pixels reduce the error by a further factor of $\sqrt{2}$, since only three of the original values are involved. What has happened is that the errors of adjacent pixels are no longer independent.

In a case like this, one could track the errors by propagating a higher-dimensional covariance matrix, where the covariance of two variables is defined as the expected value of the product of deviations from the means:

$$\sigma_{xy} = \langle (x - \bar{x})(y - \bar{y}) \rangle = \langle xy \rangle - \bar{x}\bar{y}. \tag{2.18}$$

The variance of a single variable is therefore equivalent to its own covariance:

$$\sigma_x^2 = \sigma_{xx} = \langle x^2 \rangle - \langle x \rangle^2. \tag{2.19}$$

With this information, the error on a sum $\sum_{i=1}^{n} x_i$ is calculated using

$$\sigma^2 = \sum_{i=1}^{n} \sigma_{x_i}^2 + 2 \sum_{i<j} \sigma_{x_i x_j}. \tag{2.20}$$

For more information, see an error analysis or statistics textbook, such as Taylor (1997).

Since this level of error propagation is computationally complex, reduction software usually does not track covariance. It is nevertheless important to be aware that simple variance estimates may not be exactly correct after resampling. In the ideal limit of resampling a sinc interpolant at the original rate, however, covariance is not an issue because no smoothing occurs (at any given point, the RMS distance of positively and negatively weighted sample contributions is zero).

2.6.3 *Data quality*

As well as storing variance data alongside each science image, it is useful to include 'data quality' information. A separate integer array is used to track which pixels are good or bad (and why) in the main science array. Each bit of the integer represents a Boolean yes/no flag for a particular defect, allowing for multiple problems with each pixel. For example, a value of zero typically represents a good pixel, with different bits set to one where issues such as cosmic ray contamination, saturation or high dark current are identified. Clearly, some of these result in total loss of information in a pixel, whereas others just lead to noisy values.

Data quality propagation can help significantly with masking out values at each reduction stage, given that defective samples may be impossible to identify directly from their values after processing steps such as interpolation or sky subtraction. Perhaps unfortunately, the convention as to which bits represent what defects depends on the software.

2.6.4 *File storage format*

Like other astronomical data, IFU spectra are typically stored and processed in 'flexible image transport system' (FITS) files. This standard is overseen by an IAU working group and documented at http://fits.gsfc.nasa.gov. Each single FITS file can contain multiple datasets, along with information describing the data. The most common elements are the following:

- one or more N-dimensional image arrays;
- ASCII header information, using keyword=value pairs of multiple data types (e.g. OBJECT = 'NGC1068' or EXPTIME = 120);
- one or more binary tables, using named columns (e.g. XCOORD) and mixed types to store heterogeneous data.

EXT#	EXTTYPE	EXTNAME	EXTVER	DIMENS	BITPIX	OBJECT
0						NGC1068
1	BINTABLE	MDF		32x21	8	
2	IMAGE	SCI	1	32x1022	-32	NGC1068
3	IMAGE	VAR	1	32x1022	-32	Variance
4	IMAGE	DQ	1	32x1022	32	DQ
5	IMAGE	SCI	2	32x1022	-32	NGC1068
6	IMAGE	VAR	2	32x1022	-32	Variance
7	IMAGE	DQ	2	32x1022	32	DQ

[... etc ...]

FIGURE 2.38. Example multi-extension FITS structure for part-processed image slicer data, from the edited output of the IRAF task fxheader. The full file continues until EXTVER=21, corresponding to the 21st image slice. The 'MDF' (mask definition file) extension is a binary table containing the IFU sky-to-detector mapping. Fibre or lenslet data are more likely to be stored in a 2D or 3D array in a single EXTVER.

Multiple datasets within a FITS file are kept in separate 'extensions'. First is a primary header, containing keywords applicable to the whole file, such as the object name, telescope pointing, airmass, filter and central wavelength. This is followed by one or more numbered and/or named extensions, each containing an image array or binary table, along with any header keywords specific to that particular dataset. Figure 2.38 shows an example FITS structure for image slicer data, including variance and data quality arrays and a table with details of the IFU mapping.

2.6.5 *Formats for reduced data*

Inside a FITS file or its equivalent, fully reduced IFU data can be stored in one of at least three formats for subsequent analysis. A basic option for extracted 1D spectra is to work with 'row-stacked spectra', each of which is stored as a separate row of a single 2D image. Without further image reconstruction, this method is limited to 1D spectral analysis such as velocity mapping. One then has to create a 2D spatial map (such as a velocity field) later on from the results.

An IFU datacube is a 3D image with two spatial axes and one wavelength axis, usually arranged in that order. It is therefore a natural extension of 2D image storage. As a simple array, the format is straightforward to read and manipulate in IRAF, Python, IDL or any other generic data processing environment. Generating the datacube usually requires resampling processed IFU data onto a 3D grid (unless the IFU has an already square lenslet array). It is therefore possible that a datacube could represent the original data less faithfully than a minimally processed format, due to interpolation-related artifacts or smoothing. On the other hand, it provides an opportunity for resampling to be done carefully just once, at a well-defined stage. Where interpolated data are oversampled to produce 'smooth' images for visualization, datacube file sizes can become quite large, which can lead to redundant processing later on as well as requiring more storage than interpolating 'on the fly'. By resampling close to the Nyquist rate, however, datacubes can be made at least as compact as other formats.

The Euro3D data format was developed as a standard for exchanging data between reduction and analysis software packages, without requiring spatial resampling and using a relatively sophisticated storage arrangement. This is the native format of the 'E3D'

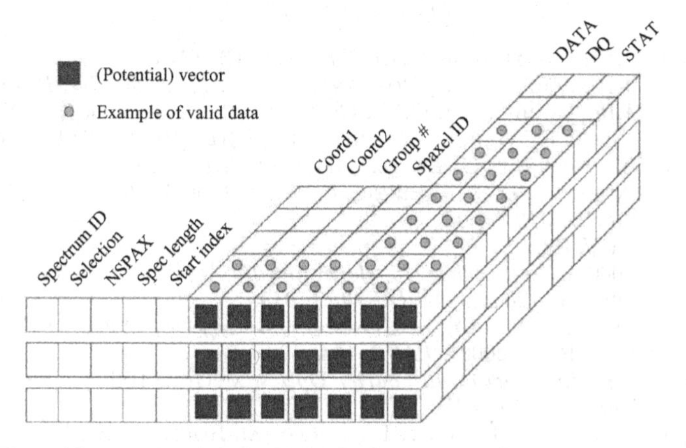

FIGURE 2.39. The table structure of the Euro3D data format (From Kissler-Patig *et al.*, 2003).

visualization tool (although the latter can also read datacubes). Both 1D or 2D spectra and the attributes describing them are stored together in a binary table (Figure 2.39). The format is closer to the raw data than a cube, avoiding resampling until needed during visualization or analysis. The file size is therefore minimal, as for row-stacked spectra. One limitation is that special libraries are needed to work with the format, although these are now publicly available in the C programming language from the Centre de Recherche Astronomique de Lyon.

2.6.6 *Summary of the data reduction process*

The reduction sequences shown here for optical and near-infrared IFU data should give an idea how to tie together the steps discussed in Section 2.5. For more specific details, it is best to refer to reduction examples or software documentation for the instrument being used, which are tailored to that instrument's particular characteristics.

Most data reduction software packages operate on FITS files, each of which has one or more image extensions. Software packages sometimes have the option to propagate error and data quality arrays alongside each image, to help understand the accuracy of results better and perhaps also optimize data combination. The final output of the process, ready for analysis, is usually a set of row-stacked spectra, a datacube or a Euro3D file. The best option amongst these depends on the analysis software, scientific application and user preference. For those working in the Euro3D framework, the data format specification, libraries and visualization tool are now available under the appropriate links at http://www.aip.de/Euro3D.

2.7 Acknowledgements

Thanks to the organizers for their kind invitation to the Winter School. My initial studies of IFU data reduction and sampling were aided by input from Jeremy Allington-Smith, Robert Content and the instrumentation group at Durham, UK. The Gemini Observatory has since given me the opportunity to commission and observe with numerous IFUs. Some of the details in these lectures are based on my work or discussions with Bryan Miller at Gemini, Frank Valdes at NOAO, the Euro3D team and other colleagues.

This work was supported by the Gemini Observatory, which is operated by the Association of Universities for Research in Astronomy, Inc., on behalf of the international Gemini partnership of Argentina, Australia, Brazil, Canada, Chile, the UK and the USA.

REFERENCES

ALLINGTON-SMITH, J.R., CONTENT, R. (1998), *PASP*, **110**, 1216

ALLINGTON-SMITH, J.R., MURRAY, G., CONTENT, R. ET AL. (2002), *PASP*, **114**, 892

ALLINGTON-SMITH, J.R., DUBBELDAM, C.M., CONTENT, R. ET AL. (2007), *MNRAS*, **376**, 785

ARRIBAS, S., MEDIAVILLA, E., GARCÍA-LORENZO, B. ET AL. (1999), *A&AS*, **136**, 189

BACON, R., COPIN, Y., MONNET, G. ET AL. (2001), *MNRAS*, **326**, 23

BERSHADY, M.A., ANDERSEN, D.R., HARKER, J. ET AL. (2004), *PASP*, **116**, 565

BEUTLER, F.J. (1966), *SIAMR*, **8**, 328

BOWEN, I.S. (1938), *ApJ*, **88**, 113

BRACEWELL, R.N. (2000), *The Fourier Transform and Its Applications*, McGraw-Hill, 3

CAPPELLARI, M., COPIN, Y. (2003), *MNRAS*, **342**, 345

CARDIEL, N., GORGAS, J., CENARRO, J., GONZALEZ, J.J. (1998), *A&AS*, **127**, 597

FRUCHTER, A.S., HOOK, R.N. (2002), *PASP*, **114**, 144

GOODMAN, J.W. (1968), *Introduction to Fourier Optics*, McGraw-Hill

HECHT, E. (1987), *Optics*, Addison-Wesley, 2

KISSLER-PATIG, M., COPIN, Y., FERRUIT, P., PÉCONTAL-ROUSSET, A., ROTH, M.M. (2003), *Euro3D data format, format definition*, http://www.aip.de/Euro3D

KOTEL'NIKOV, V.A. (1933), On the transmission capacity of 'ether' and wire in electrocommunications, *Izd. Red. Upr. Svgazi RKAA*

LAUER, T.R. (1999), *PASP*, **111**, 227

LEHMANN, T.M., GÖNNER, C., SPITZER, K. (1999), Survey: interpolation methods in medical image processing, IEEE Trans. *Medical Imaging*, **18**, 1049

MARVASTI, F.A. ED. (2001), *Nonuniform sampling: theory and practice*, Kluwer Academic/ Plenum

MOFFAT, A.F.J. (1969), *A&A*, **3**, 455

NYQUIST, H. (1928), Certain topics in telegraph transmission theory, *AIEE Trans*, **47**, 617

OKE, J.B. (1990), *AJ*, **99**, 1621

OLIVEIRA, A.C., DE OLIVEIRA, L.S., DOS SANTOS, J.B. (2005), *MNRAS*, **356**, 1079

PASQUINI, L., AVILA, G., BLECHA, A. ET AL. (2002), Installation and commissioning of FLAMES, the VLT Multifibre Facility, *Msngr*, **110**, 1

RAAB, W., POGLITSCH, A., LOONEY, L.W. (2004), FIFI LS: the far-infrared integral field spectrometer for SOFIA, *SPIE*, **5492**, 1074

RHOADS, J.E. (2000), *PASP*, **112**, 703

SHANNON, C.E. (1949), Communication in the presence of noise, *IRE Proc.*, **37**, 10

STROHMER, T. (1993), Efficient methods for digital signal and image reconstruction from nonuniform samples, PhD, University of Vienna

TAYLOR, J.R. (1997), An introduction to error analysis, the study of uncertainties in physical measurements, *University Science Books*, 2

TURNER, J.E.H. (2001), Astronomy with integral field spectroscopy: observation, data analysis and results, PhD, University of Durham, UK

VANDERRIEST, C.A. (1980), *PASP*, **92**, 858

WHITTAKER, E.T. (1915), On the functions which are represented by the expansions of the interpolation theory, *Proceedings of the Royal Society Edinburgh, Section A*, **35**, 181

WRIGHT, G.S., RIEKE, G.H., COLINA, L. ET AL. (2004), The JWST MIRI instrument concept, *SPIE*, **5487**, 653

3. 3D spectroscopic instrumentation

MATTHEW A. BERSHADY

In this chapter we review the challenges of, and opportunities for, 3D spectroscopy and how these have led to new and different approaches to sampling astronomical information. We describe and categorize existing instruments on 4 m and 10 m telescopes. Our primary focus is on grating-dispersed spectrographs. We discuss how to optimize dispersive elements, such as VPH gratings, to achieve adequate spectral resolution, high throughput, and efficient data packing to maximize spatial sampling for 3D spectroscopy. We review and compare the various coupling methods that make these spectrographs '3D', including fibres, lenslets, slicers, and filtered multi slits. We also describe Fabry–Perot (FP) and spatial-heterodyne interferometers, pointing out their advantages as field-widened systems relative to conventional, grating-dispersed spectrographs. We explore the parameter space all these instruments sample, highlighting regimes open for exploitation. Present instruments provide a foil for future development. We give an overview of plans for such future instruments on today's large telescopes, in space and in the coming era of extremely large telescopes. Currently-planned instruments open new domains but also leave significant areas of parameter space vacant, beckoning further development.

3.1 Fundamental challenges and considerations

3.1.1 *The detector limit I: six into two dimensions*

Astronomical data exist within a six-dimensional hypercube sampling two spatial dimensions, one spectral dimension, one temporal dimension, and two polarizations. In contrast, high-efficiency, panoramic digital detectors today are only two-dimensional (with some limited exceptions). The instrument builder's trick is to down-select the critical observational dimensions relevant to addressing a well-motivated subset of science problems. Here we consider the application to 3D spectroscopy at high photon count rates, where both spatial and spectral domains must be parsed on to, for example, a charge coupled device (CCD) detector, as illustrated in Figure 3.1. The choice is in how the datacube is sliced along orthogonal dimensions, since it is not easy to rotate a slice within the cube. Such 'rotation' could be accomplished via multi-fibre or multi-slicer feeds to multiple spectrographs, but to date the science motivation has not led to such a design. In practice, then, we have the extremes of single-object, cross-dispersed echelle spectrographs, to FP monochromators. The 'traditional' integral-field spectrograph (IFS) is between these two limiting domains.

In addition to balancing the trade-offs between spatial versus spectral information, there is also the issue of balancing sampling (i.e. resolution) versus coverage in either of these dimensions. Science-driven trades formulate any specific instrument design. When sampling spatial and spectral domains, not all data have equal information content. Hence one may also consider integral versus sparse sampling. Fibre-fed IFS such as Hexaflex (Arribas *et al.*, 1991) and SparsePak (Bershady *et al.*, 2004) are examples of sparse sampling in the spatial domain. Multi-exposure FP observations, multi-beam spectrographs, or notch gratings (discussed below) are examples of instruments with the capability of sparse sampling in the spectral domain.

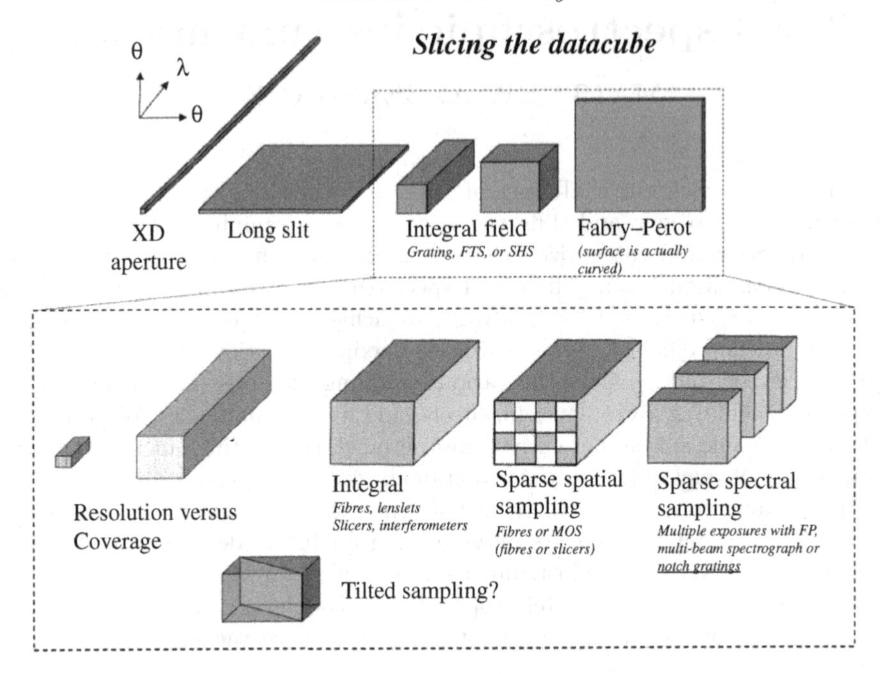

FIGURE 3.1. Sampling the datacube with equal volumes and detector elements.

3.1.2 *Merit functions*

There are a number of generic merit functions found in the instrumentation literature, in a variety of guises used, or tailored, to suit the need of comparing or contrasting the niche of specific instruments. Some useful preliminary definitions (used throughout this chapter) are:

- the spectral resolution, $R = \lambda/d\lambda$;
- the number of spectral resolution elements, N_R;
- spectral coverage $= \Delta\lambda = N_R \times d\lambda$;
- spatial resolution $d\Omega$, i.e. the sampling element on the sky (fibre, lenslet, slicer slitlet, or seeing disc);
- number of spatial resolution elements, N_Ω; and
- spatial coverage $\Omega = N_\Omega \times d\Omega$.

With these definitions, the trade-offs discussed above may be summarized by stating that $N_R \times N_\Omega$ must be roughly constant for a given detector. Another important statement is that $A \times \Omega$, or grasp, is conserved in an optical system (A is the telescope collecting area): The same instrument has the same $A \times \Omega$ on any diameter telescope with the same focal ratio, something derived from the identify $\Omega = a/f^2$, where a is the instrument focal area and f the focal length. What changes with aperture, of course, is the angular sampling. For sufficiently extended sources, angular sampling is not necessarily at a premium. Imagine, for example, dissecting nearby galaxies with a MUSE-like instrument on a 4 m- or 1 m-class telescope. (MUSE is discussed later in this chapter; Bacon *et al.*, 2004).

In addition to the basic ingredients listed above, the most common merit functions are the grasp, the specific grasp, $A \times d\Omega$ (how much is grasped within each spatial resolution element of the instrument), and etendue, $A \times \Omega \times \epsilon$, where ϵ is the total system efficiency from the top of the atmosphere to the detected photoelectron. Etendue is more fundamental than grasp since high-efficiency instruments are the true performance engines.

Despite the fact that an instrument with an unreported efficiency is much like a car *sans* fuel-gauge or speedometer, recovering ϵ from the literature is often not possible. For this reason we resort to grasp, but note that in some cases this gives an unfair comparison between instruments.

If there is no premium on spatial information then 'spectral power', $R \times N_R$, is suitable. At the opposite extreme, where spatial information is paramount, a suitable merit function is $A \times d\Omega^n \times N_\Omega = A \times d\Omega^{n-1} \times \Omega$, where $n = 1$ for high specific grasp and -1 for high resolution. In the context of 3D spectroscopy, merit functions that combine spatial and spectral power are appropriate: $\Omega \times R$, $A \times \Omega \times R \times \epsilon$, or their counterparts replacing Ω with $d\Omega$. If any information will do, $N_R \times N_\Omega$ alone gives a good synopsis of the instrument power since this effectively gives the number of resolution elements (related to detector elements) that have been effectively utilized by the instrument.

An attempt at a grand merit function can be formulated by asking the following, sweeping question: 'How many resolution elements can be coupled efficiently to the largest telescope aperture (A) covering the largest patrol field (Ω_s) for as little cost as possible?' In this case, the figure of merit may be written:

$$FOM = \epsilon \times (\Delta\lambda/\lambda) \times (\Omega/d\Omega) \times A \times \Omega_s \times \pounds^{-1} \tag{3.1}$$

$$= \epsilon \times N_R \times N_\Omega \times A \times \Omega_s \times \pounds^{-1} \tag{3.2}$$

where $\Delta\lambda$ is the sampled spectral range, and £ is the cost in the suitable local currency. To this figure of merit one may add the product $R^n \times d\Omega^m$, where $n, m = 1$ if resolution is science-critical in the spectral and spatial domains (respectively), $n, m = -1$ if coverage is science-critical, or $n, m = 0$ if resolution and coverage are science-neutral (in which case you are not trying hard enough!).

From this discussion it is clear that a suitable choice of merit function is complicated, and must be science driven. The relative evaluation of instruments cannot be done sensibly in the absence of a science-formulated FOM; the outcome of any sensible evaluation will therefore depend on the science formulation. For this reason, when we compare instruments we strategically retreat and explore the multi-dimensional space of the fundamental parameters of spatial resolution, spectral resolution, specific grasp, total grasp, spectral power, and N_R versus N_Ω.

3.1.3 *Why spectral resolution is so important*

In addition to the intrinsic merits and requirement of high spectral resolution for certain science programmes, high resolution is of general importance for improving signal-to-noise (S/N) in the red and near-infrared (NIR). For ground-based observations, terrestrial backgrounds from 0.7–2.2 microns suffer a common malady of being dominated by extremely narrow ($\mathrm{m\,s^{-1}}$) airglow lines, typically from OH molecules. Unlike the thermal IR, however, there is a cure to lower the background without going to high altitude or space. The airglow lines cluster in bands, and the lines within the bands may be separated at $R = 3000$–5000. This means that at these resolutions, while the mean background level within the spectral bandpass is constant, the median drops precipitously: more spectral resolution elements are at lower background level in interline regions. The lines themselves, however, remain unresolved until $R \sim$ a few $\times\ 10^5$, so that above $R = 4000$ one continues to increase the fraction of the spectral bandpass at low-background levels.

As an illustration, we show the terrestrial sky background in a spectral region at 0.8 microns observed by D. York and J. Lauroesch (private communication) with the KPNO (Kitt Peak National Observatory) 4 m echelle. In Figure 3.2 the sky spectra, observed at an instrumental resolution of 33,000, is degraded to illustrate the resulting

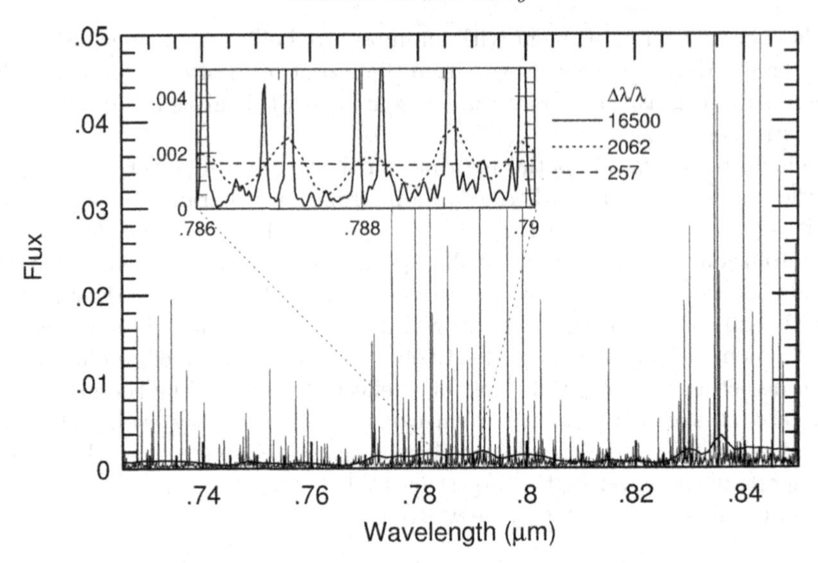

FIGURE 3.2. Night sky near 0.8 μm at $250 < R < 33{,}000$.

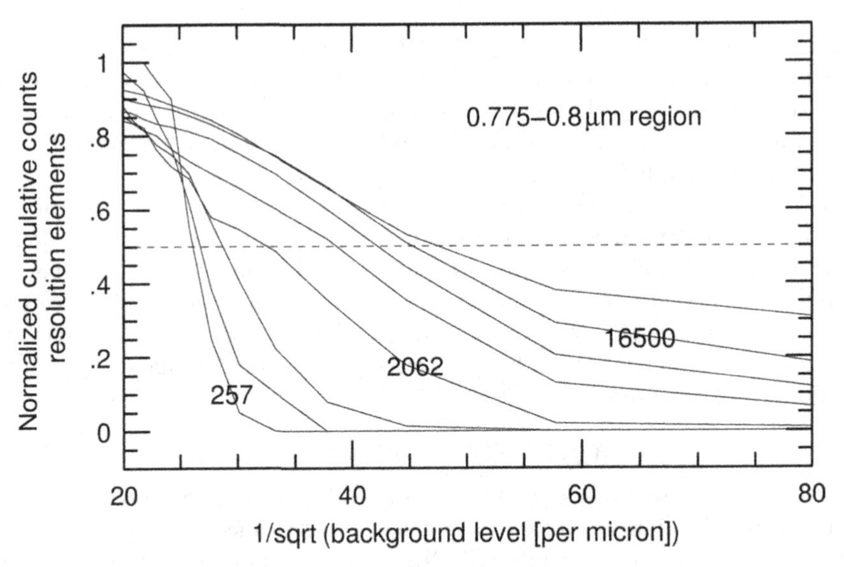

FIGURE 3.3. Cumulative distribution of resolution elements as a function of the background level proportional to S/N (increasing to the left) for $250 < R < 33{,}000$ (labelled).

change in the distribution of background levels. In Figure 3.3, the normalized, cumulative distribution of resolution elements as a function of background level is plotted for different instrumental resolutions. For background-limited measurements, the S/N is proportional to the inverse square root of the background level. Hence the median background level gives an effective scaling for sensitivity gains with spectral resolution. It can be seen that the largest changes in the median background level occur between $1000 < R < 4000$, but significant gains continue at higher resolution. The result can be qualitatively generalized to other wavelengths in the 0.7–2.2 μm regime. While the lines become more intense moving to longer wavelengths, the power spectrum (in wavelength) of the lines appears roughly independent of wavelength in this regime (compare

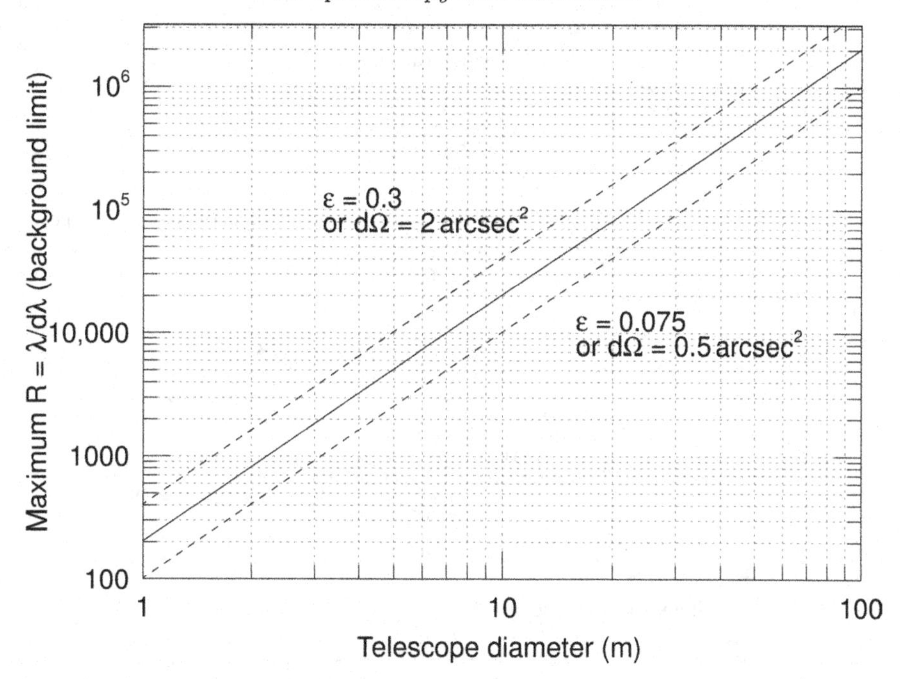

FIGURE 3.4. Maximum spectral resolution versus telescope diameter to stay background (versus detector) limited for different assumptions of instrument efficiency (ϵ) and spatial sampling ($d\Omega$). The solid line assumes $\epsilon = 0.15$ and $d\Omega = 1\,\text{arcsec}^2$.

Maihara *et al.*, 1993; Hanuschik, 2003). Note this is a qualitative assessment that should be formally quantified.

3.1.4 *The detector limit II: read noise*

Our infatuation with spectral resolution is a problem given the modern predilection for high angular resolution. After the *Hubble Space Telescope* there is no turning back! There is, however, a limit, due to detector noise, which we always want to be above. The goal is to be photon-limited (either source or background) because this is fundamental (it is the best we can do), and for practical purposes, S/N is independent of sub-exposure time and detector sampling.

To stay photon-limited in the background-limited regime puts significant constraints on the Ω–R sampling unit. The spatial and spectral sampling unit cannot be too fine for a given A and ϵ. For 8 m- and 4 m-class telescopes we calculate:

$$R/d\Omega < 16,500(D_T/9m)^2(t/1h)(\epsilon/0.15)\,\text{arcsec}^{-2},\ \text{or} \tag{3.3}$$

$$< 2500(D_T/3.5m)^2(t/1h)(\epsilon/0.15)\,\text{arcsec}^{-2} \tag{3.4}$$

where D_T is the telescope aperture diameter and t the (single destructive-read) exposure length. The general case is shown in Figure 3.4. To reach spectral resolutions well above $R = 5000$, which is advantageous for background reduction, a telescope significantly in excess of 10 m is needed for apertures significantly under 1 arcsec^{-2}.

With these considerations in mind, in the next three sections (Section 3.2–3.4) we turn to approaches and examples of existing instruments, followed by three sections (Sections 3.5–3.7) in which we summarize the range of these instruments, what parameter space is undersampled, and the prospects for future instruments. Throughout, we attempt to provide relatively complete instrument lists. No doubt, some instruments have been

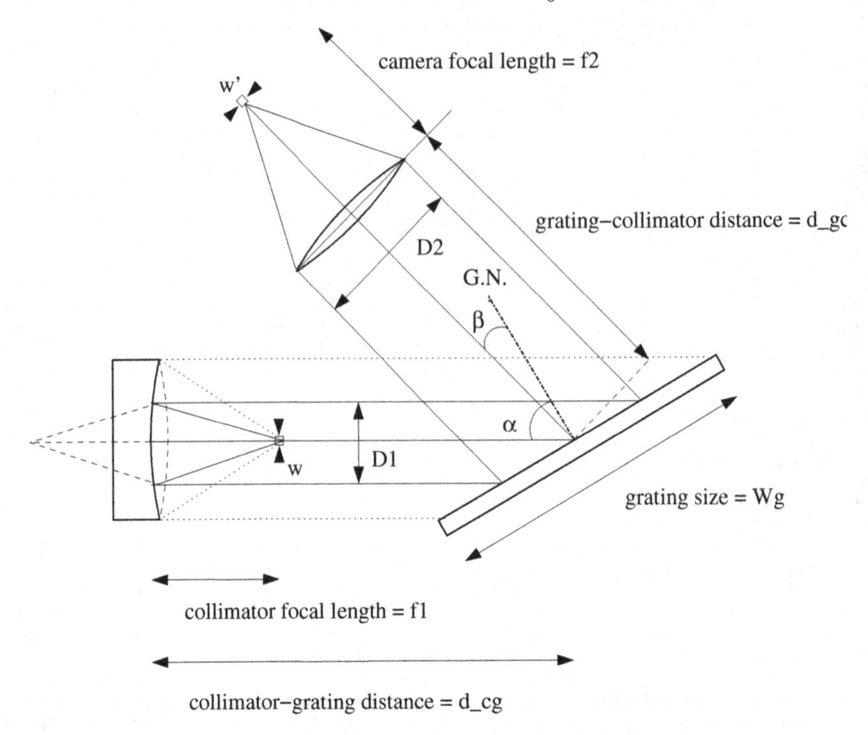

FIGURE 3.5. Basic spectrograph layout schematic for reflective/refractive collimator, reflection grating and refractive camera.

overlooked, and also the field of instrumentation advances rapidly. Reports of additional instruments or corrections are welcome.

3.2 Grating-dispersed spectrographs

Basic spectrograph theory and design can be found in most standard optics textbooks. Of particular note is the excellent monograph on astronomical optics by Schroeder (2000). In Section 3.2.1 we summarize the salient features to provide a consistent nomenclature, and to put these features into context of our discussion of 3D spectroscopy, specifically what drives consideration of merit functions that tune spatial versus spectral performance. The balance of this section includes a description of dispersive elements (Section 3.2.2), coupling methods and modes (Section 3.2.3–3.2.11), and summary considerations, including a discussion of sky-subtraction problems and solutions (Section 3.2.12).

3.2.1 *Basic spectrograph design*

In a 3D spectrographic system, there is a premium on packing spatial information on to the detector. To achieve sufficient spectral resolution at the same time requires balancing the trades between system magnification and dispersion. Starting with the grating equation, generalized for a grating immersed in medium of index n: $m\lambda = n\Lambda_g(\sin\beta + \sin\alpha)$, where Λ_g is the projected groove separation in the plane of the grating, m the order, and α and β the incident and diffracted grating angles relative to the grating normal in the medium, we can write the angular and linear dispersion as $\gamma \equiv d\beta/d\lambda = m/n\Lambda_g \cos\beta = (\sin\beta + \sin\alpha)/\lambda\cos\beta$, and $dl/d\lambda = f_2\gamma$. Figure 3.5 illustrates a basic spectrograph, defines these angles and subsequent terms.

The system magnification can be broken down into spatial and anamorphic factors. The physical entrance aperture width, w, is re-imaged on to the detector to a physical width w', demagnified by the ratio of camera to collimator focal lengths. Hence the spatial width (perpendicular to dispersion) is given as $w'_\theta = w(f_2/f_1)$. For non-imaging feeds (i.e. fibres or lenslets), it is advantageous to pack as much information as possible into a given pixel, as long as individual spatial entrance elements can be resolved. This means cameras must be as fast as possible relative to their collimators. For imaging feeds (slits or slicers), the desire to preserve and sample the spatial information retained in the slit means the choice must be science driven.

In the dispersion direction, an additional, anamorphic factor, r, arises due to the fact that grating diffraction implies incident and diffracted angles need not be the same. Hence incident and diffracted beam sizes scale as $r = D_1/D_2 = \cos\alpha/\cos\beta$. This arises because in general $A \times \Omega$ is conserved; if the beam gets larger, the angles get smaller. Another way to think of this is in terms of the definition of $r = |d\beta/d\alpha|$, and ask for a given $\mathrm{d}\alpha$ (angular slit width) what is $\mathrm{d}\beta$ such that $\mathrm{d}\lambda = 0$? This result can then be derived from the grating equation. In any case, $\beta/\alpha < 1$ implies magnification, while $\beta/\alpha > 1$ implies demagnification. The re-imaged slit width in the spectral dimension is then $w'_\lambda = rw'_\theta$. In Littrow configurations, important below, $\alpha = \beta = \delta$ (the latter being the grating blaze angle), and so there is no anamorphic factor. Since the re-imaged slit width always degrades the instrumental spectral resolution, it is always advantageous, in this sense, to have anamorphic demagnification. However, depending on the pixel sampling, optical aberrations, and slit size, w'_λ may not be the limiting factor in instrumental resolution. Anamorphic demagnification also comes at a cost: the camera must be large (larger than the collimator) to capture all of the light in the expanded beam. Demagnification never hurts resolution, but the cost should be weighed against the gains.

The spectral resolution can now be written as $R = \lambda/d\lambda$, or $R = \lambda(\gamma/r)(f_1/w)$. The term γ/r indicates that we want large dispersion, but that we can get resolution also from anamorphic demagnification. The terms f_1/w indicate that we want a long collimator at fixed camera focal length, requiring a field lens or white-pupil design to avoid vignetting.[1] Alternatively, we may rewrite the equation as $R = \lambda(\gamma/r)(D_1/\theta D_T)$, noting θ is the angle on the sky, $d\lambda = w'_\lambda/(dl/d\lambda)$, $w = f_T\theta$, and $f_1/d_1 = f_T/D_T$, where f_T/D_T refer to the effective focal ratio of whatever optics feed the spectrograph, e.g. the telescope. The combination of r and D_1 indicates that we want a larger collimator and an even larger camera. Using the grating equation we may write $R = (f_1/w)(\sin\beta + \sin\alpha)/\cos\alpha$, which, in Littrow configurations, reduces to $R = (f_1/w)2\tan\alpha$. In the latter situation it is clear that resolution can be dispersion-driven by going to large diffraction angles, α, which requires large gratings.

3.2.2 *Dispersive elements*

We distinguish here principally between reflection and transmission gratings. Transmission gratings yield much more compact spectrograph geometries. This leads to less vignetting and better performance with smaller optics.

[1] A field lens, which sits near a focus to avoid introducing power into the beam, serves to move the spatial pupil to a desirable location in the system. This is often the grating, but in general can be the location such that the overall system vignetting is minimized. A white-pupil design (Baranne, 1972; Tull *et al.*, 1995) is one that re-images a pupil placed on a grating, typically on to a second grating (e.g. a cross-disperser) or the camera objective. It is 'white' because the pupil image location is independent of wavelength even though the light is dispersed.

Reflection gratings

Reflection gratings come in three primary varieties: ruled surface-relief (SR), holographically etched SR, or volume-phase holographic (VPH). We list the pros and cons of each of these.

Ruled SR gratings have the advantage of control over the groove shape, blaze and density, which provides good efficiency in higher orders (e.g. echelle) at high dispersion. There are existing samples of masters with replicas giving up to 70% efficiency, but 50–60% efficiency is typical, with 40% as coatings degrade. Scattered light and ruling errors can be significant, and existing masters are limited in type and size. It does not appear to be possible to make larger masters with high quality at any reasonable cost.

Holographically etched SR gratings have low scattered light, the capability to achieve high line density (hence high dispersion) and large size. However, they have low efficiency ($< 50\%$) because symmetric grooves put equal power in positive and negative orders.

VPH gratings can be made to diffract in reflection (Barden *et al.*, 2000), but have not yet been well developed for astronomical use. Reflection gratings can be coupled to prisms to significantly enhance resolution via anamorphing (Wynne, 1991).

Transmission gratings

Transmission gratings are either SR or VPH and, when coupled with prisms, are referred to as grisms. SR transmission gratings and grisms are efficient at small angles and low line densities (good for low-resolution spectroscopy), but are inefficient at large angles and line densities due to groove shadowing. Transmission echelles do exist, but have 30% diffraction efficiencies or less.

VPH gratings and grisms are virtually a panacea. They are efficient over a broad range of line densities and angles. Any individual grating is also efficient over a broad range of angles (what is known as a broad 'superblaze'; see below). Peak efficiencies are as high as 90%; they are relatively inexpensive to make and likewise to customize; and they can be made to be very large (as large as your substrate and recording beam, now approaching 0.5 m). Their only disadvantages are that they have, to date, been designed for Littrow configurations.

It is worth dwelling somewhat on the theory and subsequent potential of VPH gratings. There still remain manufacturing issues of obtaining good uniformity over large areas (Tamura *et al.*, 2005), but it is reasonable to be optimistic that refinement of the process will continue at a rapid pace. Application in the NIR for cryogenic systems is also promising: coefficient thermal expansion mismatch between substrate and diffracting gelatin, potentially causing delamination, does not appear to be a concern (W. Brown, private communication, this Winter School). Blais-Ouellette *et al.* (2004) have confirmed that diffraction efficiency holds up remarkably well at 77 K, but that the effective line density changes with thermal contraction. We can expect most grating-fed spectrographs in the future will use VPH gratings alone or in combination with conventional (e.g. echelle) gratings. The capabilities of VPH gratings will open up new design opportunities, many of which will be well suited to 3D spectroscopy.

3.2.3 *VPH grating operation and design*

Diffraction arises from modulation of the index of refraction in a sealed layer of thickness d of dichromated gelatin (the material is hygroscopic), with mean optical index n_2. Typical values for n_2 are around 1.43, but the specific value depends sensitively on the modulation frequency (i.e. the line density Λ) and amplitude, Δn_2, and the specifics of the exposure and developing process. (Note that it is not currently possible to predict the precise value of n_2 from a manufacturing standpoint.) The seal is formed typically by two flat substrates, but this can be generalized to non-flat surfaces and wedges (i.e. prisms). Because this layer represents a volume ($d \gg \lambda$), the diffraction efficiency is modulated by the Bragg condition: $\alpha = \beta$. These angles are defined here with respect to the plane of the index modulations.

The wonder of VPH gratings is the ability to custom-design them. Starting with a science-driven choice of dispersion and wavelength, the grating equation and dispersion relation given the Bragg condition uniquely set the line frequency and angle, respectively, for unblazed gratings. The key to high diffraction efficiency is then to tune the gelatin thickness and index modulation amplitude such that diffraction efficiency is high in both s- and p-polarizations (the s-polarization electric vector is perpendicular to the fringes). This can be done by brute force via rigorous coupled wave calculations, or by noting that in the so-called 'Kogelnik limit' the diffraction efficiencies are periodic in these quantities (Barden *et al.*, 2000; Baldry *et al.*, 2004). The two polarizations have different periodicities, i.e. VPH gratings are in general highly polarizing, so the trick is finding the $(d, \Delta n_2)$-combination that phases one pair of s and p efficiency peaks. Thinner gel layers yield broader bandwidth over which the diffraction efficiency is high, relative to the efficiency at the Bragg condition. The thinner the layer, the larger the index modulation required to keep the efficiency high in an absolute sense. Modulations above 0.1 are very difficult to achieve, and more typical values are in the range of 0.04–0.07; gel layers are in the range of a few to a few tens of microns. In practice, because there is limited manufacturing control over the index modulation and effective depths of the gelatin exposure, gratings requiring very precise values in these parameters will be difficult to make and have large inhomogeneities. Our experience is that it is useful to understand how wavelength and resolution requirements can be relaxed to locate more robust design parameters.

Blazed VPH gratings

Nominally, the fringe plane is parallel to the substrate normal (indicated by the angle $\phi = 0$). This yields an unblazed transmission grating. Essentially all astronomical VPH gratings in use are made this way. There is concern that tilted fringes will curve with the shrinkage of the gelatin during development (Rallison and Schicker, 1992), but this concern has not been fully explored. By tilting the fringes (this is done simply by tilting the substrate during exposure in the hologram), one can enter several different interesting regimes, as illustrated concisely by Barden *et al.* (2000) in their Figure 1: small $|\phi|$ yields blazed reflection gratings, $\phi = 90\,\text{deg}$ produces unblazed reflection gratings, and large $|\phi|$ blazes the reflection gratings. 'Large' and 'small' depend on the angle of incidence, as illustrated below. The sign convention is such that positive ϕ decreases the effective incidence angle. The incident and reflected angles in the gelatin, α_2 and β_2, are related by $\alpha_2 = \beta_2 + 2\phi$, with $\alpha_2 - \phi$ being the effective diffraction angle. The grating equation, when combined with the Bragg condition yields: $m\lambda_b = 2n_2\Lambda\sin(\alpha_2 - \phi)$, where λ_b is the Bragg wavelength, and $\Lambda = \Lambda_g\cos\phi$ is the fringe spacing perpendicular to the fringes. We use Baldry *et al.*'s (2004) nomenclature; their Figure 1 is an instructive reference for this discussion.

Baldry *et al.* (2004) work out the case for no fringe tilt with flat or wedged substrates. Here we give the case of flat substrates but arbitrary ϕ. Burgh *et al.* (2007) extend this to include arbitrary fringe tilt (ϕ). The relevant angles with respect to the grating normal can be found with these equations in terms of the physical grating properties:

$$\sin\alpha = n_2\sin\alpha_2, \quad \text{and} \quad \sin\beta = n_2\sin\left[\sin^{-1}\left(\frac{\sin\alpha}{n_2}\right) - 2\phi\right] \tag{3.5}$$

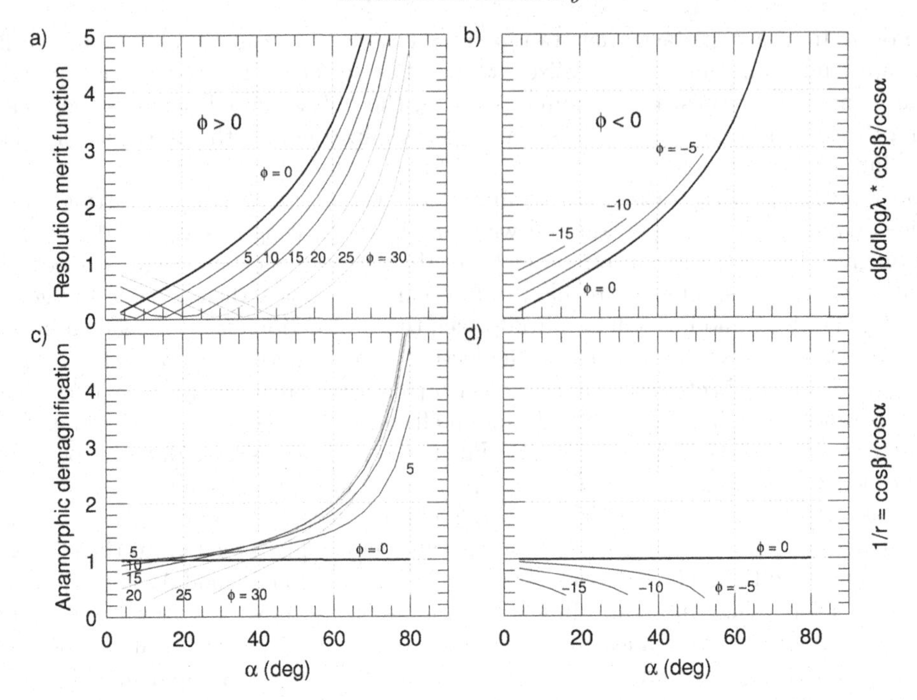

FIGURE 3.6. Resolution merit function and anamorphic factor for blazed VPH gratings with mean gel index $n_2 = 1.43$. Typical SR gratings have $1.05 < 1/r < 1.2$.

The anamorphic factor and dispersion are still defined in terms of α and β as given in Section 3.2.1. With the interrelation of these angles as given above, it is easy to show the logarithmic angular dispersion at the Bragg wavelength is:

$$d\beta/dlog\lambda = 2n_2 \cos\phi \sin\left[\sin^{-1}\left(\frac{\sin\alpha}{n_2}\right) - \phi\right] \Big/ \cos\beta. \qquad (3.6)$$

To understand the potential advantages of blazed transmission gratings, we define a resolution merit function as $\frac{1}{r}\ d\beta/dlog\lambda$, i.e. the product of the logarithmic angular dispersion and the anamorphic factor. With this function we can explore, in relative terms, whether tilting the fringes yields resolution gains. Figure 3.6 shows the anamorphic factor and the resolution merit function versus grating incidence angle for positive and negative fringe tilts. Negative fringe tilts give a small amount of increased resolution at a given α by significantly increasing dispersion, which overcomes an increase in the anamorphic magnification (*mag*). This means the detector is less efficiently used. Negative fringe tilts also limit the usable range of α for which $\beta < 90$ deg (transmission), and hence the maximum achievable resolution in transmission that can be achieved is lowered with negative fringe tilts.

With positive fringe tilts, the anamorphic demagnification increases strongly at large incidence angles, although there is little gain in going to $\phi > 15$ deg. Note that the demagnification becomes < 1 (i.e. magnification) roughly when $\alpha \sim 1.5\ \phi$. This is when the effective diffraction angle $(\alpha_2 - \phi)$, changes sign with respect to the tilted fringes (the grating remains in transmission). The overall resolution decreases with increased positive fringe tilt, but the decrease is modest for small tilt angles. Given the large increase in anamorphic demagnification relative to the modest loss in resolution, for small tilt angles there is a definite gain in information: A +5 deg tilt gives a 12% loss in the resolution

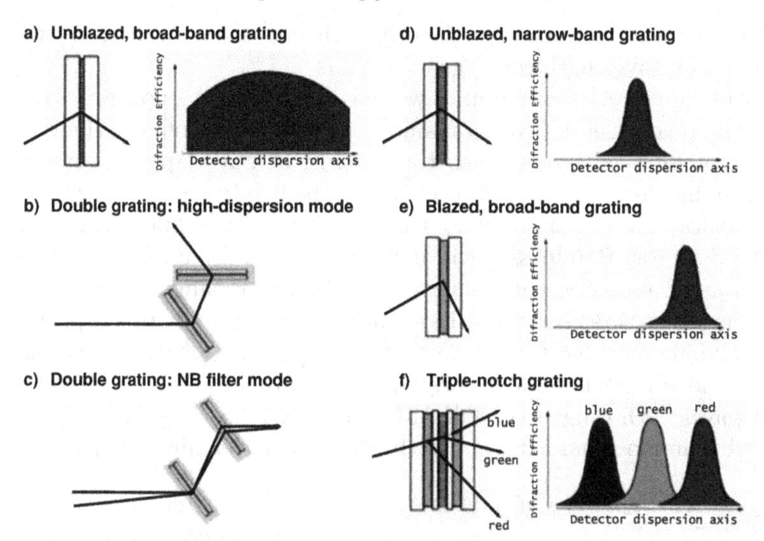

FIGURE 3.7. Novel grating modes. a) Conventional broad-band application now becoming a staple of modern spectrographs. b) Double-grating geometry yielding a net dispersion of $\sim$ the sum of the two individual grating dispersions (gratings are not necessarily identical, but angles must be adjusted accordingly). c) Double-grating geometry yielding a narrow-band filter with field-dependent bandpass given by the Bragg condition (gratings are identical) d)–f) Narrow bandpass gratings unblazed (d), blazed (e), and combined (f) to form a notch grating. Other modes are discussed in the text. Panels a), d)–f) show both the grating configuration as well as a cartoon-sketch of the diffraction efficiency as a function of location of the detector dispersion axis, labelled for the mean wavelength regime of the diffraction bandpass.

merit function at $\alpha = 60$, but a 51% gain in the anamorphic demagnification. With suitably good optics and detector sampling the demagnified image, this equates directly into an increase in the number of independent spectral resolution elements, replete with a 72% increase in spectral coverage. The loss in resolution can easily be made up by slightly increasing α (in this case, from 60 to 63 deg) and modulating Λ in the grating design to tune the wavelength. Instruments with blazed, high-angle VPH gratings with tilts of $5 < \phi < 15$ deg will allow for the high resolution needed to work between sky lines, while efficiently packing spectral elements on to the detector. This is critical in the context of 3D spectroscopy, where room must also be made for copious spatial elements.

Unusual VPH grating modes

In addition to tilted fringes, VPH gratings pose opportunities for a number of novel modes well suited to 3D spectroscopy. Figure 3.7 illustrates some of these. With very high diffraction efficiency it is now reasonable to consider combining gratings to augment the dispersion, and hence resolution. If the two gratings are kept parallel but offset along the diffraction angle, they can serve as (tunable) narrow-band filters, an alternative to etalons (e.g. Blais-Ouellette *et al.*, 2006). Barden *et al.* (2000) have explored using multiple gelatin layers with different line frequencies to select Hα and Hβ in separate bandpasses. By slightly rotating one set of lines, sufficient cross-dispersion is added to space the two spectra – one above the other – on the detector. This is well suited for spectrographs fed with widely spaced fibres or slitlets (i.e. an underfilled, conventional long-slit spectrograph), and represents an interesting trade-off in wavelength and spatial multiplex. At sufficiently high dispersion (and hence limited bandpass), the number of layers could be increased to mimic a multi-order echelle.[2] The advantage of this approach is in resolution and wavelength coverage.

An alternative approach is something we refer to as 'notch' gratings. Here, we take advantage of the relative ease (from a manufacturing standpoint) of achieving a narrow bandpass and combine gel layers tuned to different, non-overlapping wavelength bandpasses at a given incidence angle (e.g. by changing the line frequency). By also tuning the fringes with modest tilts, each bandpass can be centred on a different, non-overlapping portion of the detector. Bandpasses will have to be carefully crafted by tuning grating parameters to avoid parasitic contamination in the other bands. The figure illustrates positive and negative tilts, but the tilts could be arranged all to be positive to take advantage of the anamorphic factors described above. This offers another way to slice the datacube, one which allows for sparse spectral sampling of key spectral diagnostics over a broad wavelength range (e.g. [OII]λ3727, Hβ+[OIII]$\lambda\lambda$4959,5007 and Hα) at high dispersions, with ample room left over on the detector for significant spatial multiplex.

3.2.4 *Summary of implications for 3D spectrograph design*

The most compact spectrograph designs yield the highest-efficiency, wide-field systems needed to grapple with attaining large angular coverage for 3D spectroscopy. To also obtain high enough spectral resolution to work between the atmospheric airglow often requires significant dispersive power and anamorphic demagnification. Large anamorphic demagnification, while not free (larger camera optics are required), is well-suited to packing information on to the detector. This is particularly important in 3D applications where spatial multiplex is at a premium. VPH transmission gratings are clearly preferred because they lend themselves to compact spectrograph geometry and provide high diffraction efficiency. We have shown they can, in principle, also yield large anamorphic demagnification. With high-angle, double, and blazed VPH gratings, echelle-like resolutions can be achieved at unprecedented efficiency (75–90% in diffraction alone). Unusual modes to produce tunable narrow-band filters and notch gratings also open up the possibility for well-targeted sparse, spectral sampling.

3.2.5 *Coupling formats and methods: overview*

The essence of the 3D spectrometer lies in the coupling of the telescope focal plane to the spectrograph. We review the four principal methods: (i) direct fibres, (ii) fibres + lenslets, (iii) image-slicers, and (iv) lenslet arrays, or pupil-imaging spectroscopy. A nice, well-illustrated overview can be found in Allington-Smith and Content (1998); additional discussion of the merits and demerits of different approaches can be found in Alighieri *et al.* (2005). Here we also make an evaluation. We discuss a fifth mode not seen in the literature, which we refer to as (v) 'filtered multi slits'. Many spectrographs either have, or could easily be modified to have, this capability. We also describe (vi) multi-object configurations: a mode which will undoubtedly become more common in the future.

Throughout this discussion, we distinguish between near-field versus far-field effects. The near-field refers to the light distribution at the focal surface, e.g. fibre ends, and what is re-imaged ultimately on to the CCD. The far-field refers to the ray-bundle distribution, i.e. the cross-section intensity profile of the spectrograph beam significantly away from the focal surface. Different coupling methods offer the ability to remap near- and far-field light-bundle distributions, which can have advantages and disadvantages.

[2] A true cross-dispersed echelle-like grating would work, in principle, with two layers, rotated by 90 degrees. VPH gratings have not yet been made with high efficiency in multiple orders, but see Barden *et al.* (2000) for measurements up to order 5.

3.2.6 *Direct fibre coupling*

The simplest and oldest of methods consists of a glued bundle of bare fibres mapping the telescope to spectrograph focal surfaces. With properly doped, AR-coated fibres throughput can be at or above 95%, which can be compared to 92% reflectivity off one freshly coated aluminium surface. These have the distinct advantage of low cost and high throughput. As with all fibre-based coupling, there is a high degree of flexibility in terms of reformatting the telescope to spectrograph focal surfaces (for example, it is easy to mix sky and object fibres along slit), and the feeds can be integrated into existing long-slit, multi-object spectrographs. However, bare fibre integral field units (IFUs) are not truly integral and do not achieve higher than 60–65% fill factors (on the deleterious effects of buffer stripping of small fibres, see Oliveria *et al.*, 2005). This coupling is perhaps the most cost-effective mode for cases where near-integral sampling is satisfactory, and preservation of spatial information is not at a premium.

Information loss and stability gain with fibres

Focal ratio degradation (FRD) and azimuthal scrambling represent information loss (an entropy increase). FRD specifically results in a faster output f-ratio (Ramsey, 1988). This has an impact on spectrograph design or performance since either the system will be lossy (output cone overfills optics), or the spectrograph has to be designed for the proper feed f-ratio. PMAS (Potsdam Multiaperture Spectrophotometer; Roth *et al.*, 2005) is an excellent example of how to properly design a spectrograph to handle fast fibre-output beams. The existing WIYN (Wisconsin–Indiana–Yale) Bench spectrograph is a good example of how not to do it. In fact, it is so bad that we rebuilt it (Bershady *et al.*, 2008); we were able to recapture 60% of the light (over a factor of 2 in throughput) with no loss of spectral resolution in the highest resolution modes.

Azimuthal scrambling can help and hurt. While scrambling destroys image information, it symmetrizes the output beam, ameliorating, to some extent, the effect of a changing telescope pupil on HET-like (Hobby–Eberly Telescope) or SALT-like (Southern African Large Telescope) telescopes by homogenizing the ray bundle. Thus, the contribution of spectrograph optical aberrations to the final spectral image is more stable. (This is a far-field effect.)

Radial and azimuthal scrambling together homogenize near-field illumination, e.g. the seeing-dependent slit function is decreased. Radial scrambling and FRD are one and the same (compare Ramsey, 1988 and Barden *et al.*, 1993), so that one trades information loss for stability (similar to the trade of precision for accuracy). In practice, fibre-input beam speeds of $f/3$ (PMAS) to $f/4.5$ (HET and SALT) are desirable. However, with fast input/output f-ratios this limits possible spectrograph demagnification since it is expensive to build faster than $f/2$ for large cameras.

Telecentricity

Because azimuthal scrambling symmetrizes a beam, if the input light-cone is misaligned with the fibre axis, the output beam (f-ratio) is faster. This is not FRD. To avoid this effect, fibre telecentric alignments of under a degree are needed even for f-ratios as fast as 4–6 (Bershady *et al.*, 2004; Wynne and Worswick, 1989).

Causes of FRD

Excessive FRD in fibres is due to stress. Hectospec (Fabricant *et al.*, 2005) embodies an excellent example of how to properly treat fibres and fibre cabling (Fabricant *et al.*, 1998; see also Avila *et al.*, 2003 in the context of FLAMES on VLT). Fibre termination

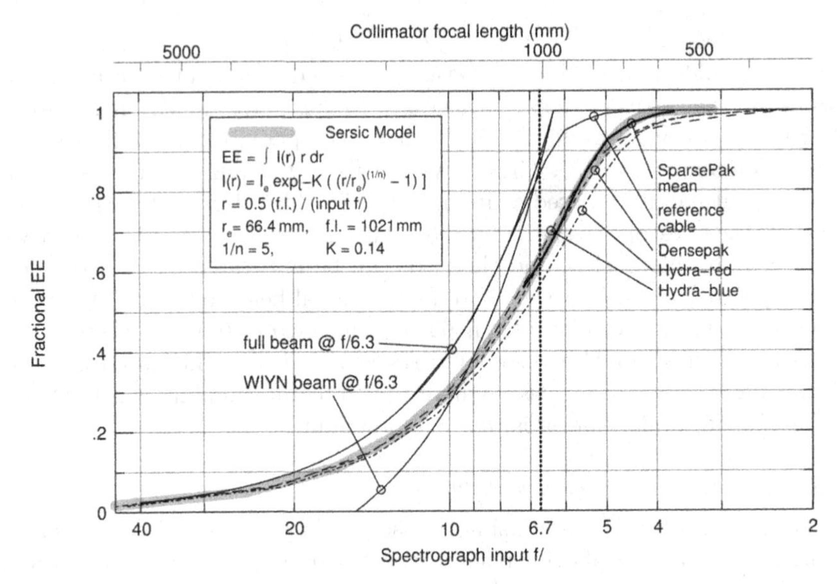

FIGURE 3.8. Output fibre irradiance (encircled energy versus beam f-ratio) for fibre cables on the WIYN Bench Spectrograph. The input beam profile is an unapodized $f/6.3$ beam with an $f/17$ central obscuration (labelled). Output beam profiles are faster, due to FRD, and are well fitted by a Sersic model of index $1/n = 5$ (S. Crawford, private communication).

and polishing can also induce stress. Bershady *et al.* (2004) discuss some other IFU-related issues in terms of buffering. However, even for perfectly handled fibres, there is internal scattering, the cause of which has long been a debate. Nelson (1988) suggested a combination of (i) Rayleigh scattering (variation in fibre refractive index); (ii) Mie scattering (fibre inhomogeneities comparable to the wavelength); (iii) stimulated Raman and Brillouin scattering (not relevant at low signal level in astronomical applications); and (iv) micro-bending. Micro-bending seems a likely culprit; it is the unsubstantiated favourite in the literature. Micro-bending models predict a wavelength-dependent FRD. While Carrasco and Parry (1994) tentatively see such an effect, neither Schmoll *et al.* (2003) nor Bershady *et al.* (2004) confirm the result. However, these studies use different measurements methods. More work is required to understand the physical cause(s) of FRD, and with this understanding, perhaps, reduce the amplitude of the effect. We find FRD produces an output fibre beam profile which can be well-modelled by a Sersic function (Figure 3.8; S. Crawford, private communication). This says something about either the scattering model or how seriously to take physical interpretations of Sersic-law profiles of galaxies.

Quality versus quantity

Fibres offer the opportunity of easily trading quality for quantity in terms of packing the spectrograph slit. Scattered light within the spectrograph, combined with fibre azimuthal scrambling means that spatial information in the telescope focal plane is coupled to all adjacent fibres in the slit. Closely packing fibres in the slit can make clean spectral extraction difficult. The WIYN Bench Spectrograph is a good example where the amplitude of scattered light is low, fibre separation is large and ghosting is negligible. This spectrograph and feeds are optimized for clean extraction with little cross-talk (1% cross-talk in the visible in optimum S/N aperture, degrading to 10% in the NIR). Information packing in the spatial dimension is modest due to fibre separation, while information packing

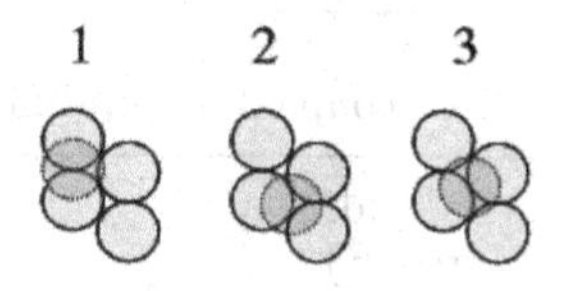

FIGURE 3.9. Critical sampling with densely packed fibres.

in the spectral dimension is high due to large anamorphic factors. Other systems have significant spectral overlap. For example, staggered slits, where fibres are separated by only their active diameter (COHSI; Kenworthy *et al.*, 1998) make it difficult to extract a clean spectrum and optimize S/N at the same time, but the spatial multiplex is increased. There is no one right answer, but definitely a decision worthy of science-based consideration.

Image reconstruction and registration

Even without lenslets, densely sampled fibres provide excellent image reconstruction on spatial scales comparable to the fibre diameter. One can achieve the theoretical sampling limit with a three-position pattern of half-fibre-diameter dithers (Figure 3.9; compare Koo *et al.*, 1994 in the context of under-sampled HST/WFPC-2 data). Even with sparse sampling, registration of the spectral datacube with broad-band images can be achieved to 10% of the fibre diameter by cross-correlating the spectral continuum with respect to broad-band images or integrated radial light profiles (Bershady *et al.*, 2005). Kelz *et al.* (2006) show how it is possible to reproduce the continuum image of UGC 463 using the PPak fibre bundle without any subsampling.

Summary of instruments

Some of the first IFUs were on the KPNO 4 m RC spectrograph: DensePak-1 followed by DensePak-2 (Barden and Wade, 1988; for other early IFUs, see also Guerin and Felenbok, 1988). The last incarnation (Barden *et al.*, 1998) was on WIYN. Conceptually, these instruments spawned SparsePak (WIYN; Bershady *et al.*, 2004) and PPak (PMAS, Calar Alto; Verheijen *et al.*, 2004; Kelz *et al.*, 2006). A more versatile single instrument suite, built for the William Herschel Telescope (WHT), is INTEGRAL (WYFOS, Wide-Field Optic Spectrograph), which offers several plate-scales and formats (Arribas *et al.*, 1998), and a sophisticated and well-thought-out mapping between telescope and spectrograph focal planes. These are all shown in Figure 3.10. GOHSS is one case of a NIR (0.9–1.8 μm) application (Lorenzetti *et al.*, 2003). VIRUS (Hill *et al.*, 2004) and APOGEE (Allende Prieto *et al.*, 2008) are the only planned future instruments. A summary of existing and future optical and NIR direct fibre-fed IFU spectrographs are listed in Table 3.1.

3.2.7 *Fibre + lenslet coupling*

The basic concept of lenslet coupling to fibres is again, as with bare fibres, to remap a 2D area in the telescope focal surface to a 1D slit at the spectrograph input focal surface. The key difference is in the fore-optics, which consists of a focal expander and lenslet array; these feed the fibre bundle. The focal expander matches to the scale of the lenslet array. Allington-Smith and Content (1998) and Ren and Allington-Smith (2002) present some technical discussion and illustration of the method. Each micro-lens in the array then forms a pupil image on the fibre input face. The pupil image is suitably smaller than the lenslet to allow the fibres to be packed behind the integral lenslet array. This reduction speeds up the input beam ($A \times \Omega$ is conserved). Given the previous discussion concerning FRD, this can be advantageous to minimize entropy increase.

TABLE 3.1. Direct fibre-coupled integral field instruments

Instrument	Tel.	D_T (m)	Ω	$d\Omega$ (arcsec^2)	N_Ω	$\Delta\lambda/\lambda$	R	N_R	ϵ
Existing optical instruments									
DensePak	WIYN	3.5	564	6.2	91	1.02	1000	1024	0.04
		3.5	564	6.2	91	0.07	13750	1024	0.04
		3.5	564	6.2	91	0.04	24000	1024	0.04
		3.5	119	1.3	91	1.02	1000	1024	0.04
		3.5	119	1.3	91	0.07	13500	1024	0.04
		3.5	119	1.3	91	0.04	24000	1024	0.04
SparsePak	WIYN	3.5	1417	17.3	82	1.02	800	819	0.07
		3.5	1417	17.3	82	0.07	11000	819	0.07
		3.5	1417	17.3	82	0.03	24000	819	0.07
PPak	CA	3.5	2070	5.64	367	0.15	7800	1183	0.15
INTEGRAL	WHT	4.2	32.6	0.159	205	0.22	2350	515	...
		4.2	32.6	0.159	205	0.94	550	515	...
		4.2	139.3	0.64	219	0.22	2350	515	...
		4.2	139.3	0.64	219	0.94	550	515	...
		4.2	773	5.73	135	0.07	2350	300	...
		4.2	773	5.73	135	0.90	550	300	...
Future optical instruments									
VIRUS	HET	9.2	32604	1.0	32604	0.505	811.	410	0.16
Existing near-infrared instruments									
GOHSS	TNG	3.6	44.2	1.77	25	0.12	4380.	512	0.13
Future near-infrared instruments									
APOGEE	SDSS	2.1	942	3.14	300	0.10	23500	2350	0.22

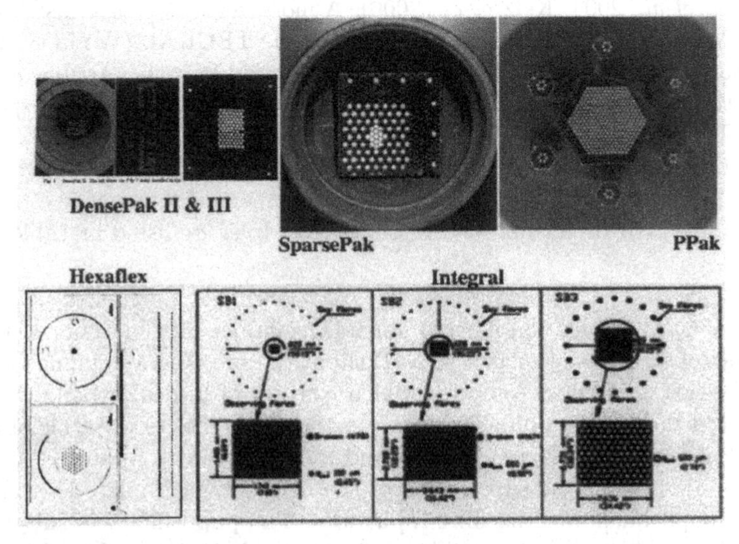

FIGURE 3.10. Direct-fibre IFUs on optical spectrographs. The top row shows the legacy started by S. Barden with DensePak-1 and DensePak, leading to SparsePak, PPaK on the KPNO 4 m, WIYN, and Calar Alto, respectively. The bottom row shows Hexaflex and INTEGRAL on WHT with their multiple, selectable bundles and ample sky-fibres.

At the output stage, the option exists to reform the (now azimuthally scrambled) slit image with an output micro-lens linear array, or to use bare fibres. Without lenslets, the input f-ratio to the spectrograph will be faster, which means there is less possibility for geometric demagnification via a substantially faster camera. In this case the spectrograph also re-images the fibre-scrambled telescope pupil: the image varies with telescope illumination, while the ray bundle distribution (far-field) varies with the telescope image.

The positive attributes of lenslet-fed fibre arrays are: (i) improved filling factors to near unity; and (ii) control of input and output fibre f-ratio. The latter permits effective coupling of a slow telescope f-ratio to fibre input at a fast, non-lossy beam speed and likewise permits effective coupling of fibre output to spectrograph. The negative attributes of this coupling method include (iii) increased scattered light (from the lenslet array); (iv) lower throughput (due to surface reflection, scattering, and misalignment). For example, typical lenslet + fibre units yield only 60–70% throughput (Allington-Smith *et al.*, 2002). When there is a science premium on truly integral field sampling, the above two factors do not outweigh the filling factor improvements. Finally, there is the more subtle effect of whether or not to use output lenslets. Aside from the matter of f-ratio coupling, there is the issue of whether swapping the near- and far-field patterns is desirable for controlling systematics in the spectral image. It amounts to assessing whether the spectrograph is 'seeing-limited', i.e. limited by spatial changes in the light distribution within the slit image formed by the fibre and lenslet, or aberration limited.

Prime examples of optical instruments on 8 m-class telescopes include VIMOS (Le Fevre *et al.*, 2003), GMOS (Gemini-N,S, Allington-Smith *et al.*, 2002), and FLAMES/GIRAFFE in ARGUS or multi-object IFU modes (Avila *et al.*, 2003).[3] Typical characteristics of these devices are fine spatial sampling (well under an arcsec) and modest spectral resolution. ARGUS is an exception, achieving resolutions as high as 39,000. Its multi-object mode is also unique, and powerful (see later discussion). On 4–6 m-class telescopes there are PMAS (Roth *et al.*, 2005), Spiral+AAOmega (Kenworthy *et al.*, 2001; Saunders *et al.*, 2004), MPFS (Afanasiev *et al.*, 1990)[4] and IMACS-IFU (Schmoll *et al.*, 2004).[5] Compared to most direct-fibre IFUs on comparable telescopes, these instruments also have finer spatial sampling.

NIR instruments include SMIRFS (Haynes *et al.*, 1999), and COHSI, which is a precursor – in some regards – to CIRPASS (Parry *et al.*, 2004). An interesting application of flared fibres is discussed by Thatte *et al.* (2000) for cryogenic systems.

A summary of existing and future optical and NIR lenslet + fibre coupled IFU spectrographs is listed in Table 3.2. While it may seem surprising that no future instruments appear to be planned, we will discuss one possible instrument for the 30 m Telescope (TMT) below.

3.2.8 *Slicer coupling*

Image-slicers have been around for a long time, primarily serving the high-resolution community, e.g. to slice a large fibre into a thin, relatively short slit to feed cross-dispersed echelles (for one recent example, see Tull *et al.*, 1995) Extending the concept into a 3D mode follows the same basic notion, which can be thought of as deflecting slices of the telescope image plane both along and perpendicular to the slice through a pair of reflections. These reflections have power to reform the focal-plane image. Given the

[3] See also www.eso.org/instruments/flames/inst/Giraffe.html.
[4] See also www.sao.ru/hq/lsfvo/devices/mpfs/mpfs_main.html.
[5] See also www.ociw.edu/instrumentation/imacs.

TABLE 3.2. Fibre+lenslet coupled integral field instruments

Instrument	Tel. Method	D_T (m)	Ω	$d\Omega$ (arcsec^2)	N_Ω	$\Delta\lambda/\lambda$	R	N_R	ϵ
			Existing optical instruments						
PMAS	Calar Alto	3.5	64.	0.5	256	0.11	9400	1000	0.15
		3.5	64.	0.5	256	0.52	1930	1000	0.15
		3.5	144.	0.75	256	0.11	9400	1000	0.15
		3.5	144.	0.75	256	0.52	1930	1000	0.15
		3.5	256.	1.0	256	0.11	9400	1000	0.15
		3.5	256.	1.0	256	0.52	1930	1000	0.15
SPIRAL	AAT	3.9	251.	0.49	512	0.29	1700	495	0.25
		3.9	251.	0.49	512	0.07	7500	495	0.25
MPFS	SAO	6.0	256.	1.0	256	0.12	8800	1024	0.045
		6.0	64.	0.25	256	0.47	2200	1024	0.045
IMACS-IFU	Magellan	6.5	62.0	0.031	2000	0.61	2500	4096	0.19
		6.5	37.7	0.031	1200	0.31	7500	2340	0.17
GMOS	Gemini	8.0	49.6	0.04	1500	0.21	3450	730	···
		8.0	49.6	0.04	1500	0.32	2300	730	···
		8.0	49.6	0.04	1500	0.82	890	730	···
		8.0	24.8	0.04	750	0.42	3450	1460	···
		8.0	49.6	0.04	1500	0.64	2300	1460	···
		8.0	49.6	0.04	1500	1.00	890	1460	···
VIMOS	VLT	8.0	2916.	0.45	6400	0.6	250	150	···
		8.0	698.	0.11	6400	0.6	250	150	···
		8.0	729.	0.45	1600	0.2	2500	500	···
		8.0	174.5	0.11	1600	0.2	2500	500	···
ARGUS/IFU	VLT	8.0	83.9	0.27	315	0.105	11000	1155	···
		8.0	83.9	0.27	315	0.042	39000	1625	···
ARGUS	VLT	8.0	27.7	0.09	315	0.105	11000	1155	···
		8.0	27.7	0.09	315	0.042	39000	1625	···
			Future optical instruments						
			none						
			Existing near-infrared instruments						
COHSI	UKIRT	3.8	8.5	0.85	100	0.26	500.	128	···
SMIRFS	UKIRT	3.8	24.2	0.34	72	0.023	5500.	128	···
CIRPASS	Gemini	8.0	54.5	0.13	490	0.41	2500.	1024	···
		8.0	54.5	0.13	490	0.085	12000.	1024	···
		8.0	27.0	0.06	490	0.41	2500.	1024	···
		8.0	27.0	0.06	490	0.085	12000.	1024	···
			Future near-infrared instruments						
			none						

deflections, the slices are re-aligned end to end as in a long slit, which then feeds a conventional spectrograph.

The latest incarnation is the so-called 'Advanced Image Slicer' (AIS) concept: a three-element system, introduced and nicely illustrated by Allington-Smith *et al.* (2004). In short, the slicer mirrors at the telescope focal plane divide it into strips and have power to place the telescope pupil on the next slicer element. This is desirable to keep these elements small and the slicer compact. The second element is an array of pupil mirrors

TABLE 3.3. Slicer coupled integral field instruments

Instrument	Tel.	D_T (m)	Ω	$d\Omega$ (arcsec^2)	N_Ω	$\Delta\lambda/\lambda$	R	N_R	ϵ
				Existing optical instruments					
ESI[a]	Keck	10.0	22.8	1.28	18	0.95	3500	3325	0.14
		10.0	15.0	0.56	27	0.95	5200	4950	0.14
		10.0	10.0	0.25	40	0.95	7800	7410	0.14
		10.0	8.4	0.09	93	0.95	13000	12350	0.14
				Future optical instruments					
WiFeS	ANU	2.3	775.	1.	775	1.03	3000	3090	...
		2.3	775.	1.	775	0.44	7000	3090	...
MUSE[a]	VLT	8.0	3600	0.04	9e4	0.67	3000	2000	0.24
				Existing near-infrared instruments					
UIST	UKIRT	3.8	19.8	0.06	344	0.15	3500	512	...
PIFS	Palomar	5.0	51.8	0.45	115	0.23	550	128	0.22
		5.0	51.8	0.45	115	0.10	1300	128	0.22
NIFS[a]	Gemini	8.0	9.0	0.01	900	0.19	5300	1007	...
GNIRS[a]	Gemini	8.0	15.4	0.023	684	0.301	1700	512	...
		8.0	15.4	0.023	684	0.087	5900	512	...
SPIFFI	VLT	8.0	0.54	0.006	1024	0.34	3000	1024	0.3
		8.0	10.2	0.001	1024	0.34	3000	1024	0.3
		8.0	64.0	0.06	1024	0.34	3000	1024	0.3
				Future near-infrared instruments					
KMOS[a]	VLT	8.0	188.0	0.04	4204	0.28	3600	1000	...
FISICA[a]	GTC	10.4	72.0	0.53	136	0.79	1300	1024	...

[a] Advanced image slicer design.

(one per slice), which reformat the slices into a pseudo-slit, where they form an image of the sky. A tertiary field lens (a lenslet for each slice) controls the location of the pupil stop in the spectrograph. This is critical for efficient use of the spectrograph. All-mirror designs exist for the NIR (FISICA, Eikenberry, 2004b), taking advantage of lower scattering at longer wavelengths to machine monolithic elements. Catadioptric designs exist for the optical (MUSE, Henault *et al.*, 2004). Here the pupil lenses replace pupil mirrors, which aids the geometric layout of the spectrograph system.

The salient features of image slicers are (i) they are the only IFU mode to preserve all spatial information. All other coupling modes destroy spatial information within the sampling element, either by fibre scrambling or pupil imaging (below). (ii) Image slicers are also the most compact at reformatting the focal plane on to the detector. (iii) They can be used in cryogenic systems and at long wavelengths where fibres do not transmit (although lenslet arrays also accomplish this; see next section). There are some disadvantages, including (iv) scattered light from the slicing mirrors (diamond-turned optics cannot be used in the optical), and (v) a lack of reformatting freedom. The latter is perhaps less of a concern given that the image is being preserved. However, for possible multi-object modes, particular attention must be paid to the design of the required relay optics to avoid efficiency losses.

We summarize the existing and planned instruments in Table 3.3. The length of the list, particularly in the planned instruments, marks a sea change over the last few years away from fibre + lenslet coupling. While slicers originated for NIR instruments – starting with the now-defunct MPE-3D (Thatte *et al.*, 1994) – the list of planned optical slicers is extensive. Existing NIR instruments include PIFS (Palomar Integral Field Spectrograph; Murphy *et al.*, 1999) and UIST (UKIRT Imager-Spectrometer; Ramsay Howat *et al.*, 2006)[6] on 4 m-class telescopes; NIFS (Near-Infrared Integral Field Spectrograph; McGregor *et al.*, 2003), GNIRS (Gemini Near Infrared Spectrograph; Allington-Smith *et al.*, 2004) and SPIFFI (Spectrometer for Infrared Faint Field Imaging; Eisenhauer *et al.*, 2003; Iserlohe *et al.*, 2004), on 8 m-class telescopes. SINFONI[7] on VLT (Bonnet *et al.*, 2004) in particular has shown the power of NIR adaptive optics (AO) coupled to an image slicer at moderate spectral resolution achieving 20–30% throughput. Future NIR instruments include KMOS (Sharples *et al.*, 2004) – a multi-object system discussed below – and FISICA. Below we also discuss three planned NIR instruments for space.

While the only existing optical instrument is ESI (Echellette Spectrograph and Imager; Sheinis *et al.*, 2002, 2006), future optical instruments include WiFeS (Wide-Field Spectrograph; Dopita *et al.*, 2004), SWIFT (Short Wavelength Integral Field Spectrograph; Goodsall *et al.*, this workshop) and MUSE (Multi-Unit Spectroscopic Explorer; Bacon *et al.*, 2004 and references therein). ESI is unique in being the only cross-dispersed IFU system. While the number of spatial elements is modest, ESI has enormous spectral multiplex (at medium spectral resolution and good efficiency), the largest of any instrument planned or in existence.

3.2.9 *Direct lenslet coupling*

This is the most significant departure in grating-dispersed 3D spectroscopy, and therefore the most interesting. The basic concept consists of pupil-imaging spectroscopy using lenslets. The same type of lenslet array used in the fibre + lenslet mode creates a pupil image from each lenslet, which again is smaller than the size of the lenslet. Here, the array of pupil-images forms the spectrograph input focal surface, or object; no fibres or slicers reformat the telescope focal plane into long-slit; the two-dimensional array of pupil-images is preserved. However, the pupil image does not preserve the spatial information within the lenslet field. These pupil images are dispersed, and then re-imaged at the output spectrograph image surface.

Because direct lenslet injection preserves the 2D spatial data format, this type of instrument typically offers more spatial coverage or sampling at the expense of spectral information. The extent of the spectrum from each pupil image must be truncated to prevent overlap between pupil images. From the instrument design perspective, what is gained is significant: The spectrograph field of view grows linearly with Ω, instead of as Ω^2 as it must in a long-slit spectrograph, where the 2D spatial information must be reformatted into a 1D slit. Hence, this mode is best suited to instruments with the largest Ω or N_Ω.

Lenslet-coupled instruments have excellent spatial fill factors, identical to fibre + lenslet systems, and comparable to slicers. Because this is achieved with fewer optical elements and no fibres, there is no information loss via FRD, and overall the system efficiency can be very high. As with fibre + lenslet coupling, concerns about scattered light from lenslets apply here too. Unlike fibre-coupled modes, there is no control over spatial reformatting. The spectra can be well packed on to the detector but, as noted above, the bandpass must be crafted to prevent overlap for a given spectral dispersion,

[6] See www.jach.hawaii.edu/UKIRT/instruments/uist/uist.html for sensitivities.

[7] SPIFFI + MACAO; Spectrograph for Integral Field Observations in the Near Infrared = Spectrograph for Infrared Faint Field Imaging + Multi-Application Curvature Adaptive Optics.

TABLE 3.4. Lenslet-coupled integral field instruments

Instrument	Tel.	D_T (m)	Ω	$d\Omega$ (arcsec^2)	N_Ω	$\Delta\lambda/\lambda$	R	N_R	ϵ
			Existing optical instruments						
SAURON	WHT	4.2	1353	0.88	1577	0.11	1213.	128	0.147
		4.2	99	0.07	1577	0.10	1475.	150	0.147
OASIS	WHT	4.2	1.92	0.002	1100	0.50	1000.	400	...
		4.2	31.0	0.026	1100	0.50	1000.	400	...
		4.2	180.	0.17	1100	0.50	1000.	400	...
			Future optical instruments						
			none						
			Existing near-infrared instruments						
OSIRIS	Keck	10.4	1.2	0.02	3000	0.12	3400.	400	...
		10.4	30.	0.10	3000	0.12	3400.	400	...
		10.4	0.3	0.02	1019	0.47	3400.	1600	...
		10.4	7.5	0.10	1019	0.47	3400.	1600	...
			Future near-infrared instruments						
			none						

i.e. there is limited spectral coverage at a given resolution. Spectral extraction is critical to minimize crosstalk while maximizing S/N.

Existing optical systems (SAURON, Bacon *et al.*, 2001; OASIS, McDermid *et al.*, 2004) have relatively low dispersion due to grism limitations, although the grisms allow for very compact, undeviated systems. Grating-dispersed systems do exists in the NIR (OSIRIS, Larkin *et al.*, 2003). Future systems with VPH grisms and gratings will have even higher efficiency; the coupling mode is well suited to articulated camera spectrographs. The systems summarized in Table 3.4 are designed to exploit superb image quality with fine spatial sampling (OASIS and OSIRIS are coupled to AO). While they cannot take advantage of high dispersion without becoming read-noise limited, systems with larger specific grasp could be optimized for high spectral resolution.

3.2.10 *Filtered multi-slit (FMS) coupling*

The notion of direct lenslet coupling motivates a poor person's alternative which returns the riches of preserving spatial information. The concept is to use a conventional, multi-object imaging spectrograph with a narrow-band filter, and a slit mask of multi slits in a grid pattern with grid spacing tailored to the desired dispersion of the grating. This is illustrated in Figure 3.11. Spatial multiplexing is increased via filtering. While this only offers sparse spatial sampling, it preserves spatial information (unlike any other mode except slicing) and can easily be adapted to existing spectrographs.

The notion of filtering to increase spatial multiplex has been used for multi-object spectroscopy, e.g. Yee *et al.* (1996) in the context of redshift surveys using MOS on CFHT (Le Fevre *et al.*, 1994). Likewise, fibre + lenslet-coupled IFUs, such as VIMOS and GMOS, use filtering as an option to prevent spectral overlap in configurations with multiple parallel pseudo-slits; this is designed to permit trade-offs in spatial versus spectral coverage.

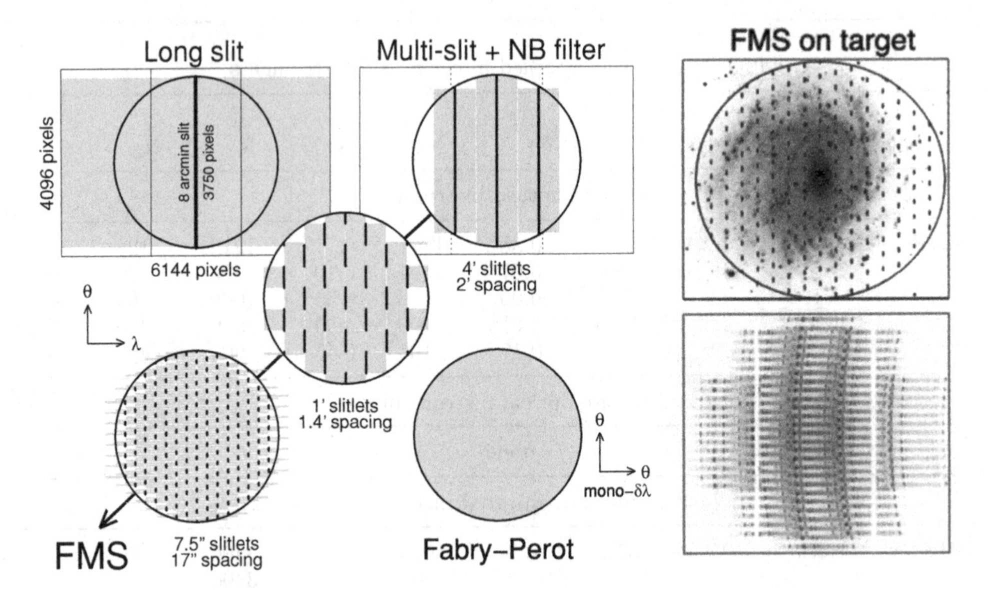

FIGURE 3.11. Filtered multi-slit schematic for SALT's Robert Stobie Spectrograph (RSS). The five panels on the left left show the progression from long slit, to filtered long slit, to two different grids of filtered multi slits. Both are tuned to a high dispersion, 10 nm bandpass, and achieve a spatial multiplex gain of 3 over a pure long slit, with a ×6 loss in bandpass. Higher spatial multiplex (two to 10 times) is achieved at lower spectral resolution. The RSS Fabry–Perot mode is shown for reference. The two rightmost panels show an overlay on a nearby, face-on galaxy and some on-telescope calibration data for that slit-mask.

What is described here is more like the multi-object mode, but instead uses a regular grid of slitlets. This is well-suited, for example, to observing single, extended sources.

An example of this type of instrument is the SALT Robert Stobie Spectrograph (RSS), a prime focus imaging spectrograph with an 8 arcmin field of view, articulating camera, VPH grating suite, dual Fabry–Perot etalons, and $R = 50$ order-blocking filters (Burgh *et al.*, 2003; Kobulnicky *et al.*, 2003). The latter can be used with the multi-slit masks to gain a factor of 3 in spatial multiplex at the highest spectral resolutions ($R = 10{,}000$ with a 10 nm bandpass). At lower resolutions (and fixed bandpass), the slit-packing can be increased by factors of 2 to 10, such that the gain in spatial multiplex is comparable to the loss of a factor of 5–6 in spectral multiplex in this particular case (the system is designed for large spectral multiplex). Even at high spectral resolution what is gained – beyond the spatial multiplex – is the ability to gain 2D spatial mapping in a single exposure. On balance, what is lost and gained is comparable from a purely information standpoint, and hence the choice is, as always, science driven. For the study of nearby galaxy kinematics, this is an outstanding approach.

3.2.11 *Multi-object configurations*

Multi-object 3D spectroscopy is a major path for future instrumentation, although it already exists today in one fabulous instrument: FLAMES/GIRAFFE. Here we are talking about instruments with multiple, independently positionable IFUs. Returning to our so-called 'grand' merit function, it is for just these types of instruments that Ω_s is relevant.

The most obvious way to feed such an instrument is with fibre or fibre + lenslet bundles (e.g. FLAMES/GIRAFFE). Fibre-based systems provide flexibility for spatial positioning, but for cryogenic NIR instruments lenslets or slicers may be required. This necessitates relay optics, which are more mechanically challenging to design and build, and the introduction of additional surfaces that lead to lowered throughput. Sharples *et al.* (2004) have considered the multiple deployable slicer design for KMOS. It is also possible to implement direct lenslet coupling (pupil imaging), as demonstrated by the MUSE concept (Henault *et al.*, 2004), albeit in the context of splitting up a monolithic field into chunks fed to separate spectrographs.

3.2.12 *Summary of considerations*

The various coupling methods discussed above present different opportunities for down-selecting information, and packing three into two dimensions in ways that trade quality versus quantity.

Information selection and reformatting

Fibre + lenslet, slicers and lenslet modes yield comparable spatial, telescope focal-surface sampling, while pure fibre systems have at best 65% integral coverage. Fibre-based systems, however, offer the most flexibility in reformatting telescope-to-spectrograph focal surfaces. Slicers and FMS preserve full spatial information, but only slicers preserve full integral spatial information. As a result of this coherency, slicers can give the most efficient packing on the detector. In terms of spectral information, lenslets and FMS have limited sampling, but other coupling modes all essentially feed long-slit spectrographs and are therefore comparable.

Coverage versus purity

Scattered-light and cross-talk limit signal purity, but to avoid their deleterious effects requires less efficient use of the detector by, for example, broader spacing of fibres in the pseudo-slit, or band-limiting filters, thereby limiting coverage in either or both spatial and spectral dimensions. The trade-off optimization should be science driven. Within this context, pure fibre systems and FMS minimize scattered light, although fibre azimuthal scrambling broadens potential cross-talk between spatial channels of the spectrograph. Slicer systems, again by virtue of the spatial coherency of each slice, are able to utilize detector real-estate while maintaining signal purity.

Sky subtraction

There are four primary issues concerning, and that are root causes of, sky subtraction problems in spectroscopy: (i) low dispersion (sky lines contribute overwhelming shot-noise); (ii) aberrations and non-locality (skyline profiles vary with field angle – spectral and spatial – and time); (iii) stability (instrument); and (iv) under-sampling (compound problems of field-dependent aberrations and flexure). All of these conditions are further compounded if there is fringing on the CCD.

The solutions to these problems are instrumental, observational and algorithmic, i.e. in the approach to the data analysis. The instrumental solution involves having a well-sampled, high-resolution and stable system (you get what you pay for). Fibre-based systems offer the most mapping flexibility, which is critical for spectrographs with aberration-limited spectral image-quality. Pupil imaging (lenslets with or without fibres) may offer advantages for HET/SALT-style telescopes, again if sky subtraction is spectrograph aberration-limited.

The observational approach includes (i) beam-switching, where object and sky exposures are interleaved, and (ii) nod-and-shuffle, where charge is shuffled on the detector in concert with telescope nods between object and sky positions. Both of these approaches have a 50% efficiency in either on-source exposure or in the number of sources that can be observed (the on-detector source packing fraction).

An algorithmic approach entails aberration modelling, which is well suited to any of the coupling methods that feed a spectrograph in a pseudo long slit. The question is to what extent data analysis can compensate for instrumental limitations and avoid inefficient observational protocol.

Some examples exist of telescope-time-efficient sky-subtraction algorithms; these are solutions that do not require beam-switching or nod-and-shuffle. For example, Lissandrini *et al.* (1994) identify flux and wavelength calibration, as well as scattered light, as the dominant problems in their fibre-fed spectroscopic data. They use sky lines for second-order flux calibration (after flat fields), model scattered light from neighbouring fibres and map image distortions in pixel space to obtain accurate wavelength calibration. The improvement is dramatic. Bershady *et al.* (2005) show that higher-order aberrations are important; wavelength calibration is critical, but so too is line shape. They describe a recipe for subtracting the continuum and fitting each spectral channel with a low-order polynomial in the spatial dimension of the datacube. The algorithm works spectacularly well for sources with narrow line emission with significant spectral channel offsets (e.g. high internal dispersion as in a rapidly rotating galaxy, or intrinsically large velocity range as in a redshift survey) and well-sampled data. For other instruments or sources (poor sampling, low dispersion, broad lines, small velocity range), if aberrations are significant, more dedicated sky fibres are needed. On balance, the optical stability of the instrument is critical.

Are these post-facto algorithmic solutions 100% efficient? Not quite. One still needs to sample sky, but, as derived in Bershady *et al.* (2004), the fraction of spatial elements devoted to sky is relatively low (under 10%, and falling below 3% when the number of spatial resolution elements exceeds 1000). So here is a case where, with a stable spectrograph, considerable efficiency may be gained by employing the right processing algorithm. Consequently, fibre-fed, bench-mounted spectrographs offer the greatest opportunities to realize these gains. Regardless of spectrograph type and feed, attention to modelling optical aberrations is critical for good sky subtraction (Kelson, 2003; Viton and Milliard, 2003).

3.3 Interferometry I: Fabry–Perot interferometry

Fabry–Perot interferometry (FPI) provides a powerful tool for 3D spectroscopy because FPI is field-widening relative to grating-dispersed systems. That is to say, higher spectral resolution can be achieved with FPI for a given instrument beam size and entrance aperture. This has long been recognized in astronomy. Unfortunately, the breadth of applications of FPI to sample the datacube has been under-utilized in astronomy. Astronomical applications almost exclusively use FPs as monochromators, i.e. field-dependent, tunable filters. This allows for a premium on spatial multiplex at the loss of all spectral multiplex at a given spatial field angle. Multi-order spectral multiplex can be regained via additional grating dispersion, as noted below; however, in astronomical applications, this is largely a concept (with one exception). Further, it is also possible to use FPs for spectroscopy. In this mode, FPI yields the converse trade in spatial versus spectral multiplex. There is again only one example of such an existing instrument. In short, FPI to date has offered two (orthogonal) extremes in sampling the datacube. The third

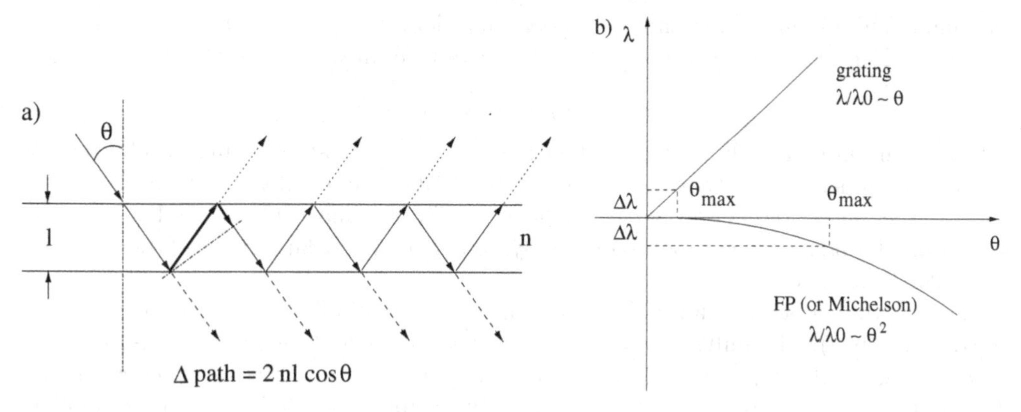

FIGURE 3.12. Basic concept of etalon interference (a), and FP versus grating spectral resolution as a function of angular aperture, θ (b).

dimension (bandpass or field sampling on the sky) has been gained via the temporal domain, i.e. multiple observations. In this sense FPI has not yet been implemented for truly 3D spectroscopy.

The basic principles of FPI, in the context of astronomical monochromators, can be found in Geake *et al.* (1959), Vaughan (1967) and many other references. We summarize the salient aspects to highlight here the field-widened capabilities (we are indebted to R. Reynolds for the structure of the formal development). We discuss and give examples of the two different FPI applications noted above and sketch how one might balance spatial and spectral multiplex in future 3D instruments.

3.3.1 *Basic concepts and field widening*

Etalons (high-precision flat glass plates) are parallel-spaced by some distance l, filled with gas of refractive index n and coated to have high reflectivity. Light incident at some angle θ produces internal reflections, with transmission when the added path ($\Delta\text{path} = 2nl\cos\theta$) between reflections yields positive interference (Figure 3.12a). The ratio of transmitted to incident intensity, I_t/I_i, is given approximately by an Airy function with peaks ($I_t = I_i$) when $\Delta\text{path} = m\lambda$, where m is the order. Given the geometry, this yields an angular dependence to the transmitted wavelength: $\lambda = (2nl/m)\cos\theta$. This can be compared to the grating equation (Littrow configurations for simplicity), where $\lambda = 2n\Lambda_g/m\sin\theta$. At small angles, this means that the instrument entrance aperture can be larger in angle for an FP compared to a grating spectrograph for the same $\delta\lambda$, as illustrated in Figure 3.12b. In other words, an FP system is field-widened for the same spectral resolution (see also Roesler, 1974; Thorne, 1988).

The central wavelength of the FP is controlled by tuning the gap (l) or pressure (index n). The free spectral range is given by the spacing between Airy function peaks in wavelength: $Q = 1/2nl\cos\theta$. Order-blocking filters are needed to suppress other orders. Double etalons suppress the Lorentian wings in the Airy function. The resolution, which is the full width at half maximum of the Airy formula peak, is given by: $R = \lambda/\delta\lambda = 2l\cos\theta N_{\Re}/\lambda = mN_{\Re}$, where $N_{\Re}$ is the reflective finesse defined as $N_{\Re} = \pi\Re^{1/2}/(1-\Re)$, and $\Re$ is the reflectivity. The finesse is equivalent to roughly the number of back and forth reflections; a typical value is $\sim$30 (for a more detailed discussion, see Tanaka *et al.*, 1985). This implies that the spectral resolution, R, is roughly the total path difference divided by the wavelength. High spectral resolution requires high finesse or high order, with the gap size tuned for the desired wavelength. This also achieves high contrast between the maximum and minimum transmittance between orders: $I_{max}/I_{min} = (1+\Re)^2/(1-\Re) = 1 + \frac{4}{\pi^2}N_{\Re}^2$. Herbst and Beckwidth (1988) provide a nice illustration of these quantities.

3.3.2 *FP monochromators*

FPs are conventionally thought of as being used with collimated beams (a mini-review of such instruments from that era is presented by Bland and Tully, 1989). In this case, there is the classic radial wavelength dependence in the image plane. At low spectral resolution the bandpass can be made nearly constant over a large field of view (Jones *et al.*, 2002), as follows.

One way to characterize an etalon is by the size of its 'bull's eye', or Jacquinot spot (Jacquinot, 1954). The bull's eye refers to the physical angle θ such that $\lambda_0/|\lambda_0 - \lambda_\theta| < R$, and is given by $\theta_{max} = \cos^{-1}(1 - 1/R) \simeq \sqrt{2/R}$. This quantity is independent of the telescope and is a property of the etalon. By coupling to a telescope, it is possible to modify the angular scale (α) sampled on the sky by the bull's eye. Since $A \times \Omega$ is conserved, $\alpha = \theta D_e/D_T$, where D_e is the etalon diameter and D_T the telescope diameter.

FPs can, however, be used in converging (or diverging) beams, even near a focus (Bland-Hawthorn *et al.*, 2001). Some examples include the optical FP on the 3.6 m CFHT, when used with the AO Bonnette (AOB)[8] and the future F2T2, a near-infrared double-etalon system for FLAMINGOS-2 (8 m Gemini; Eikenberry *et al.*, 2004a; Scott *et al.*, 2006). Image information is preserved by sampling the beam at a downstream focus, but the spectral resolution is lowered (for a given finesse) at any spatial location because each field angle on the sky is mapped into a range of physical angles through the etalon. The degradation is not particularly severe for lower-finesse etalons or very slow beams. The FLAMINGOS-2 multi-conjugate adapative optics (MCAO) focus for F2T2 is $f/30$, and the AOB F-P beam is $f/40$. If the total angular field of view is much smaller than the beam angle, or the focus is made telecentric, the bandpass is constant across field angles on the sky, and the system forms a highly uniform tunable filter. The AOB optics are not telecentric; this produces a radial degradation in the resolution.

3.3.3 *FP spectrometers*

Alternatively, the full spectral information can be extracted at the loss of the spatial information by placing the etalons at or near a telecentric focus and sampling the pupil in a collimated beam. The Wisconsin Hα Mapper (WHAM; Reynolds *et al.*, 1998) is the only astronomical example of this type of instrument. In this instance, the light is collimated after it passes through the etalons, never refocused, and a detector is placed at the pupil formed by the collimator. Field position on the detector contains spectral information: each radius corresponds to a different wavelength. This is similar to the monochromator application, except in this case each radial location on the detector has a superposition from all spatial locations on the sky within the instrument entrance aperture.

3.3.4 *3D FP spectrophotometers*

Grating-dispersed FPI

Arguably the most interesting FP monochromator mode is to eliminate the order-blocking filters and grating-disperse the output beam to separate the orders on to the detector to increase the spectral multiplex. See, for example, le Coarer *et al.*'s (1995) description of PYTHEAS. Baldry and Bland-Hawthorn (2000) work out a particularly compelling case

[8] For an earlier incarnation on this telescope, see www.cfht.hawaii.edu/Instruments/Spectroscopy/Fabry-Perot/ and Joncas and Roy (1984).

for a cross-dispersed echelle system. The gain in spectral multiplex does not necessarily cost spatial multiplex. In practice, some FPs are in spectrographs where they underfill the detector and usable field in the image plane (e.g. RSS and F2T2). If the dispersion is significantly greater than the etalon resolution, then, in addition to spectral multiplex, this mode adds band-limited slitless spectroscopy in each FP order.

Pupil-imaging FPI

The above discussion frames the notion that detection downstream of an etalon at the pupil of a collimated beam provides spectral information but no spatial information, while detection at a focal surface provides the complement. A simple ray-trace shows that between these two locations spectral and spatial information are mixed. By using pupil imaging at the system input via a lenslet array (Section 3.2.9), detection at an intermediate surface in a converging beam can separate spatial and spectral information. Although this has never been done, in principle this could balance spatial and spectral multiplex and allow for true 3D spectroscopy in future, field-widened instruments.

3.3.5 *Sky stability*

Because spectral channels are not observed simultaneously in monochromatic modes, atmospheric changes must be calibrated (in the context of TAURUS, see, for example, Atherton *et al.*, 1982). Field stars may suffice if they are sufficiently featureless over the scanned wavelength range. Built-in calibration is desirable, which can be achieved, for example, via a diachronic feeding a monitoring camera. This capability is designed for a new generation of instruments (e.g. ARIES, T. Williams, private communication).

3.3.6 *Examples of instruments*

Two extremes in FP instrumentation are highlighted by the RSS imaging FP (Williams *et al.*, 2002) and the WHAM non-imaging FP. Both have 150 mm etalons, but the RSS system is coupled to a 9.2 m telescope with an 8 arcmin field of view, 0.2 arcsec sampling and spectral resolutions of 500, 1250, 5000, and 12,500. In contrast, WHAM is coupled to a 0.6 m telescope, with a 1 deg field of view and angular resolution, spectral resolution of $R = 25{,}000$, and spectral coverage of about 166 resolution elements for one spatial element.

There are a large number of existing FP monochromators (also know as tunable filters), indicated even by the following incomplete list. Optical systems include, but are not limited to:

- PUMA (2.1 m OAN-SPM; Rosado *et al.*, 1995);
- RFP (1 m and 4 m CTIO; e.g. Sluit and Williams, 2006);
- CIGALE (3.6 m ESO [European Southern Observatory] and 1.9 m OHP; Boulesteix *et al.*, 1984);
- FaNTOmM (1.6 m OMM, 1.9 m OHP, and 3.6 m CFHT; Canada–France–Hawaii Telescope; Hernandez *et al.*, 2003);
- Goddard FP (3.5 m APO; Gelderman *et al.*, 1995);
- SCORPIO FP (6 m SAO; Afanasiev and Moiseev, 2005);
- IMACS FP (Inamori Magellan Cassegrain Spectrograph FP; 6.5 m Magellan; Dressler *et al.*, 2006);

as well as the above-mentioned CFHT FP etalons, which can be used with the AOB as well as the MOS (Multi-Object Imaging Spectrograph) and SIS systems. The most widely cited system is TTF/TAURUS-II (3.9 m AAT; Anglo-Australian Telescope; 4.2 m WHT; Gordon *et al.*, 2000 and references therein).

Existing infrared instruments include:

- NIC-FPS (3.5 m Arc; Hearty *et al.*, 2004);
- GriF (3.6 m CFHT; Clenet *et al.*, 2002);
- PUMILA (2.1 m OAN-SPM; Rosado *et al.*, 1998);
- UFTI (3.8 m UKIRT; United Kingdom Infrared Telescope; Roche *et al.*, 2003); and
- NaCo (8 m VLT; Hartung *et al.*, 2004; Iserlohe *et al.*, 2004).

GriF, NaCo and F2T2 are AO-fed. By virtue of their use in collimated beams, many of the FP systems are designed to be transportable between instruments (i.e. spectrographs or focal reducers) and telescopes. Future instruments include the optical OSIRIS (Optical System for Imaging and Low/intermediate-Resolution Integrated Spectroscopy; 10.4 m GTC; Gran Telescopio CANARIAS) and near-infrared FGS-TF (6.5 m JWST; James Webb Space Telescope; Davila *et al.*, 2004) and F2T2 (above). These systems span a wide range of wavelength, spectral and spatial resolution. One attribute they have in common is a spectral multiplex of unity.

3.4 Interferometry II: spatial heterodyne spectroscopy

A spatial heterodyne spectrometer (SHS) is a Michelson interferometer with gratings replacing the mirrors. The principles of operation are described and illustrated by Harlander *et al.* (1992); this is a paper that is well worth careful study.[9] Briefly, each grating diffracts light at wavelength-dependent angles. Because of the 90 degree fold between the two beams, the wavefronts at a given wavelength are tilted with respect to each other after beam recombination. This tilting produces a sinusoidal interference pattern with a frequency dependent on the tilt angle. The degree of tilt is a function of wavelength, simply due to the grating diffraction, and hence the interference pattern frequency records the wavelength information.

It is easiest to conceptualize this in terms of two identical gratings (as illustrated by Harlander *et al.*, 1992 in their figures 2 and 3), but in principle the gratings do not need to be the same. Wavefronts produce interference patterns with frequencies set by wavelength, with the central wavelength producing no interference. Hence, the signal is heterodyned about the frequency of the central wavelength. Resolution is set by the grating aperture diameter because this sets the wavelength (i.e. angular tilt) that minimally departs from the central wavelength which can produce the first (lowest) frequency for interference. Bandwidth is set by the length of the detector, i.e. how many frequencies can be sampled depends on the number of pixels.

The advantage of an SHS over a Michelson is that no stepping is required to gain the full spectral information, but the field of view is reduced. The SHS can be fed with a long slit or lenslet array, although with the latter a band-limiting filter is needed (as with a conventional dispersed spectrograph). As with a Michelson, however, field-widening is possible via prisms. In the SHS application, the prisms give gratings the geometric appearance of being more perpendicular to the optical axis; hence, larger field angles are mapped within the beam deviation producing the lowest-order interference fringe. Cross-dispersion is possible (by tilting one of the gratings about the optical axis), but the same fundamental limits apply concerning 3D information formatted into a 2D detector.

One of the problems with the standard Michelson or SHS interferometer is that their geometry throws out half the light right from the start. Non-lossy geometries are possible. Harlander *et al.* (1992) give an example of working off-axis on the collimating mirror (see their figure 5). This is a perfect application for holographic gratings. Transmission-grating

[9] The presentation here benefited from discussion with J. Harlander, A. Sheinis, R. Reynolds, F. Roesler and E. Merkowitz.

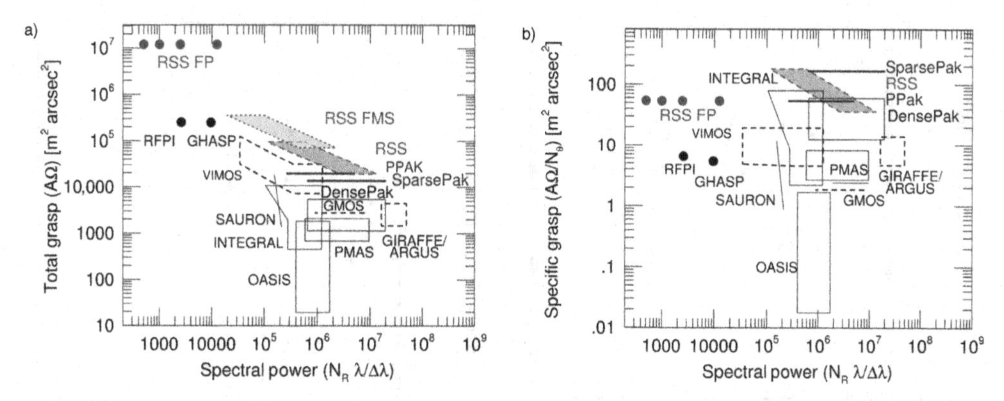

FIGURE 3.13. Total and specific grasp versus spectral power for a range of instruments on 4-m- and 10-m-class telescopes (solid and dashed lines, respectively) partially updated from Bershady *et al.* (2005). See text for comments on instrument efficiency.

geometries would eliminate the need to go off-axis and probably allow for larger field. Another approach is a Mach–Zender-style interferometer (Douglas *et al.*, 1990). The latter requires twice the detector real estate for the same number of spectral resolution elements.

The primary advantage of an SHS is that it allows for very high spectral resolution for a given solid angle relative to a conventional, grating-dispersed spectrograph. The SHS is field-widened like an FP. This means the SHS can be built for low cost even on large telescopes because the optics are small.

However, because the signal is in the form of an interferogram, there is what is known as the 'multiplex disadvantage.' This can be expressed as the S/N performance of the SHS relative to a grating spectrograph: $S/N_{\mathrm{SHS}} = S/N_{\mathrm{GS}}(f/2)^{1/2}(S_{\mathrm{SHS}}/S_{\mathrm{GS}})^{1/2}$, where S/N_{SHS} and S/N_{GS} are the signal to noise in SHS and grating spectrometer, respectively, S_{SHS} and S_{GS} are the total photon signal, respectively, and f is the fraction of total signal in a given spectral channel ($f < 1$, and decreases with bandwidth). In words, this means that an SHS loses competitiveness with grating-dispersed spectrographs when the bandpass is large. This has implications for design and use. Clearly, one must make S_{SHS} and f as large as possible. The small compact optics of an SHS system lend themselves to efficiency optimization. To make f as large as possible, one must choose a small bandwidth (but more than a Fabry–Perot monochromator) and remove OH lines via pre-filtering, or by selecting bandpasses between them. Returning to Figure 3.1, SHS is between an FP monochromator and other IFS methods and will therefore have application to a broad range of science programmes that seek high spectral resolution over a limited bandpass with good spatial coverage.

3.5 Summary of existing instruments

Here, we explore the sampled parameter space in spatial versus spectral information, as well as coverage versus resolution, starting with grasp and spectral power (Figure 3.13). Recall that, because reliable, consistent measurements of efficiency are unavailable for most instruments, we use grasp instead of etendue (warning: we really want etendue). Note, however, that there is a factor of 6 range in the known efficiencies of instruments tabulated in this chapter. Further, note that there are two ways of viewing the specific grasp. From the perspective of staying photon-limited at high spectral resolution, high specific grasp is important. The 'flip side' is that low specific grasp implies high angular resolution.

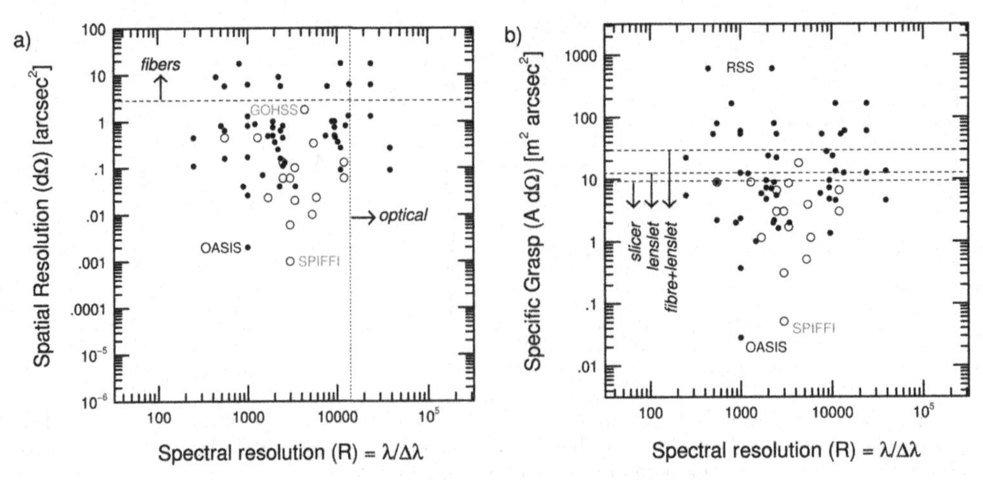

FIGURE 3.14. Spatial resolution (a) and specific grasp (b) versus spectral power for all instruments in Tables 3.1–3.4, highlighting differences between optical (filled symbols) and NIR (open symbols), as well as between different coupling methods (labelled).

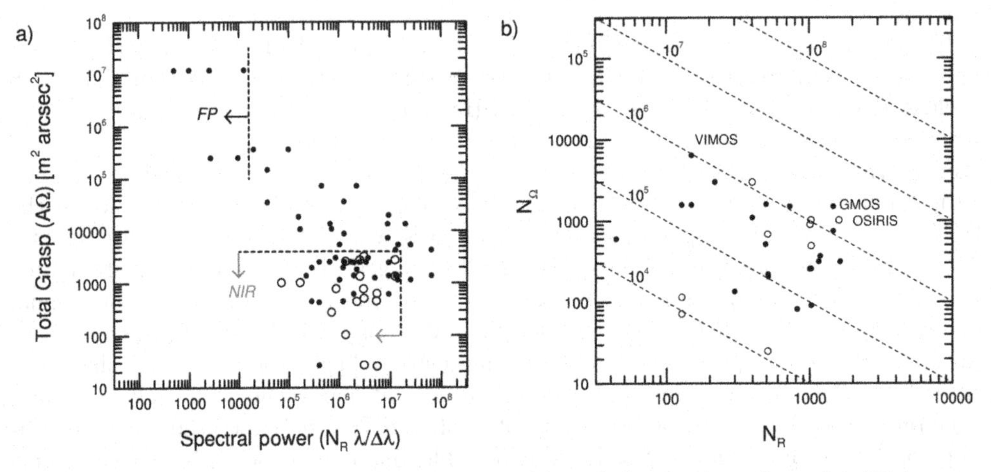

FIGURE 3.15. Total grasp versus spectral power (a) and the number of spatial (N_Ω) versus spectral (N_R) resolution elements for all instruments in Tables 3.1–3.4, highlighting differences between optical (filled symbols) and NIR (open symbols). Dashed lines are at constant (labelled) total information ($N_R \times N_\Omega$).

Figure 3.14 shows that spatial resolution is higher in NIR instruments, while spectral resolution is higher in optical instruments. Fibre IFUs have the largest specific grasp, reflected in the bifurcation seen in spatial resolution, i.e. fibre-fed instruments have large footprints per element ($d\Omega$). There is a trend of decreasing specific grasp going from fibre +lenslet, lenslet, and finally to slicers. ESI has unusually large $A \times d\Omega$ for a slicer; RSS in FMS mode has the highest specific grasp overall.

Figures 3.14 and 3.15 together show that optical and near-infrared instruments trade spatial resolution for grasp; there are no high-grasp NIR instruments; the highest spectral power instruments are optical. Optical and near-infrared instruments sample comparable total information, with optical instruments sampling a broader range of trades between spatial versus spectral information. Older NIR instruments clearly suffer from being detector-size limited. IMACS-IFU stands out as having a significantly larger number of total information elements, $N_R \times N_\Omega$, and in this sense is on par with future-generation instruments.

3.6 The extended-source domain

One area of extragalactic science is clearly under-sampled by existing instrumentation, namely high spectral resolution yet low surface brightness 3D spectroscopy of extended sources. The scientific impetus is for detailed nebular studies (ionization, density, metallicity, abundances) of not only compact HII regions, but to extend such study to the diffuse, ionized gas. Likewise, a significant fraction of the stellar light in galaxies is in extended distributions at low surface-brightness, i.e. below the night sky background. The kinematic and chemical properties of these stars is largely unknown outside of resolved populations in the Local Group. Stellar kinematics of galaxies on spatially resolved scales are required to dissect the mass distribution and detailed dynamics of disc, bulge and halo components. This information is effectively the Rosetta Stone for deciphering how galaxies have assembled.

One concern with most existing IFU spectrographs is their focus on very fine spatial sampling. Referring back to Figure 3.4, on telescopes as small as only 10 m, this severely limits the spectral resolution that can be achieved at sub-arcsec sampling in the photon-limited regime. For example, FLAMES/GIRAFFE is unusual in its high spectral resolutions of 10–40,000. Each IFU unit is 2×3 arcsec with 20 rectangular microlenses sampling only 0.52×0.52 arcsec; this is equivalent to a 1 arcsec fibre on 3.5 m telescope. The instrument is very close to the photon detector-limited divide. The IMACS-IFU should be in a similar domain at its high spectral-resolution limit.

There is no question that FLAMES/GIRAFFE has proven spectacular for emission line work, particularly if line emission is clumpy and unresolved, e.g. ionized gas kinematics of distant galaxies (Flores *et al.*, 2004). The need for high angular resolution in the distant-galaxy kinematic game is paramount. Even with $\sim$0.5 arcsec resolution, *Hubble Space Telescope* images are needed to super-resolve the IFU data (Flores *et al.*, 2004). It will be difficult, however, to use this same facility to study diffuse gas or the stellar continuum in resolved sources. Furthermore, resolved structures at high redshift are all at apparently low surface brightness because of cosmological dimming. To stay photon-limited, observing in the low surface brightness regime requires lower spectral resolution, larger apertures ($d\Omega$) or larger telescopes. Will this be addressed by future instrumentation?

3.7 Future instruments

The next generation of instruments will compete on both space-based platforms such as the *James Webb Space Telescope* (JWST) and ground-based telescopes reaching 30 m or more in diameter. Why build these bigger telescopes? The argument of simply collecting more photons is compelling but not sufficient. New facilities, which come at increasingly greater cost, must yield gains above the linear increase in area. Such 'windfalls' may include overcoming detector noise, the diffraction limit (at long wavelengths), backgrounds (in the case of space-borne platforms) or critical combinations thereof.

A discussion of backgrounds and the relative merits and niches of 8 m-class space-borne telescopes such as JWST, and large 30 m-class ground-based telescopes was vetted in the early planning stages of what was once known as 'MAXimum Aperture Telescopes', or MAXAT. Gillett and Mountain (1998)[10] pointed out that a cooled spacecraft has significantly lower background in the infrared compared to the ground, even at high spectral resolution. This contrast is dramatic for $\lambda > 2.5\,\mu\mathrm{m}$, i.e., in the thermal IR. However, they calculated that above $R \sim 1000$, 8 m-class space telescopes are detector-limited at any wavelength, assuming 0.05 arcsec apertures, a generous system throughput and realistic detectors. They constructed a competitiveness criteria that assumed diffraction-limited performance for stellar imaging or spectroscopy. Compared to JWST, they concluded that ground-based telescopes can be competitive at $\lambda < 2.5\,\mu\mathrm{m}$ for imaging if $D_T > 20\,\mathrm{m}$, and for diffraction-limited spectroscopy at $R > 1000$ for $D_T > 8\,\mathrm{m}$ at any wavelength. These considerations have been influential in the planning and design of future-generation instruments in the era of JWST.

3.7.1 *Ground-based instruments on 10 m telescopes*

Given rapid growth in 3D spectroscopy, we expect many new and retro-fitted systems in the coming years. We outline three instruments – MUSE, VIRUS and KMOS – because they highlight the common themes of object and instrument multiplexing. (US scientists will note that two of these systems are on the VLT.) Object multiplexing is a departure for 3D instrumentation; instrument multiplexing is a departure overall. The basic parameters of these three instruments are summarized in Tables 3.1 and 3.3.

Both MUSE and VIRUS offer unprecedented spatial sampling. MUSE provides a truly integral 1 arcmin^2 area, sampled at the 0.2 arcsec scale, accomplished via image slicing (AIS-type). The most significant portions of the system are the slicers, which must perform well (with little scattered light) in the optical, and the field-partitioning between a bank of 24 identical spectrographs. In comparison, the individual spectrographs are modest, albeit high-efficiency, articulated VPH-grating systems.

VIRUS uses the same notion of a replicated spectrograph unit (also articulated VPH-grating systems), but in this case fed by bare fibres on a much coarser scale (25× larger $d\Omega$). The field sampling is sparse; hence, this instrument follows directly in the path of bare-fibre IFUs. What stands out in the VIRUS design is the 'ultra-cheap' notion of the spectrograph unit, i.e. by building many replicated units, costs are lowered by economies of scale. Such a demonstration has important implications for future large-telescope instrument design (how large must the replication scale be to manifest significant economy?). If the full replication of 132 spectrographs is accomplished, this will be by far the widest-field (largest grasp) IFU in existence, probably for years to come. To achieve the VIRUS-132 goal requires an all-new wide-field prime-focus spherical-aberration corrector yielding a 16 arcmin science-grade field for the HET; this is a significant optomechanical challenge in itself.

MUSE, in contrast to VIRUS, has 9× less total grasp, but almost 3× more spatial elements (N_Ω). In other words, both stand out as remarkable in spatial sampling in their own way. The differences in spatial resolution versus coverage between MUSE and VIRUS lies in their respective science themes. VIRUS is designed as a precision-cosmology engine to measure the baryon oscillations by detecting $z \sim 3$ Lyα-emitters, and using their distribution as a density tracer. These sources are relatively rare (in surface-density to a given detected flux), although the exact flux density relation is still uncertain. Rather low spectral resolution ($R < 1000$) is needed, since only line identification (in the blue where backgrounds are low) and redshifts are required.

[10] See also the AURA MAXAT Final Report (1999), www.gemini.edu/science/maxat.

TABLE 3.5. Future TMT integral field instruments

Instrument	Coupling	D_T (m)	Ω (arcsec^2)	$d\Omega$ (arcsec^2)	N_Ω	$\Delta\lambda/\lambda$	R	N_R	ϵ
IRMOS	slicer	30.	40.	0.01	4000	0.25	2000	500	...
IRMOS	slicer	30.	40.	0.01	4000	0.25	10000	2500	...
IRIS	slicer	30.	0.26	1.6e-5	16384	0.05	4000	200	...
IRIS	slicer	30.	1.33	8.1e-5	16384	0.05	4000	200	...
IRIS	slicer	30.	7.93	4.8e-4	16384	0.05	4000	200	...
IRIS	slicer	30.	41.0	2.4e-4	16384	0.05	4000	200	...
WFOS	fibre+lens	30.	810.	0.56	1440	1.37	5000	6850	0.3

MUSE, in contrast, is designed to probe the detailed internal properties (dynamics, stellar populations) of galaxy populations over a wide range in redshift and in a representative cosmological volume. The aim of this instrument is essentially to enable spectroscopic versions of many 'Hubble Deep Fields', each with sufficient spectral and spatial information to extract kinematics and line-diagnostics of many thousands of $z < 1$ galaxies.

KMOS has much the same science goals as MUSE, with the key distinction of pushing to higher redshifts by using the NIR to capture the optical rest-frame. By pushing to higher redshift to gain temporal leverage on the galaxy formation and evolution process the source distribution becomes apparently fainter and the NIR backgrounds are higher. Consequently, on the same size telescope, one is forced to look at intrinsically more luminous and hence rarer objects. Therefore, the KMOS design moves away from the notion of a monolithic integral-field, to a 24-probe system in a large, 7.5 arcmin diameter patrol field. Each probe spans a $2.8 \times 2.8\,\mathrm{arcsec}^2$ area sampled at $0.2 \times 0.2\,\mathrm{arcsec}^2$. While a multi-IFU instrument already exists (again on the VLT) in the optical with FLAMES/GIRAFFE, the extension to the NIR using slicers with twice the number of probes will be a significant technical achievement.

What is missing from this suite of remarkable instruments is a design which pushes forward a significant increase in spectral sampling (spectral power) or specific grasp. For example, none of these instruments offers over $R = 4000$ and $N_R = 2000$. This means, for example, that advances in the study of low surface brightness, dynamically cold ($\sigma < 80\,\mathrm{km\ s}^{-1}$) systems or nebular regions will require additional instrument innovation.

3.7.2 *Ground-based instruments on 30–50 m-class telescopes*

We summarize some of the specific examples of TMT 3D spectroscopic instrumentation in Table 3.5, based on D. Crampton's overview (Crampton and Simard, 2006).[11] TMT instrument design is largely driven by AO capabilities, where the salient point is that there are many 'flavours' of AO, with associated levels of difficulty and risk (inversely proportional to their performance in either image quality or field of view, or both). The IFU-capable TMT instruments include, in order of decreasing AO requirements: (i) IRMOS, a NIR multi-object integral-field spectrograph fed by the multi-object adaptive object system (MOAO), capable of 20 positional, 5 arcsec compensated patches within a 5 arcmin patrol field; (ii) IRIS, a NIR imager and integral field spectrograph working at the diffraction limit, fed by the narrow-field facility AO system (NFIRAOS);

[11] See also www.tmt.org/observatory/index.html.

TABLE 3.6. Future space-based integral field instruments

Instrument	Coupling	Tel.	D_T (m)	Ω	$d\Omega$ (arcsec^2)	N_Ω	$\Delta\lambda/\lambda$	R	N_R	ϵ
FGS-TF	FP	JWST	6.5	38088.	0.018	2.10e7	0.01	100	1	···
NIRSpec	AIS	JWST	6.5	9.	0.0056	1600	0.34	3000	1024	···
MIRI	AIS	JWST	6.5	51.8	0.30	173	1.48	2800	4096	···
SNAP-IFU	AIS	SNAP	2.	9.0	0.022	400	1.95	100	195	0.44

and (iii) WFOS, an optical wide-field, seeing-limited spectrograph with potential for a modest-grasp IFU with good spectral power and spectral resolution ($R < 6000$).

With the exception of WFOS, instrument design is driven by AO considerations because of the enormous physical size of the image (which scales with mirror diameter for a constant f-ratio). WFOS is necessarily a monster. The AO-driven focus is suitable for scientific studies of un- or under-resolved sources (stars, planets, sub-kpc scales in distant galaxies), and excellent science cases have been developed. Of this excellence there is no doubt. Of concern is that once wedded to the notion of a very large telescope with no clear path to building affordable, comparably monstrous instruments, one is forced down a path, ab initio, of considering only science enabled by high angular resolution. It is not surprising to note that WFOS – the one non-AO corrected instrument – stands out as also the one TMT instrument concept that breaks into the high specific-grasp domain at modest spectral-resolution domain (Figure 3.16). Indeed, as seen in Figure 3.17, WFOS breaks new ground in terms of its total grasp with a higher spectral power than any existing instrument (save ESI). To optimize low surface brightness studies, other paths will need to be forged to push to higher spectral power and resolution at comparably high grasp.

These same trends are also being played out for instrument design for extremely large telescopes (ELTs) (e.g. Eisenhauer *et al.*, 2000; Russell *et al.*, 2004). We have focused on TMT because of the more mature stage of this telescope's planning. No doubt ELT's complement of instruments will open up exciting new capabilities, as demonstrated by the superb forefront instrumentation on the VLT. The TMT instrumentation programme, like that of the European ELT, is evolving rapidly. What is presented here is a snapshot circa late 2005.

3.7.3 *Space-borne instruments*

We summarize the planned 3D spectroscopic instruments for JWST and SNAP (Super Nova Acceleration Probe; Aldering *et al.*, 2002) in Table 3.6. (There are other missions, which include IFUs, also in the planning stages.) On JWST, in remarkable contrast to HST, three of the four instruments have 3D spectroscopic modes in the near- and mid-infrared: (i) FGS-TF (of which F2T2 is the ground-based analogue) delivers a 2.3×2.3 arcmin^2 field at $R \sim 100$, with two cameras covering 1.2 to 4.8 μm; (ii) Near-Infrared Spectrograph (NIRSpec) (Prieto *et al.*, 2004) has a 3×3 arcsec^2 IFU using and AIS with 40 3×0.075 arcsec^2 slices, covering 0.8–5 μm at $R = 3000$; (iii) Mid-Infrared Instrument (MIRI) (Wright *et al.*, 2004) has 4 simultaneous image-slicers at $R \sim 3000$ feeding four wave-bands between 5 and 28 μm. Each samples 4.6×5.5 arcsec^2 (increasing by a factor of two between the bluest and the reddest channel) with a 0.37 arcsec slit width (changing by a factor of 4 between the bluest and the reddest channel). Quoted numbers represent mean values over all channels.

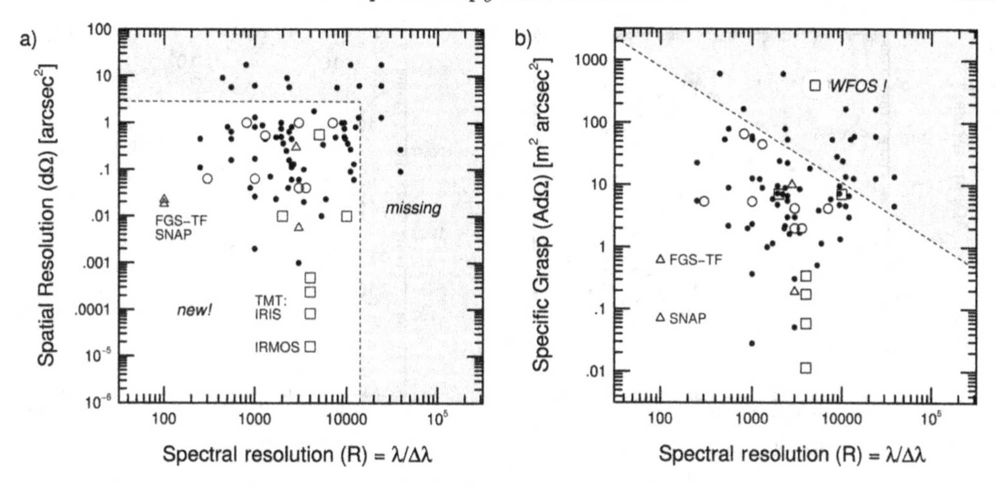

FIGURE 3.16. Spatial resolution (a) and specific grasp (b) versus spectral power including future instruments in Tables 3.5 and 3.6: existing (filled circles), future ground-based 10 m-class telescopes (open circles), future TMT (open squares), and future space-borne (open triangles).

The SNAP IFU (Ealet *et al.*, 2003) is designed to identify SNe type out to $z \sim 1.7$. As such, it is unique in being a dual optical–NIR system (0.35–1.7 microns), with a 3×3 field using AIS, but very low spectral resolution ($R = 100$). With its very high expected efficiency, co-added datasets should yield superb, spatially resolved spectrophotometry of galaxies on 1–2 kpc scales.

Overall, future space-borne capabilities can be characterized as having $3 \times 3\,\mathrm{arcsec}^2$ fields mapped with AIS technology with 0.15 arcsec sampling, lower spatial resolution than TMT. Spectral resolution is in the $100 < R < 3000$ range, again lower than TMT. This is consistent with their being competitive in performance relative to TMT-class instruments, given Gillett and Mountain's (1998) argument. However, there are no large-grasp systems that take full advantage of the low backgrounds of space. There are no high- or even medium-resolution spectrographs to couple, competitively, to such large angular apertures. Nonetheless, barring past fiascos, the space-borne missions offer the guarantee of superlative image quality and low backgrounds extending into the mid-IR, while ground-based observatories face the intense challenge of developing advanced AO systems.

3.7.4 *Summary of future instruments*

While Figure 3.16 shows some of the areas not accessed by currently planned future instrumentation, at the same time clearly great strides are planned for accessing new domains in spatial resolution from the ground. This is encouraging because only with the largest apertures can we stay photon-limited at moderate spectral resolution. JWST instruments present the unique ability to work at more modest spectral resolution and still remain source-photon limited. Space-borne instruments, overall, will also provide the most-stable and best-characterized point-spread functions, a premium for high angular-resolution spectrophotometry.

Figure 3.17 reveals where new instruments open up new territory: both in added grasp at high spectral power and simply in more resolution elements ($N_R \times N_\Omega$). Of particular note is the thrust towards instruments with many thousands of spectral resolution elements. These gains are seen for both ground-based and space-borne instruments, on both 10 m- and 30 m-class telescopes.

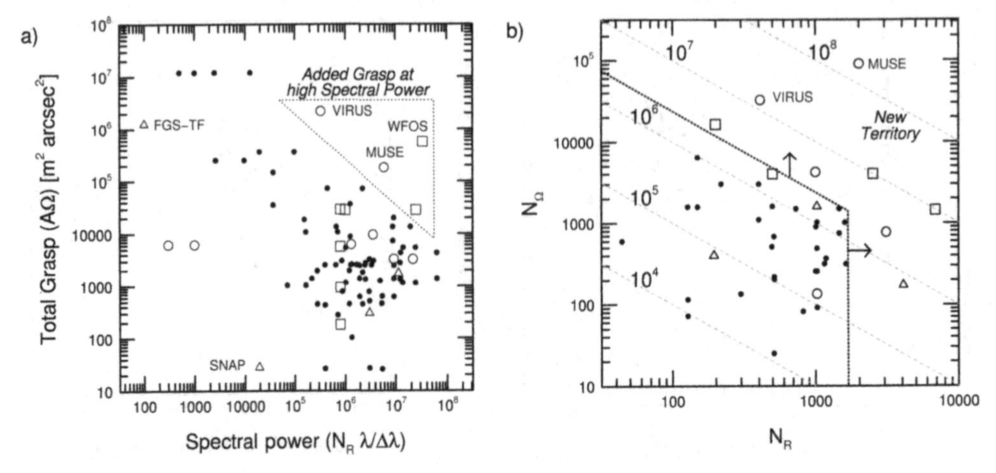

FIGURE 3.17. Total grasp versus spectral power (a) and the number of spatial (N_Ω) versus spectral (N_R) resolution element (b) including future instruments in Tables 3.5 and 3.6; symbols as in previous figure.

These gains are made with conceptually conventional grating-dispersed systems or FP monochromators. Clearly, there is opportunity for less conventional, field-widened instruments (such as the interferometric concepts discussed above), which can amplify both grasp and spectral power or spectral resolution. Given the relative novelty of these approaches, they present higher risk, but potentially higher return, and are best suited for ground-based development.

Acknowledgements

We would like to thank the Instituto de Astrofísica de Canarias (IAC) and Winter School organizers, the University of Toronto Department of Astronomy and Astrophysics for their gracious hospitality during a sabbatical year where this work was done, and the NSF for their financial support of this research (AST-0307417 and AST-0607516).

REFERENCES

AFANASIEV V.L., MOISEEV, A.V. (2005), *Astr. Lett.*, **31**, 193

AFANASIEV, V.L, DODONOV, S.N., SIL'CHENKO, O.K., VLASYUK, V.V. (1990), *SAO preprint*, **N54**

ALDERING, G., AKERLOF, C.W., AMANULLAH, R. ET AL. (2002), *Proc. SPIE*, **4835**, 146

ALIGHIERI, S.S. (2005), in the Proceedings of the ESO Workshop on *Science Perspectives for 3D Spectroscopy*, (Garching, Germany)

ALLENDE PRIETO, C., MAJEWSKI, S.R., SCHIAVON, R. ET AL. (2008) *AN*, **329**, 1081

ALLINGTON-SMITH, J., CONTENT, R. (1998), *PASP*, **110**, 1216

ALLINGTON-SMITH, J., MURRAY, G., CONTENT, R. ET AL. (2002), *PASP*, **114**, 892

ALLINGTON-SMITH, J., DUBBELDAM, C.M., CONTENT, R. ET AL. (2004), *Proc. SPIE*, **5492**, 701

ARRIBAS, S., MEDIAVILLA, E., RASILLA, J.L. (1991), *ApJ*, **369**, 260

ARRIBAS, S., DEL BURGO, C., CARTER, D. ET AL. (1998), *ASPCS*, **152**, 149

ATHERTON, P.D., TAYLOR, K., PIKE, C.D., HARMER, C.F.W., PARKER, N.M., HOOK, R.N. (1982), **201**, 661

AVILA, G., GUINOUARD, I., JOCOU, L., GUILLON, F., BALSAMO, F. (2003), *Proc. SPIE*, **4841**, 997

BACON, R., COPIN, Y., MONNET, G. ET AL. (2001), *MNRAS*, **326**, 23

BACON, R., BAUER, S.-M., BOWER, R. ET AL. (2004), *Proc. SPIE*, **5492**, 1145

BALDRY, I.K., BLAND-HAWTHORN, J. (2000), *PASP*, **112**, 1112
BALDRY, I.K., BLAND-HAWTHORN, J., ROBERTSON, J.G. (2004), *PASP*, **116**, 403
BARANNE, A. (1972), in *ESO/CERN Conference on Auxiliary Instrumentation for Large Telescopes*, ed. S. Lautse & A. Reiz (Geneva), 227
BARDEN, S.C., WADE (1988), *ASPCS*, **3**, 113
BARDEN, S.C., ELSTON, R., ARMANDROFF, T., PRYOR, C. (1993), *ASPCS*, **37**, 223
BARDEN, S.C., SAWYER, D.G., HONNEYCUTT, R.K. (1998), *Proc. SPIE*, **3355**, 892
BARDEN, S.C., ARNS, J.A., COLBURN, W.S., WILLIAMS, J.B. (2000), *PASP*, **112**, 809
BERSHADY, M.A., ANDERSEN, D.R., HARKER, J., RAMSEY, L.W., VERHEIJENN, M.A.W. (2004), *PASP*, **116**, 565
BERSHADY, M.A., ANDERSEN, D.R., VERHEIJEN, M.A.W., WESTFALL, K.B., CRAWFORD, S.M., SWATERS, R.A. (2005), *ApJS*, **156**, 311
BERSHADY, M., BARDEN, S., BLANCHO, P.-A. ET AL. (2008) *Proc. SPIE*, **7014**, 70140H-1
BLAIS-OUELLETTE, S., GUZMAN, D., ELGAMIL, A., RALLISON, R. (2004), *Proc. SPIE*, **5494**, 278
BLAIS-OUELLETTE, S., DAIGLE, O., TAYLOR, K. (2006), *Proc. SPIE*, **6269**, 174
BLAND, J., TULLY, R.B. (1989), *AJ*, **98**, 723
BLAND-HAWTHORN, J., VAN BREUGEL, W., GILLINGHAM, P.R., BALDRY, I.K. (2001), *ApJ*, **563**, 611
BONNET, H. ET AL. (2004), *Msngr*, **117**, 17
BOULESTEIX, J., GEORGELIN, Y., MARCELIN, M., MONNET, G. (1984), *Proc. SPIE*, **445**, 37
BURGH, E.B., NORDSIECK, K.H., KOBULNICKY, H.A., WILLIAMS, T.B., O'DONOGHUE, D., SMITH, M.P., PERCIVAL, J.W. (2003), *Proc. SPIE*, **4841**, 1463
BURGH, E.B., BERSHADY, M.A., WESTFALL, K.B., NORDSIECK, K.H. (2007), *PASP*, **119**, 1069
CARRASCO, E., PARRY, I.R. (1994), *MNRAS*, **271**, 1
CLENET, Y., LE COARER, E., JONCAS, G. ET AL. (2002), *PASP*, **114**, 563
CRAMPTON, D., SIMARD, L. (2006), *Proc. SPIE*, **6269**, 59
DAVILA, P.S., BOS, B.J., CONTRERAS, J. ET AL. (2004), *Proc. SPIE*, **5487**, 611
DOPITA, M.A., WALDRON, L.E., MCGREGOR, P. ET AL. (2004), *Proc. SPIE*, **5492**, 262
DOUGLAS, N.G., BUTCHER, H.R., MELIS, M.A. (1990), *Ap&SS*, **171**, 307
DRESSLER, A., HARE, T., BIGELOW, B.C., OSIP, D.J. (2006), *Proc. SPIE*, **6269**, 13
EALET, A., PRIETO, E., BONISSENT, A. ET AL. (2003), *Proc. SPIE*, **4850**, 1169
EIKENBERRY, S.S., ELSTON, R., RAINES, S.N. ET AL. (2004a), *Proc. SPIE*, **5492**, 1196
EIKENBERRY, S.S., ELSTON, R., GUZMAN, R. ET AL. (2004b), *Proc. SPIE*, **5492**, 1264
EISENHAUER, F., TECZA, M., THATTE, N., MENGEL, S., HOFMANN, R., GENZEL, R. (2000), *Proceedings of the Backaskog Workshop on Extremely Large Telescopes*, ed. T. Andersen, A. Ardeberg, R. Gilmozzi (Lund/ESO), **57**, 292
EISENHAUER, F., ABUTER, R., BICKERT, K. ET AL. (2003), *Proc. SPIE*, **4841**, 1548
FABRICANT, D.G., HERTZ, E.N., SZENTGYORGYI, A.H. ET AL. (1998), *Proc. SPIE*, **3355**, 285
FABRICANT, D.G., FATA, R., ROLL, J. ET AL. (2005), *PASP*, **117**, 1411
FLORES, H., PEUCH, M., HAMMER, F., GARRIDO, O., HERNANDEZ, O. (2004), *A&A*, **420**, L31
GEAKE, J.E., RING, J., WOOLF, N.J. (1959), *MNRAS*, **119**, 161
GELDERMAN, R., WOODGATE, B.E., BROWN, L.W. (1995), *ASPC*, **71**, 89
GILLETT, F., MOUNTAIN, M. (1998), *ASPC*, **133**, 42
GORDON, S., KORIBALSKI, B., HOUGHTON, S., JONES, K. (2000), *MNRAS*, **315**, 248
GUERIN, J., FELENBOK, P. (1988), *ASPCS*, **3**, 52
HANUSCHIK, R.W. (2003), *A&A*, **407**, 1157
HARLANDER, J., REYNOLDS, R.J., ROSELER, F.L. (1992), *ApJ*, 730
HARTUNG, M., LIDMAN, C., AGEORGES, N., MARCO, O., KASPER, M., CLENET, Y. (2004), *Proc. SPIE*, **5492**, 1531
HAYNES, R., LEE, D., ALLINGTON-SMITH, J. ET AL. (1999), *PASP*, **111**, 1451
HEARTY, F.R., MORSE, J., BELAND, S. ET AL. (2004), *Proc. SPIE*, **5492**, 1623
HENAULT, F.R., BACON, R., CONTENT, R. ET AL. (2004), *Proc. SPIE*, **5249**, 134
HERBST, T.M., BECKWIDTH, S. (1988), *PASP*, **100**, 635
HERNANDEZ, O., GACH, J.-L., CARIGNAN, C., BOULESTEIX, J. (2003), *Proc. SPIE*, **4841**, 1472
HILL, G.J., MACQUEEN, P.J., TEJADA, C. ET AL. (2004), *Proc. SPIE*, **5492**, 251
ISERLOHE, C., TECZA, M., EISENHAUER, F. ET AL. (2004), *Proc. SPIE*, **5492**, 1123

Jacquinot, P. (1954), *JOSAA*, **44**, 761
Joncas, G., Roy, J.R. (1984), *PASP*, **96**, 263
Jones, D.H., Shopbell, P.L., Bland-Hawthorn, J. (2002), *MNRAS*, **329**, 759
Kelson, D. (2003), *PASP*, **115**, 688
Kelz, A., Verheijen, M.A.W., Roth, M.M. et al. (2006), *PASP*, **118**, 119
Kenworthy, M.A., Parry, I.R., Ennico, K.A. (1998), *ASPCS*, **152**, 300
Kenworthy, M.A., Parry, I.R., Taylor, K. (2001), *PASP*, **113**, 215
Kobulnicky, H.A., Norsieck, K.H., Burgh, E.B., Smith, M.P., Percival, J.W., Williams, T.B., O'Donoghue, D. (2003), *Proc. SPIE*, **4841**, 1634
Koo, D.C., Bershady, M.A., Wirth, G.D., Stanford, S.A., Majewski, S.R. (1994), *ApJ*, **427**, L9
Larkin, J.E., Quirrenbach, A., Krabbe, A. et al. (2003), *Proc. SPIE*, **4841**, 1600
Le Coarer, E., Bensammar, S., Comte, G., Gach, J.L., Georgelin, Y. (1995), *A&AS*, **111**, 359
Le Fevre, O., Crampton, D., Felenbok, P., Monnet, G. (1994), *A&A*, **282**, 325
Le Fevre, O., Saisse, M., Mancini, D. et al. (2003), *Proc. SPIE*, **4841**, 1671
Lissandrini, C., Cristiani, S., La Franca, F. (1994), *PASP*, **106**, 1157
Lorenzetti, D., Cortecchia, F., D'Alessio, F. et al. (2003), *Proc. SPIE*, **4841**, 94
Maihara, T., Iwamuro, F., Yamashita, T. et al. (1993), *PASP*, **105**, 940
McDermid, R., Bacon, R., Adam, G., Benn, C., Cappellari, M. (2004), *Proc. SPIE*, **5492**, 822
McGregor, P.J., Hart, J., Conroy, P.G. et al. (2003), *Proc. SPIE*, **4841**, 1581
Murphy, T.W., Matthews, K., Soifer, B.T. (1999), *PASP*, **111**, 1176
Nelson, G.W. (1988), *ASPCS*, **3**, 2
Oliveira, A.C., de Oliveira, L.S., dos Santos, J.B. (2005), *MNRAS*, **356**, 1079
Parry, I., Bunker, A., Dean, A. et al. (2004), *Proc. SPIE*, **5492**, 1135
Prieto, E., Ferruit, P., Cuby, J.-G., Blanc, P.-E., Le Fevre, O. (2004), *Proc. SPIE*, **5487**, 777
Rallison, R.D., Schicker, S.R. (1992), *Proc. SPIE*, **1667**, 266
Ramsay Howat, S.K., Todd, S., Wells, M., Hastings, P. (2006), *New Astronomy Reviews*, **50**, 3513
Ramsey, L.W. (1988), *ASPCS*, **3**, 26
Ren, D. & Allington-Smith, J. (2002), *PASP*, **114**, 866
Reynolds, R.J., Tufte, S.L., Haffner, L.M., Jaehnig, K., Percival, J.W. (1998), *PASA*, **15**, 14
Roche, P.F., Lucas, P.W., Mackay, C.D. et al. (2003), *SPIE*, **4841**, 901
Roesler, F.L. (1974), *Methods in Experimental Physics*, 12A, Academic Press (San Diego), Chapter 12
Rosado, M., Langarica, R., Bernal, A. et al. (1995), *RMxAC*, **3**, 263
Rosado, M., Cruz-Gonzales, I., Salas, L. et al. (1998), *SPIE*, **3354**, 1111
Roth, M.M., Kelz, A., Fechner, T. et al. (2005), *PASP*, **117**, 620
Russell, A.P.G., Monnet, G., Quirrenbach, A. et al. (2004), *Proc. SPIE*, **5492**, 1796
Saunders, W., Bridges, T., Gillingham, P. et al. (2004), *Proc. SPIE*, **5492**, 389
Schmoll, J., Roth, M.M., Laux, U. (2003), *PASP*, **115**, 854
Schmoll, J., Dodsworth, G.N., Content, R., Allington-Smith, J.R. (2004), *Proc. SPIE*, **5492**, 624
Schroeder, D. (2000), *Astronomical Optics*, Academic Press (San Diego), 2nd Edition
Scott, A., Javed, M., Abraham, R. et al. (2006), *Proc. SPIE*, **6269**, 176
Sharples, R.M., Bender, R., Lehnert, M.D. et al. (2004), *Proc. SPIE*, **5492**, 1179
Sheinis, A.I., Bolte, M., Epps, H.W. et al. (2002), *PASP*, **114**, 851
Sheinis, A.I. et al. (2006), *Proc. SPIE*, (astro-ph/0606176)
Sluit, A.P.N., Williams, T.B. (2006), *AJ*, **131**, 2089
Tamura, N., Murray, G.J., Sharples, R.M., Robertson, D.J., Allington-Smith, J.R. (2005), *OExpr*, **13**, 4125 (astro-ph/0509913)
Tanaka, M., Yamashita, T., Sato, S., Okuda, H. (1985), *PASP*, **97**, 1020
Thatte, N.A., Weitzel, L., Cameron, M. et al. (1994), *Proc. SPIE*, **2224**, 279
Thatte, N., Anders, S., Eisenhauer, F. et al. (2000), *ASPCS*, **195**, 206
Thorne, A.P. (1988), *Spectrophysics*, Chapman & Hall (London)
Tull, R.G., MacQueen, P.J., Sneden, C., Lambert, D.L. (1995), *PASP*, **107**, 251

Vaughan, A.H. (1967), *ARA&A*, **5**, 139
Verheijen, M.A.W., Bershady, M.A., Andersen, D.R. et al. (2004), *AN*, **325**, 151
Viton, M., Milliard, B. (2003), *PASP*, **115**, 243
Williams, T.B., Nordsieck, K.H., Reynolds, R.J., Burgh, E.B. (2002), *ASPC*, **282**, 441
Wright, G.S., Rieke, G.H., Colina, L. et al. (2004), *SPIE*, **5487**, 653
Wynne, C.G. (1991), *MNRAS*, **250**, 796
Wynne, C.G., Worswick, S.P. (1989), *MNRAS*, **237**, 239
Yee, H.K.C., Ellingson, E., Carlberg, R.G. (1996), *ApJS*, **102**, 269

4. Analysis of 3D data

PIERRE FERRUIT

4.1 Introduction

4.1.1 *Presentation and scope*

This chapter contains the proceedings of the course on analysis of 3D spectrographic data given as part of the XVII Canary Island Winter School of Astrophysics. It provides an overview of some basic and generic analysis techniques for 3D spectrographic data.

It includes a description of an arbitrary selection of tasks with, whenever possible, examples on real data and a lot of discussion about noise and errors. To illustrate the examples, we will make a heavy use of tools that are part of the XOasis software developed at the Centre de Recherche Astrophysique de Lyon (CRAL) and of 3D datasets obtained using the TIGER and OASIS instruments.

This course is not limited to pure 3D analysis techniques as the core of the analysis of 3D datasets is either identical or similar to what is done for regular spectrographic data.

It has some obvious caveats and limitations:

- It is not exhaustive but contains a rather arbitrary selection of tasks and tools that we have considered as unavoidable.
- The methods and examples are biased toward extragalactic astronomy.
- It is limited to the data analysis techniques used in visible and near-infrared (NIR) astronomy.
- It does not address those used in the radio and X-ray communities (long-time users of 3D spectrography).

4.1.2 *Data analysis*

Before starting, we need to define better what data analysis is and, in particular, where it starts and stops in the process leading from raw data to ready-to-publish information (see Figure 4.1). In a general way, data analysis is considered to follow the data reduction (processing), which is instrument specific. The analysis will therefore be performed on radiometrically calibrated datacubes with the instrument signature removed. On output, it will produce physical quantities (e.g. fluxes, velocities, abundances, etc.).

Note that this is in fact a simplified view as the boundaries between data processing and analysis are not that well defined. Figure 4.2 shows that many data analysis tools are in fact at the boundary between data processing and analysis. It also shows that visualization tools play an important role in the analysis.

4.2 Data, noises, biases and artifacts

4.2.1 *Where is the signal?*

In a dataset, the 'signal' (i.e. what the astronomer will be really interested in) is never alone. It will be accompanied by undesired backgrounds, noises, artifacts, etc. A significant part of the work to be performed while analysing the data consists in carefully extracting the signal and evaluating the associated level of noise. The signal to noise ratio of a dataset is a key element for its analysis as it will indicate how real, trustworthy are features detected in a datacube.

In this context, much of the emphasis in this document is on the fact that simply extracting the signal from a dataset is not enough. One always needs to keep track of the

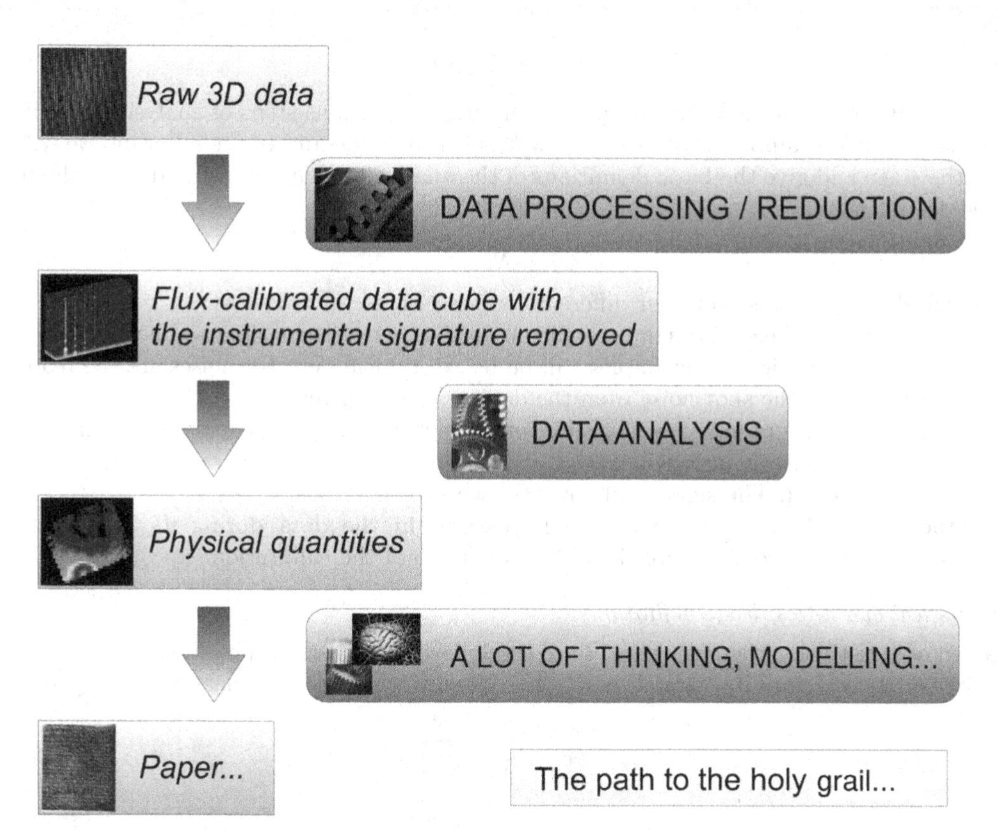

FIGURE 4.1. Flow chart showing where we locate the data analysis steps in the process leading from raw 3D dataset to published data.

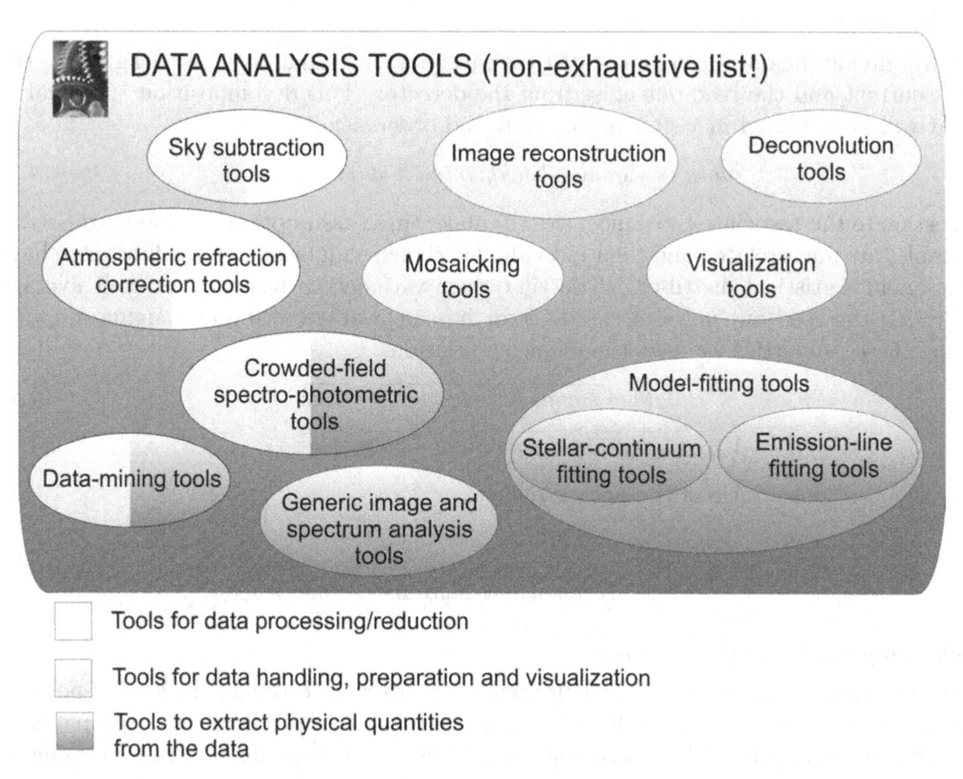

FIGURE 4.2. Sketch showing a (non-exhaustive) set of tasks, and an unsuccessful attempt to sort them between data processing and analysis tasks.

associated noises and how they propagate through the various steps of analysis. We will use very simple examples to illustrate how this can be done and in the following we will set the stage and give the basic definitions of the quantities that will be used throughout this document.

4.2.2 *Noises*

We will distinguish between four different sources of noise in a dataset:

- the instrument, e.g. through its detector. In the case of the visible and near-infrared detectors on which our examples will be based, typical detector noises are the read-out noise and the shot noise from the detector dark current.
- undesired backgrounds like sky emission. Their shot noise will remain even if they have been perfectly subtracted.
- the signal itself. The signal will have its own shot noise.
- the data reduction. Noise is also introduced in the data during its reduction/ processing, due to the finite signal to noise ratio of the calibration exposures.

Statistical quantities: basic definitions

Given a statistical variable x, we will use the following statistical quantities: its average $\overline{x}$, its variance V_x and its 'sigma' $\sigma_x = \sqrt{V_x}$. We have the following classical relations:

$$\overline{x} = \langle x \rangle = \frac{1}{N} \times \sum_{i=1}^{N} x_i \tag{4.1}$$

$$V_x = \langle (x - \overline{x})^2 \rangle = \frac{1}{N} \times \sum_{i=1}^{N} (x_i - \overline{x})^2 \tag{4.2}$$

Application to the noise in a dataset

We will divide the data into four statistical variables: the signal itself, a background, the dark current and the read-out noise from the detector. This decomposition is typical of what will be obtained in visible or near-infrared observations.

$$data = signal + background + dark + read \tag{4.3}$$

To estimate the associated variance, we will make three assumptions: the four statistical variables are uncorrelated and their individual variances can be added; the first three have Poissonian statistical distributions so that their variance corresponds to their average ($V_x = \overline{x}$); the read-out noise of the detector has an average of 0 and a sigma $\sigma_{readout}$. Using these properties we have the following relations:

$$\overline{data} = \overline{signal} + \overline{background} + \overline{dark} \tag{4.4}$$

$$V_{data} = \overline{signal} + \overline{background} + \overline{dark} + \sigma_{readout}^2 \tag{4.5}$$

The signal to noise ratio of the data is defined as:

$$\frac{S}{N} = \frac{\overline{signal}}{\sqrt{V_{data}}} = \frac{\overline{signal}}{\sqrt{\overline{signal} + \overline{background} + \overline{dark} + \sigma_{readout}^2}} \tag{4.6}$$

Noise propagation: basic definitions

If the relations obtained in the previous section are useful to estimate the level of noise in the raw (unprocessed) data, it will be very important to understand how noise propagates, i.e. what is gained or lost in terms of signal to noise when applying a data processing or analysis step. If we have a statistical variable x that is a function of two other statistical variables y and z ($x = f(y, z)$), we have the following classical relations (that can be easily generalized to a function of n statistical variables):

$$x_i - \overline{x} = (y_i - \overline{y}) \times \left(\frac{\partial f}{\partial y}\right)_{\overline{y}\,\overline{z}} + (z_i - \overline{z}) \times \left(\frac{\partial f}{\partial z}\right)_{\overline{y}\,\overline{z}} \tag{4.7}$$

$$\overline{x} = \overline{y} \times \left(\frac{\partial f}{\partial y}\right)_{\overline{y}\,\overline{z}} + \overline{z} \times \left(\frac{\partial f}{\partial z}\right)_{\overline{y}\,\overline{z}} \tag{4.8}$$

$$V_x = V_y \left(\frac{\partial f}{\partial y}\right)^2_{\overline{y}\,\overline{z}} + 2 \times COV_{yz} \left(\frac{\partial f}{\partial y}\right)_{\overline{y}\,\overline{z}} \left(\frac{\partial f}{\partial z}\right)_{\overline{y}\,\overline{z}} + V_z \left(\frac{\partial f}{\partial z}\right)^2_{\overline{y}\,\overline{z}} \tag{4.9}$$

with

$$COV_{yz} = \langle (y - \overline{y}) \times (z - \overline{z}) \rangle = \frac{1}{N} \times \sum_{i=1}^{N} (y_i - \overline{y}) \times (z_i - \overline{z}) \tag{4.10}$$

If the statistical variables are uncorrelated, then the covariance will be zero.

Noise propagation: example

We will apply the noise propagation relations provided in the previous section to the very simple case where we want to average spectrally the data of a given spectrum of a datacube. This case corresponds typically to what happens when a datacube is averaged over a given wavelength range to create an image (see Section 4.3.1). From the signal to noise ratio in the individual spectral pixels, we will compute the resulting signal to noise ratio for the averaged data.

Assuming that the spectra are in flux per unit of wavelength, we need to take into account the size of the spectral pixels when computing the average. We therefore perform the following operation (for each spectrum in the datacube):

$$F = \frac{1}{\Delta\lambda} \times \sum_{i=1}^{N} F_i \times \delta\lambda_i \;\; with \;\; \Delta\lambda = \sum_{i=1}^{N} \delta\lambda_i \tag{4.11}$$

Using the relations listed in the previous section, we can derive the average of the statistical variable F as a function of those of the F_i:

$$\overline{F} = \frac{1}{\Delta\lambda} \times \sum_{i=1}^{N} \overline{F}_i \times \delta\lambda_i \;\; with \;\; \Delta\lambda = \sum_{i=1}^{N} \delta\lambda_i \tag{4.12}$$

Here we have assumed that the size of the spectral pixels was perfectly known. If, in addition, we assume that the F_i are uncorrelated (i.e. their covariance is zero), we have:

$$V = \sum_{i=1}^{N} V_i \times \left(\frac{\delta\lambda_i}{\Delta\lambda}\right)^2 \tag{4.13}$$

$$\left(\frac{S}{N}\right) = \frac{\overline{F}}{\sqrt{V}} = \frac{\frac{1}{\Delta\lambda} \times \sum_{i=1}^{N} \overline{F}_i \times \delta\lambda_i}{\sqrt{\sum_{i=1}^{N} V_i \times \left(\frac{\delta\lambda_i}{\Delta\lambda}\right)^2}} \tag{4.14}$$

We can immediately see that the contribution of each spectral pixel to the final variance is weighted by the square of the size of the spectral pixel.

If we now assume that we have a constant spectral pixel size ($\delta\lambda_i = cte = \delta\lambda$; $\Delta\lambda = N \times \delta\lambda$) and that we are limited by the photon noise of the object ($V_i = \overline{F}_i$), this yields the following simplified relations:

$$\overline{F} = \frac{1}{N}\sum_{i=1}^{N}\overline{F}_i, \quad V = \frac{1}{N^2}\sum_{i=1}^{N}\overline{F}_i \quad and \quad \frac{S}{N} = \sqrt{N} \times \sqrt{\overline{F}} \tag{4.15}$$

We find the classical Poissonian statistics gain in signal to noise ratio proportional to the square root of N and this example shows how the signal to noise can be tracked when performing an operation on the data.

Noise propagation: comments

The example of noise propagation given in the previous section aimed primarily at showing how to track noise during a data-processing or analysis step. It also shows that things can easily get complicated (we had to make a number of simplifying assumptions to obtain a simple relation). We also point out to the reader that an additional (significant) source of complexity is avoided as, in most cases, we only keep track of the variance of the noise and not of the covariance terms. In particular, resampling operations usually introduce correlations between data. As a consequence, averaging performed further down the analysis may not yield the expected increase in signal to noise ratio due to the covariance terms. It is worth noticing that some data formats allow the storage of variance or sigma of a dataset. This is the case of the Euro3D data format (Kissler-Patig *et al.*, 2003), which carries the noise level information for each spectral pixel of each spectrum in the datacube. We are not aware of 3D data formats providing the user with a built-in way of keeping track of the complete covariance matrix (of size n^2 where n is the number of spectral pixels in the datacube).

4.2.3 *Biases or systematic errors*

If noise is unavoidable, at least its importance relative to the signal will decrease most of the time when a large number of independent measurements are put together. This is not the case when an error affects all the measurements in the same way. We quite often refer to this type of errors as systematic errors. One typical example is the error on the spectrophotometric calibration of an instrument. As the same calibration is applied to all measurements, they are no longer independent, and they are all impacted in the same way. This introduces a systematic error.

Biases are another type of error sources that will affect a complete set of measurements. They are often present when trying to derive properties of a class of objects through the observation of a sample. In this case, the way the sample is selected is extremely important and limitations on which object can be observed may introduce biases in the results. These latter error sources, which can impact the results in a very indirect way, are usually very difficult to detect and eliminate.

Systematic error: example

In this section, we give a simple example of systematic error and how it can be avoided. We are placing ourselves in the case of a deep observation of an object made of a sequence of N exposures of the same field of view. First, we assume that the object is always located at the same position within the field of view of our instrument, i.e. it falls on the same spaxels, and the corresponding spectra fall on the same detector pixels each time. For the sake of simplicity, we will assume that our object corresponds to a single spectrum per exposure.

During the processing of the dataset, the object spectrum of each exposure is spectrophotometrically calibrated by applying a wavelength-dependent factor (which has a finite accuracy). As the object is always falling at the same location within the instrument field of view, this factor is the same for all exposures. Once the N spectra are calibrated, they are then averaged to increase the signal to noise.

To simplify the computation, we will assume that all the exposures are independent and that the variance of the measurements is the same for all N exposures (i.e. $V_i(\lambda) = V_0(\lambda)$). We have:

$$F(\lambda) = \frac{1}{N} \times \sum_{i=1}^{N} F_i(\lambda) = \frac{1}{N} \times \sum_{i=1}^{N} \gamma(\lambda) \times \Gamma_i(\lambda) \tag{4.16}$$

where $F(\lambda)$ is the final averaged spectrum and $\gamma(\lambda)$ is the spectrophotometric correction factor that is applied to all the uncalibrated spectra of each exposure $\Gamma_i(\lambda)$.

The relation giving the variance of $F(\lambda)$ as a function of the variance of $\gamma(\lambda)$ and of the variance of the uncalibrated data is as follows:

$$V(\lambda) = \sum_{i=1}^{N} V_i(\lambda) \times \left(\frac{\gamma(\lambda)}{N}\right)^2 + \frac{V_\gamma(\lambda)}{N^2} \times \left(\sum_{i=1}^{N} \overline{\Gamma}_i(\lambda)\right)^2 \tag{4.17}$$

Using our simplifying assumptions, this yields:

$$V(\lambda) = \underbrace{\frac{1}{N} \times (\gamma(\lambda))^2 \times V_0(\lambda)}_{\text{term decreasing as 1/N}} + \underbrace{\frac{V_\gamma(\lambda)}{(\gamma(\lambda))^2} \times \left(\overline{F}(\lambda)\right)^2}_{\text{constant term}} \tag{4.18}$$

This shows that the spectrophotometric calibration of the data has introduced a systematic error term that does not decrease when the datasets are averaged. This means that after a certain limit, increasing the number of exposures N will not help to increase the signal to noise ratio. This latter becomes limited by the constant term in the variance.

This constant, systematic error term is due to the fact that the same spectrophotometric correction factor has been used for the spectra of all exposures, introducing a correlated source of noise in the data. A common way to solve this problem is to place the object at different locations in each exposure. In this case, the correction factor is now exposure dependent and becomes $\gamma_i(\lambda)$. We will however assume that its variance is uniform over the instrument field of view, i.e. $V_{\gamma_i}(\lambda) = V_\gamma(\lambda)$. The initial relation now becomes:

$$F(\lambda) = \frac{1}{N} \times \sum_{i=1}^{N} F_i(\lambda) = \frac{1}{N} \times \sum_{i=1}^{N} \gamma_i(\lambda) \times \Gamma_i(\lambda) \tag{4.19}$$

And this yields:

$$V(\lambda) = \underbrace{\frac{1}{N^2} \times \sum_{i=1}^{N} V_i(\lambda) \left(\gamma_i(\lambda)\right)^2}_{\text{term decreasing as 1/N}} + \underbrace{\frac{V_\gamma(\lambda)}{N^2} \times \sum_{i=1}^{N} \left(\overline{\Gamma}_i(\lambda)\right)^2}_{\text{term decreasing as 1/N}} \tag{4.20}$$

The constant term in the variance has now disappeared and has been replaced by a term decreasing as $1/N$. By changing the observation strategy, it has been possible to avoid introducing a systematic error term. This illustrates how keeping track of the noise can help in devising better observation, processing or analysis strategies.

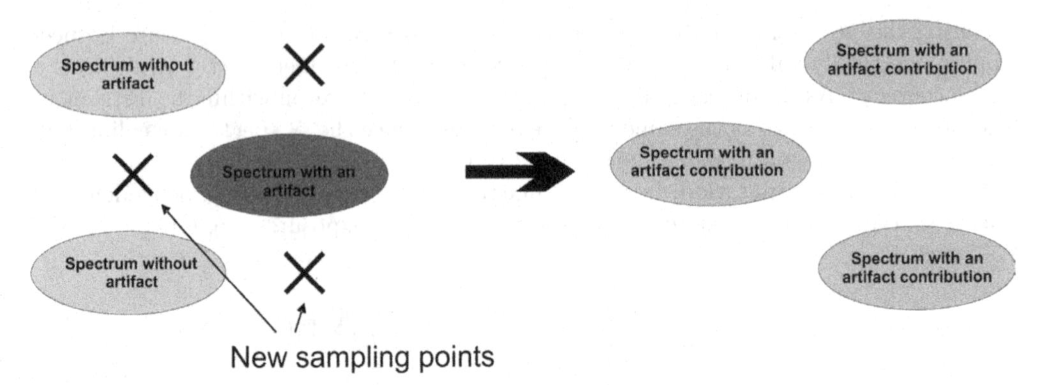

FIGURE 4.3. Sketch illustrating how an artifact can be spread over neighbouring spaxels when resampling spatially a dataset.

4.2.4 *Artifacts*

Artifacts are one last item in our list of features that will affect data. Typical sources of artifacts in visible and near-infrared datasets are:

- the detector (bad columns, bad pixels, etc.);
- the cosmic ray impacts on the detector and the residuals of their removal; and
- the residuals from the sky subtraction (see Section 4.4).

As for the noise, it is important to keep track of pixels affected by artifacts. In the Euro3D format, this is done through the use of a data quality flag for each spectral pixel of each spectrum of the datacube (see Section 4 of Kissler-Patig *et al.*, 2003).

The artifacts and the resampling of data

When resampling the data (spectrally and/or spatially), artifacts can be spread over several spaxels and/or spectral pixels. This is one of the main reasons why resampling of data is quite often seen as evil by many astronomers (another reason being that it introduces correlation between one pixel/spaxel and another). This is illustrated in Figure 4.3. Unfortunately, resampling of data in wavelength is usually quite unavoidable during the data processing (wavelength calibration). This is not the case of spatial resampling that can quite often be avoided (and usually is) until the very last stages of the data processing/analysis.

4.3 Image reconstruction and datacube visualization

4.3.1 *From a cube to an image*

Displaying images and spectra extracted from a datacube remains one of the most efficient ways (or at least one of the most used methods) of exploring a datacube. However, the reconstruction of an image by collapsing a datacube along the wavelength axis is not completely straightforward. The main reason for that is the fact that many datacubes do not have a regular spatial sampling.

The spatial sampling provided by the spectrograph can be irregular in different ways:

- Many 3D spectrographs have sampling patterns that are not a regular grid (e.g. in the case of circular or hexagonal spaxels).
- The sampling of the field of view can be sparse.
- There can be holes in the field of view coverage even when the sampling is not intrinsically sparse (observation that is the sum of non-contiguous fields of view, rejected spaxels, etc.).

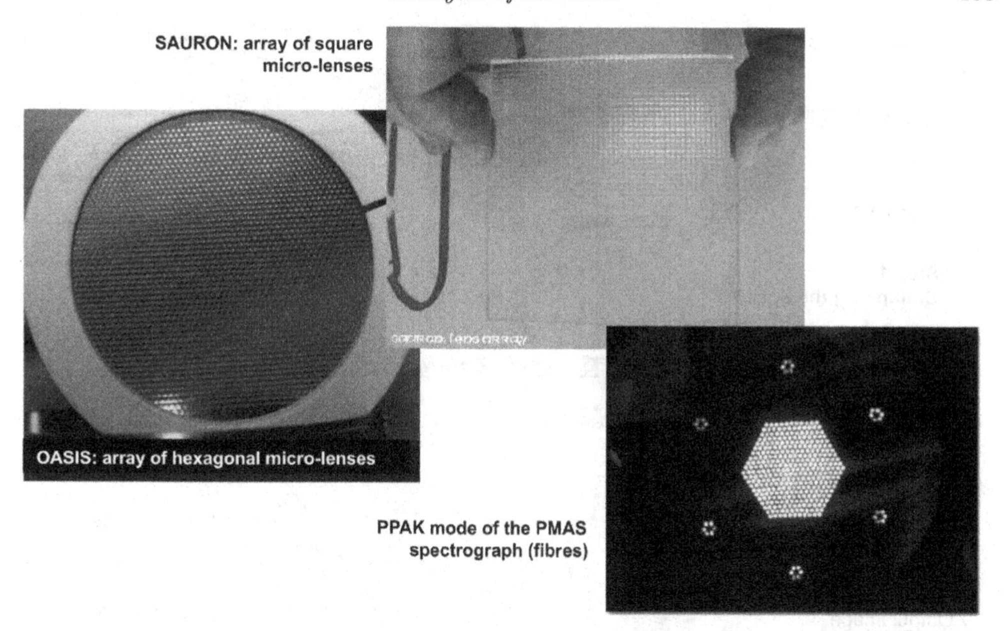

FIGURE 4.4. Images showing the location of the spaxels within the field of view of three different 3D spectrographs (SAURON, OASIS and PMAS in its PPAK mode).

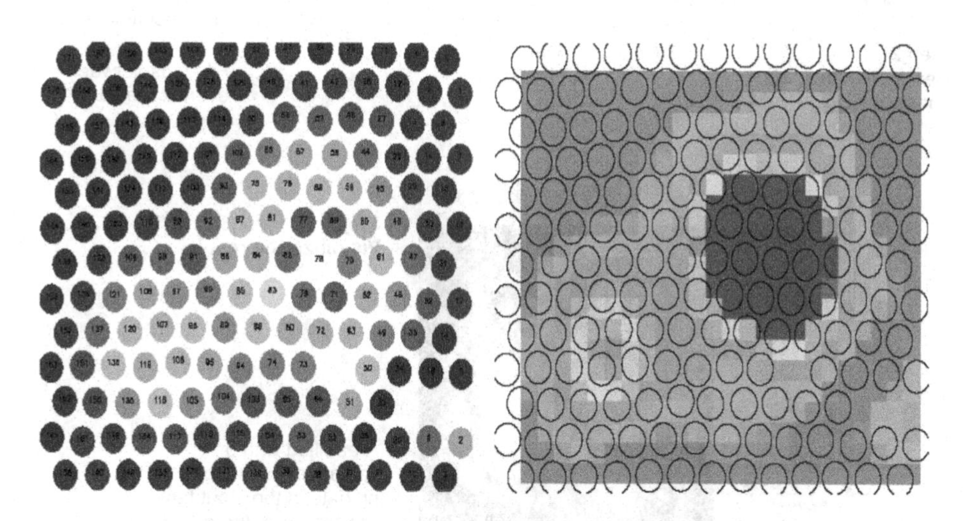

FIGURE 4.5. Example from the image reconstruction and visualization of a datacube without (left panel) and with (right panel) spatial resampling. The figure has been extracted from Sánchez (2004) and corresponds to a dataset of the INTEGRAL 3D spectrograph, which has circular, non-contiguous spaxels.

Examples of the spatial distribution of spaxels for various 3D spectrographs are shown in Figure 4.4.

When reconstructing an image from a datacube, two different approaches can be used depending on the adopted philosophy about the resampling of data. To avoid resampling, a special visualization tool like the E3D software (see tutorial on the use of E3D in this book) can be used that will take into account the location and shape of the spaxels and display them without resampling (see left panel of Figure 4.5).

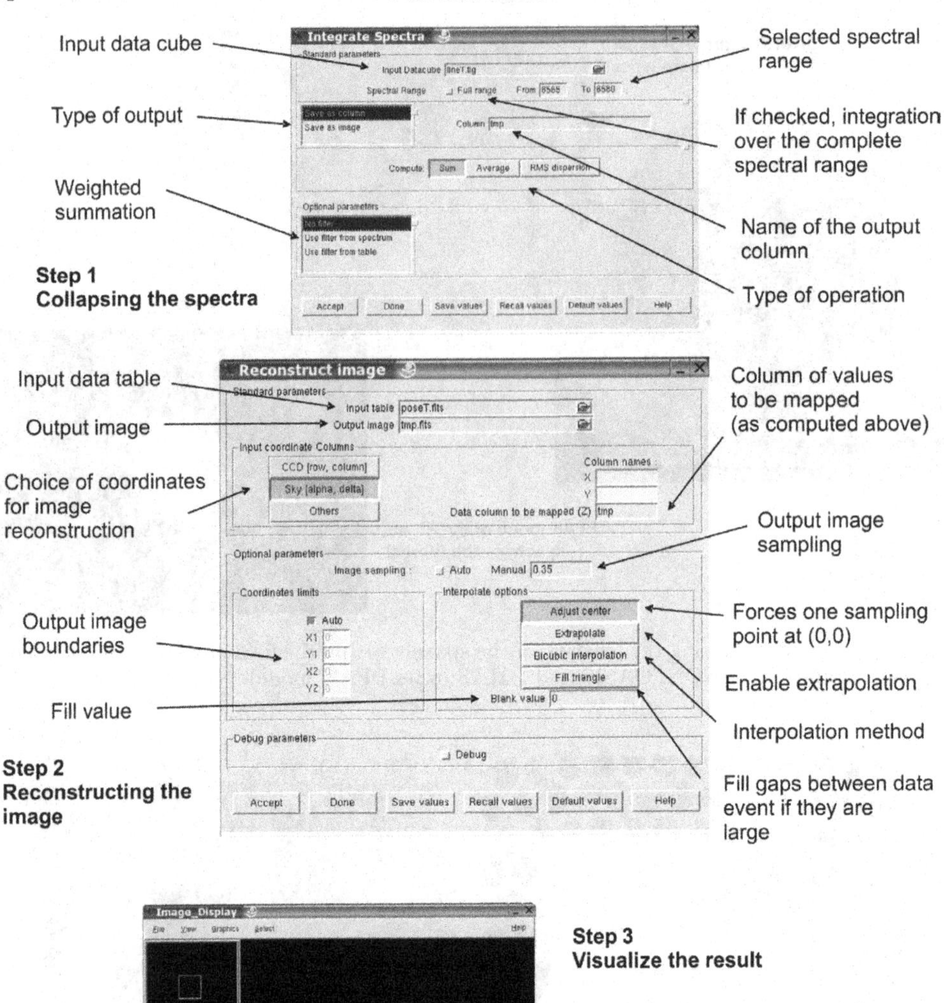

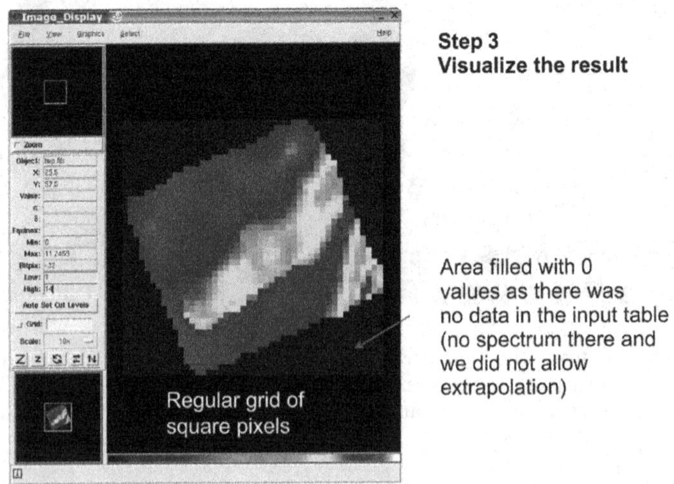

FIGURE 4.6. Example of image reconstruction procedure using the XOasis software.

The right panel of Figure 4.5 shows the results obtained for the same dataset when resampling and interpolation of the data are used (the black circles superimposed on the image are the shape and position of the spaxels). In Figure 4.6, we show screenshots of the graphical user interface (GUI) associated to the image reconstruction tools of the

XOasis software suite. It clearly shows the steps to be followed from a datacube to an image:

- The first step is the 'collapsing' of the spectra along the wavelength direction (upper panel of Figure 4.6). The user defines a wavelength range (that can be the whole spectrum) over which the spectral pixels will be averaged or summed (possibly using weighted summation). As an output, a table is produced that – for each spaxel of the datacube – contains its position and the value obtained when collapsing the corresponding spectrum.
- The second step (central panel of Figure 4.6) is the actual image reconstruction that will produce an image with square pixels and a regular sampling from the table produced during the first step. A triangulation will first be performed (looking for the nearest neighbours of each new sampling point in the table) and will be followed by the interpolation itself. The user is provided with the choice of allowing extrapolation of data and/or filling in the large gaps between spaxels by interpolation. In the example we have chosen extrapolation was not allowed, and the output image values outside the original field of view are filled with zero values (see bottom panel of Figure 4.6).

In much software (e.g. E3D) the user is also provided with a choice of interpolation methods.

4.3.2 *Datacube visualization*

The question of datacube visualization itself will only be briefly assessed in this document as it is the subject of a dedicated tutorial in this book. The art is, of course, to be able to display a 3D datacube on a 2D display. As mentioned at the beginning of this chapter, the simplest and most common solution is to reduce the number of dimensions of the data to be displayed by either collapsing the datacube along the wavelength axis (image representation; from 3D to 2D), or extracting individual spectral by collapsing the datacube along the spatial directions (spectrum representation; 3D to 2D or 1D). The E3D software is a very good example of how to navigate between spectrum and image representations of a datacube.

In the radio community, the side-to-side display of velocity slices (channel maps corresponding to an image representation) is commonly used. It is also possible to display position–velocity diagrams in which only one spatial axis of the data is preserved and displayed together with the wavelength information. This representation is familiar to the users of long-slit spectrographs.

However, none of these visualization schemes is actually a true 3D visualization method. An example of a 'real' 3D visualization developed and used by J. Gerssen is given in Figure 4.7. It has been used for the visualization of a Lyα datacube (see Wilman *et al.*, 2005). This type of visualization is best used when displaying well-defined features like emission lines. It is usually difficult to extract quantitative information from these views.

4.4 Getting rid of atmospheric effects and undesired backgrounds

Remaining at the borderline between data processing and analysis, we will now focus on the removal of atmospheric effects and undesired backgrounds.

4.4.1 *Sky emission, zodiacal light and thermal background*

If we do not consider instrument-related backgrounds like stray light or thermal emission from the optics in the infrared (IR), the most common (and unavoidable) background source for ground-based observations is the atmosphere.

FIGURE 4.7. Snapshot of datacube visualization using the VolView tools extracted from the work of J. Gerssen.

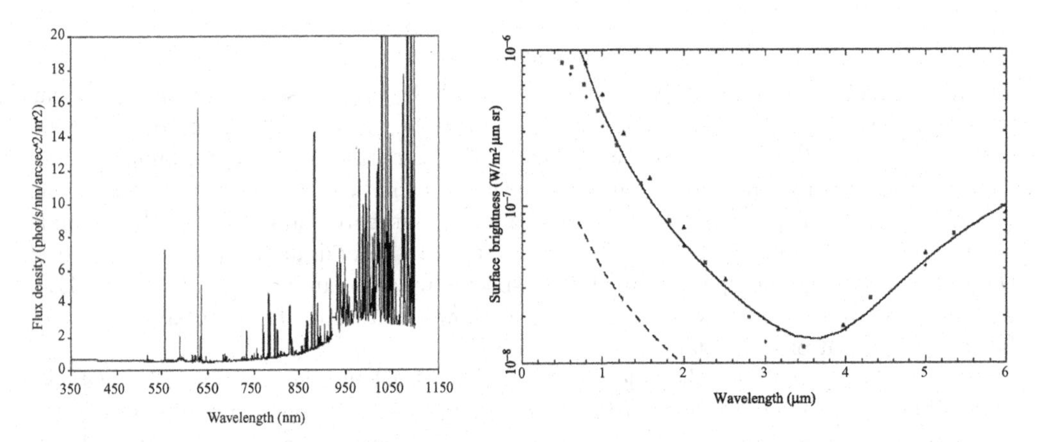

FIGURE 4.8. Spectra of the sky emission in the visible (left panel, from the GEMINI observatory) and of the zodiacal light (right panel).

Obviously, all the atmospheric background will disappear in the case of observations in space. The dominant external background then becomes the (much fainter) zodiacal light originating from the dust localized in the ecliptic plane of the Solar System. It is a mix of scattering of solar light (toward the blue) and thermal emission from the dust (toward the red). Typical sky and zodiacal light spectra are shown in Figure 4.8.

4.4.2 *Subtracting a background*

In order to get the signal out of the data, it will be necessary to subtract these backgrounds. However, even a perfect background subtraction will not remove the associated noise. In the following, we give a very simple example to illustrate the impact of the background noise on the final signal to noise ratio.

We will assume that we have data T obtained at the location of an object as well as data T_i obtained specifically for background subtraction at a set of N different locations.

We will have:

$$T = S + B + d \text{ and } T_i = B_i + d_i \tag{4.21}$$

where S is the signal, B and B_i the background and d and d_i the detector dark current. The associated variances are (assuming a 'shot-noise' type of contributions for the signal, background and dark current):

$$V_T = \overline{S} + \overline{B} + \overline{d} + \sigma^2 \text{ and } V_T^i = \overline{B}^i + \overline{d}^i + \sigma^2 \tag{4.22}$$

where σ is the detector read-out noise term.

The first step toward isolating the signal S is to subtract the dark current (that could be considered as an instrumental background). This will be performed using a dark current map and we will assume that the relative accuracy of this map is uniform ($\delta d/\overline{d} = \delta d^i/\overline{d}^i = cte$), as is the dark current rate ($\overline{d} = \overline{d}^i$). This yields the following relation for the variance associated to the dark current subtraction:

$$V_d = V_{d^i} = \overline{d} \times \left(\frac{\delta d}{\overline{d}}\right)^2 \tag{4.23}$$

And we have:

$$(T - d) = S + B \text{ and } V_{T-d} = \overline{S} + \overline{B} + \left(1 + \frac{\delta d^2}{\overline{d}^2}\right) \times \overline{d} + \sigma^2 \tag{4.24}$$

$$(T - d)^i = B^i \text{ and } V_{T-d}^i = \overline{S}^i + \overline{B}^i + \left(1 + \frac{\delta d^2}{\overline{d}^2}\right) \times \overline{d} + \sigma^2 \tag{4.25}$$

The second step is the background subtraction. For this purpose, we will average all the background-only measurements $B^i = (T - d)^i$ and subtract the result to $S + B = (T - d)$. Assuming that the background is uniform over the field of view ($\overline{B} = \overline{B}^i = cte$), we have:

$$S = (T - d) - \frac{1}{N} \times \sum_{i=1}^{N} B^i = (T - d) - \frac{1}{N} \times \sum_{i=1}^{N} (T - d)^i \tag{4.26}$$

$$V = \overline{S} + \left(1 + \frac{1}{N}\right) \times \left(\overline{B} + \left(1 + \frac{\delta d^2}{\overline{d}^2}\right) \times \overline{d} + \sigma^2\right) \tag{4.27}$$

The multiplicative factor $1 + 1/N$ witnesses the increase in variance associated with the background subtraction. For low N values, this latter can have a significant impact on the signal to noise ratio of an observation.

4.4.3 *Where things can get even more complicated*

Beware of instrumental effects

Obtaining the background and object spectra using different optical paths, as is often the case, can be a problem. This is mainly due to two different effects:

- The instrument spectral point-spread function (PSF) can differ from one spaxel to the other. As an example, the spectral PSF of an integral field unit (IFU) quite often varies within its field of view.
- The wavelength calibration of a datacube is not perfect and there may be shifts between spaxel spectra.

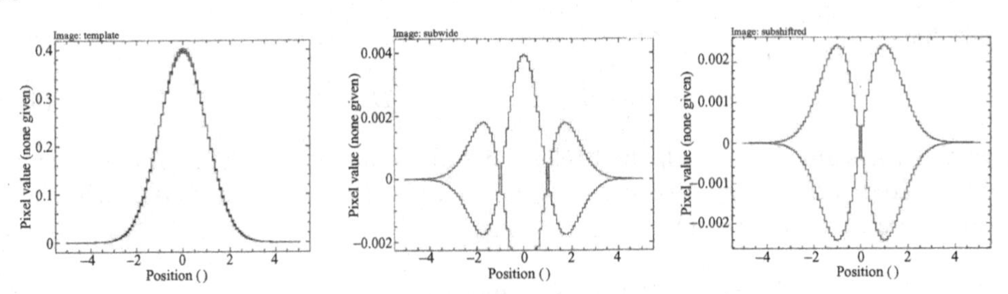

FIGURE 4.9. Left panel: plot of the nominal Gaussian line profile with the plots of the narrow/broader and blueshifted/redshifted profiles superimposed. Central panel: residuals from the subtraction of the narrower/broader profiles from the nominal one. Right panel: residual from the subtraction of the blueshifted/redshifted profiles from the nominal one.

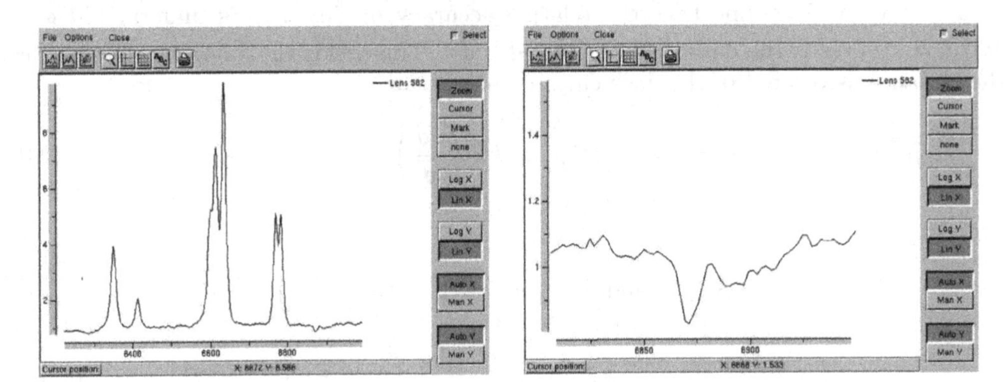

FIGURE 4.10. Left panel: spectrum of an active galaxy in the 630–700 nm range displaying the prominent emission lines from the object and an atmospheric absorption feature between 680 and 690 nm. Right panel: zoom on the atmospheric absorption feature (which is made of several lines).

In the case of the subtraction of a background with a continuum spectrum, these effects will have little impact. They can however be a source of artifacts when subtracting a background with spectrally unresolved emission lines (i.e. the sky).

To illustrate this effect, we have performed a simple simulation. We have produced an artificial emission line with a Gaussian profile (centred on zero and with a sigma of unity; see Figure 4.9). We have then introduced very small changes in the Gaussian profile parameters (changes in sigma of 1%; shifts of the centroid of the line by 1% of the sigma of the Gaussian profile) and created the corresponding spectra. A plot with all three profiles overlapped is displayed in Figure 4.9, and it shows that the lines are barely distinguishable.

However, if we now subtract the lines with the modified Gaussian profile to the original one, we see that we end up with negative and positive residuals that can reach 0.5–1.0% of the peak intensity of the line (see Figure 4.9). Although the amplitude of the residuals seems small, this can be a major issue when the background is stronger than the object, as it is quite often the case for deep exposures and in the near-infrared.

The case of atmospheric absorption features

So far, we have only addressed the subtraction of additive backgrounds. In the case of ground-based observations, the sky background can indeed be considered as an additive background, but it is also associated with absorption features (see Figure 4.10). They can be a nuisance and are quite difficult to correct for.

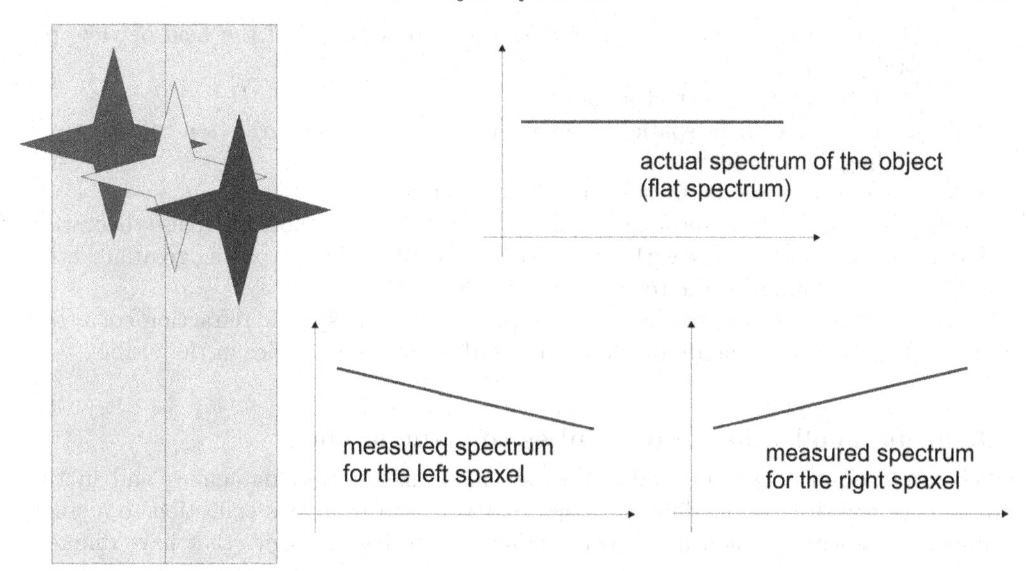

FIGURE 4.11. Sketch illustrating how the spectrum of a point source can be distorted due to the wavelength-dependent shift induced by atmospheric refraction.

One common way to determine the correction for this absorption is to calibrate it using a featureless spectrum such as the spectrum of the spectrophotometric standard star used for the radiometric calibration of the instrument (this latter being quite often selected to have a spectrum as featureless as possible). The steps are as follows:

- Assume the stellar continuum is flat at the location of the absorption feature.
- Fit the absorption line and determine the relative absorption (with respect to the continuum level).
- Apply the corresponding multiplicative correction to all the spectra of the datacube.

A major caveat is the fact that the importance of this absorption feature depends strongly on the parameters of the observation (in particular the time and the zenithal angle), and they can be quite different for the spectrophotometric star and the object itself. It is therefore not usually possible to apply the correction directly, and one needs to scale it to the object airmass and hope that time variation of the absorption between the two observations has remained small.

Atmospheric refraction

One last atmospheric effect is atmospheric refraction. It will cause the image of an object to be shifted as a function of wavelength. The effect decreases as wavelength increases and it is usually negligible starting in the near-infrared.

In slit spectroscopy it will cause the object to gradually move outside of the slit as wavelength changes. In 3D spectroscopy, there is no longer a risk of 'losing' the object (as long as the field of view is large enough) as it will just 'move' within the field of view as a function of wavelength. Reconstructing monochromatic images at different wavelengths and comparing them will easily show the shift. However, if it is not corrected, this shift will distort the spectra of the object, changing its slope from one spaxel to the other, as illustrated in Figure 4.11.

A detailed description of the impact of atmospheric refraction and how to correct for it can be found in Arribas *et al.* (1999). Typically, the steps of the correction procedure are the following (and they are only possible when using a 3D spectrograph):

- Divide the datacube in monochromatic 'slices' (one image of the field of view per spectral pixel).
- Shift all slices by the correct amount.
- Resample all the slices spatially using a common sampling grid (see comments on resampling data in Section 4.2.4).
- Reconstruct a datacube by stacking the corrected slices.

One difficulty can be the determination of the correct shift. It is possible to use theoretical values or to use a point source within the field of view (if available), paying great attention not to measure shifts intrinsic to the source itself.

Note that some instruments are now equipped with atmospheric refraction correctors (ADC). This is almost mandatory when using 3D + adaptive optics in the visible.

4.5 From a collection of datacubes to a single one

A complete observation quite often consists in a collection of datacubes and in this chapter we will discuss the different steps that will lead from this collection to a single datacube. We will distinguish between different observing strategies that have different goals:

- Obtaining a collection of observations of the same field of view with only a small pointing offset between each exposure in order to improve the spatial sampling of the datacube, i.e. the so-called 'drizzling' techniques.
- Obtaining a collection of observations of the same field of view to go as deep as possible (to improve the final signal to noise ratio).
- Obtaining observations of a collection of neighbouring (although sometimes non-contiguous or overlapping) sub-fields in order to increase the observed area on the sky (mosaicking technique).

Note that we will not address the question of datacubes obtained on the same field of view but within different spectral ranges.

4.5.1 *Drizzling*

The drizzling technique has been advertised a lot within the astronomical community through its use for HST (Hubble Space Telescope) images. It aims at improving the sampling of the image and can be very useful for a diffraction-limited instrument operating over a large wavelength range. In this case, the size of the PSF will change quickly with wavelength, and quite often the instruments cannot sample it correctly at all wavelengths and the PSF ends up undersampled in the blue; see Hook and Fruchter (2000) for a description of the technique and its application to images.

So far it has only been applied to images but it can be naturally extended to 3D datacubes (after all, they can be considered as stacks of monochromatic slices). However, we are not aware of any drizzling software adapted to datacubes.

4.5.2 *The deep-field approach*

In the deep-field approach, multiple exposures of a single field of view are obtained, aiming at pushing the detection limit of an instrument. Systematic dithering of the exposures is usually performed in order to facilitate the elimination of artifacts and limit the introduction of systematic errors. The observation strategy for datacubes is, in essence, similar to the one used for the imaging deep-fields.

The basic minimum steps needed to co-add (merge) the deep-field datacubes are: get the datacubes; register them with respect to each other, normalize them with respect to each other; co-add (merge) them.

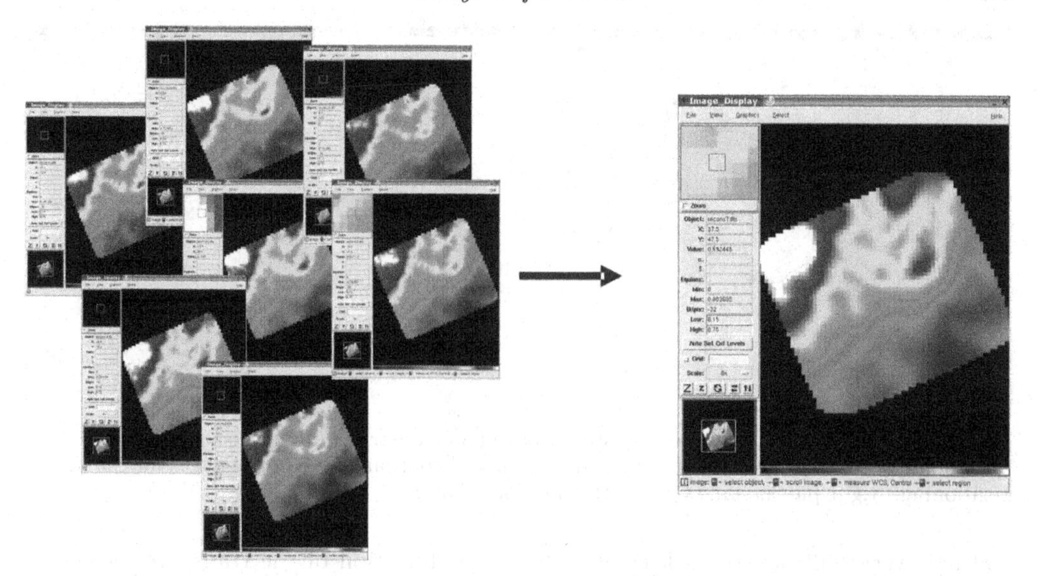

FIGURE 4.12. Reconstructed images for the 7 individual NGC 4258 datacubes (to the left) obtained with the OASIS/CFHT instrument and for the final datacube (to the right) obtained by co-adding (merging) them.

We will discuss some of these points in the following subsections and we will systematically use some OASIS/CFHT (Optically Adaptive System for Imaging Spectroscopy / Canada–France–Hawaii Telescope) observations of the galaxy NGC 4258 as an example (see Figure 4.12).

Relative positioning

Before trying to co-add the datacubes, it is necessary to determine their relative positioning. The easy way is of course to know the offset beforehand. This is rarely possible, especially as many observations have sub-arcsecond samplings and it is necessary to reach accuracies of a fraction of a spaxel size. Quite often, this information must therefore be obtained from the data itself.

If a 'sharp' morphological feature (e.g. the nucleus of a galaxy, a star, etc.) is available within the field of view it is possible to use the position of this source (as derived, for example, from centroiding or Gaussian profile fitting) as a reference. If no 'sharp' morphological feature is available, it is also possible to use contour plots of reconstructed images to get the relative positioning of two datacubes. This is illustrated in Figure 4.13. The difficulties are usually as follows:

- As it is, with inspection of the overlap being done by eye, this is a very subjective method.
- Changes in observing conditions (e.g. a seeing change) will mess up with the contour superposition.
- Noise in the individual exposures does not help.

Relative normalization

After registering the datacubes with respect to each other, we also have to check their relative normalization, i.e. tracking differences in radiometric response of the complete system (including the atmosphere if relevant) from one observation to another. In most cases, a dedicated spectrophotometric calibration will not be available for each exposure

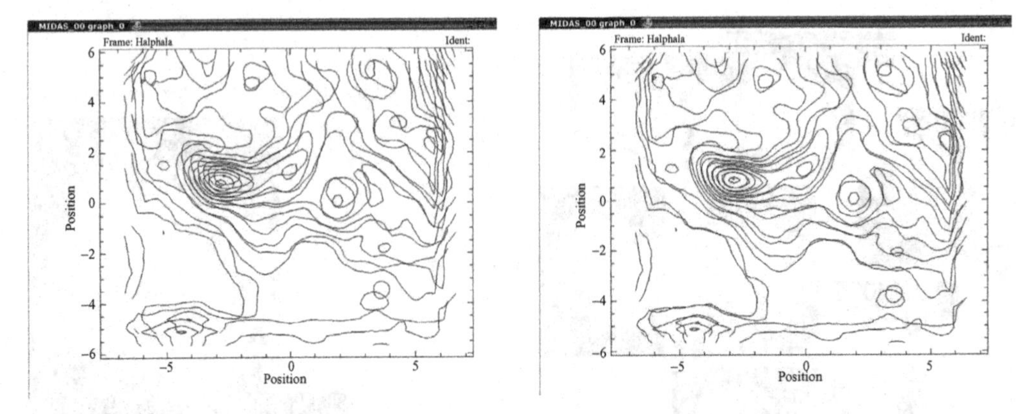

FIGURE 4.13. Example of the use of contour plots of a reconstructed image to derive the relative positioning of two datacubes. Left panel: superimposed contours of the two images for the initial positioning. Right panel: same but for the final positioning.

and it is very difficult to track changes like a modification of the transparency of the atmosphere. In the case of the deep-field approach, getting the relative normalization between the datacubes is straightforward as they all cover the same field of view and their flux or surface brightness levels can easily be compared.

Ready to co-add?

At this stage, we should have N datacubes that have been registered and normalized with respect to each other and we should be ready to co-add (merge) them. However, before starting the process we need to answer the following questions:

- How should we co-add them?
- Is it optimal in terms of signal to noise (which is the driver for the deep field approach)?

In order to answer these questions, we are going to use a very simple and quite artificial example. It will show that performing a weighted summation when co-adding the datacubes can help a lot. This allows us to take into account the fact that the datacubes can show up differences in signal to noise ratio and reflects the quite obvious fact that it is not desirable to give the same weight to low and high signal to noise datacubes.

We will assume that the calibration of the datacubes has introduced a division by the integration times (to get the spectra in unit of power, i.e. energy per unit of time) and that we are in photon noise regime. This yields:

$$S^i_{calibrated} = \frac{1}{t^i} \times S^i \text{ and } V^i = \overline{S}^i \text{ and } \left(\frac{S}{N}\right)^i = \sqrt{S^i} \tag{4.28}$$

The variance of a calibrated datacube is:

$$V^i_{calibrated} = V^i \times \left(\frac{\partial V^i_{calibrated}}{\partial V^i}\right)^2 = \overline{S}^i \times \left(\frac{1}{t^i}\right)^2 = \frac{1}{t^i} \times \overline{S}^i_{calibrated} \tag{4.29}$$

If we co-add (merge) the datacubes using a straight-forward average (no weights), we have:

$$S_{calibrated} = \frac{1}{N} \times \sum_{i=1}^{N} S^i_{calibrated} \tag{4.30}$$

If we assume that all the calibrated signals have the same average value $\overline{S}_{calibrated} = \overline{S}^i_{calibrated}$ (further simplification), this yields the following relations:

$$V_{calibrated} = \frac{1}{N^2} \times \sum_{i=1}^{N} V^i_{calibrated} = \frac{\overline{S}_{calibrated}}{N^2} \times \sum_{i=1}^{N} \frac{1}{t^i} \tag{4.31}$$

$$\left(\frac{S}{N}\right)_{calibrated} = \frac{\overline{S}_{calibrated}}{\sqrt{V_{calibrated}}} = \underbrace{\frac{1}{\sqrt{\sum_{i=1}^{N} \frac{1}{t^i}}}}_{source\ of\ trouble} \times N \times \sqrt{\overline{S}_{calibrated}} \tag{4.32}$$

The sum of the inverse of the integration times that appear in the equation for the signal to noise ratio is going to be a source of trouble. To see that, we are going to compare two practical cases.

First, if all the integration times are the same ($t^i = t = cte$):

$$\left(\frac{S}{N}\right)_{calibrated} = \frac{N}{\sqrt{\frac{N}{t}}} \times \sqrt{\overline{S}_{calibrated}} = \underbrace{\sqrt{t} \times \sqrt{N}}_{usual\ factor} \times \sqrt{\overline{S}_{calibrated}} \tag{4.33}$$

Second, if we have two exposures with different integration times ($t^{i=1} = t$ and $t^{i=2} = t/2$):

$$\left(\frac{S}{N}\right)_{calibrated} = \frac{2}{\sqrt{\frac{3}{t}}} \times \sqrt{\overline{S}_{calibrated}} = \underbrace{\sqrt{t} \times \frac{2}{\sqrt{3}}}_{smaller\ factor} \times \sqrt{\overline{S}_{calibrated}} \tag{4.34}$$

We can see that for exposures of different integration times, we do not have the expected gain in signal to noise ratio (expected to be proportional to the square root of the cumulated integration time). This is due to the contribution of the short exposure time exposures. The solution is to put more weight on long exposures than on short ones. With t_{tot} being the sum of all individual integration times (cumulated integration time) and using a normalized weight t^i/t_{tot}, we have:

$$S^{weighted}_{calibrated} = \frac{1}{t_{tot}} \times \sum_{i=1}^{N} t^i \times S^i_{calibrated} = \frac{1}{t_{tot}} \times \sum_{i=1}^{N} S^i \tag{4.35}$$

$$V^{weighted}_{calibrated} = \frac{1}{t^2_{tot}} \times \sum_{i=1}^{N} V^i = \frac{1}{t^2_{tot}} \times \sum_{i=1}^{N} S^i \tag{4.36}$$

$$\left(\frac{S}{N}\right)^{weighted}_{calibrated} = \frac{\overline{S}^{weighted}_{calibrated}}{\sqrt{V^{weighted}_{calibrated}}} = \sqrt{t_{tot}} \times \sqrt{\overline{S}_{calibrated}} \tag{4.37}$$

We have obtained the classical factor of square root of the cumulated integration time. This simple and naive example outlines the fact that it is necessary to assess the impact of each processing/analysis step on the signal to noise of the data. In this case the simple fact of having calibrated data has forced us to modify the way the datacubes are co-added (if data had still been in counts, a direct average would have worked). This explains as an example why the HST images are never in physical units (per unit of time) by default. It is simply to allow the user to perform direct summations on them without having to worry about weights.

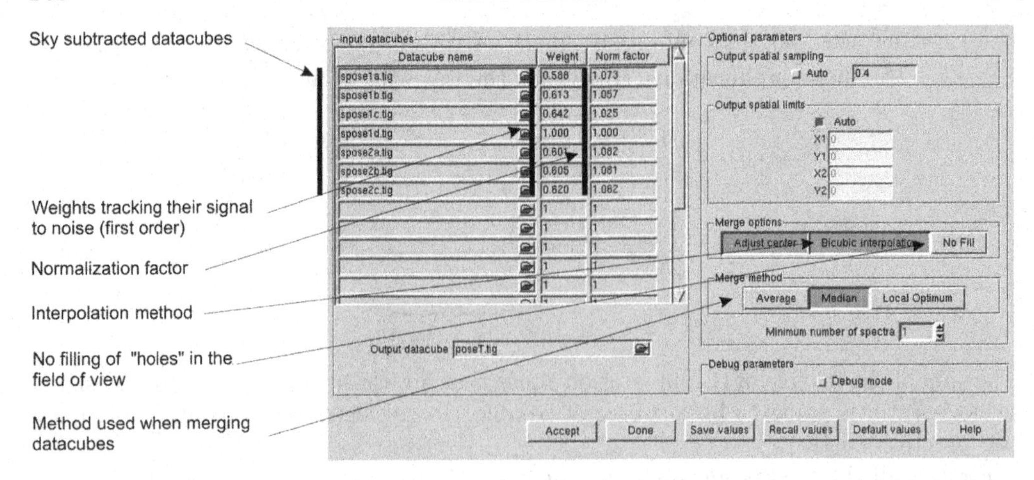

FIGURE 4.14. Example of graphical user interface (from the XOasis software) used as a front-end for a datacube merging tool. It can be seen that this program allows use of a weighted summation (with the weights set by default to t^i/t_{tot}).

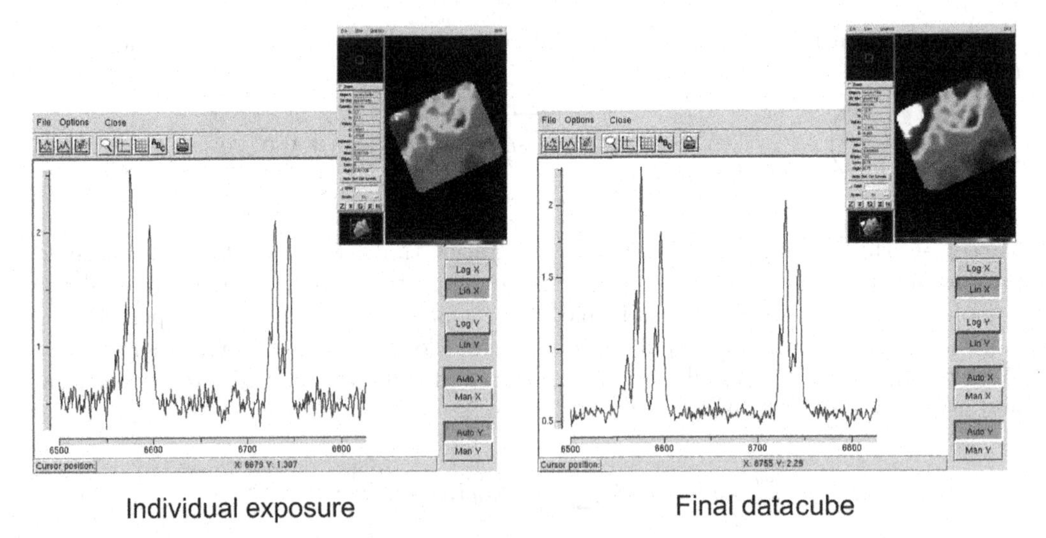

FIGURE 4.15. Spectra extracted from an individual datacube (left panel) and from the final merged datacube (right panel) at the same location in the object.

The (very simple) procedure we have described is in fact quite similar (although less sophisticated) than the ones used in optimal summation techniques.

A simple illustrated example

To finish on an example, we show a snapshot of the graphical interface of the 'merge datacube' task of the XOasis software in Figure 4.14. Note that in this case we have simultaneously corrected for atmospheric refraction. Coupling this correction with the merging of the datacubes allows for a limiting of the number of spatial resamplings of the data. In Figure 4.15, we also display a comparison between a spectrum extracted from an individual datacube and a spectrum extracted from the final, merged datacube. The gain in signal to noise ratio is obvious.

Back to the deep-field approach and having a look at what is ahead of us

In the previous sections, we have used simple and fairly artificial examples to illustrate how to co-add the exposures for a deep-field type of observation. A real example of 3D spectrographic deep-field can be found in Wilman *et al.* (2005). It has been performed using the SAURON (Spectrographic Areal Unit for Research on Optical Nebulae) instrument, with a cumulative integration time of more than 15 hours (combination of 30 individual exposures). The main difficulties that have been identified are:

- having a very good and deep knowledge of the instrument (well-calibrated and -characterized instrument);
- eliminating the backgrounds and systematic errors;
- coping with changes of image quality and atmospheric transmission from one exposure to another. Note that how to deal with changes in image quality is an entire subject in itself and is not addressed here;
- coping with the intrinsic stability of the telescope plus instrument system. This is critical for this type of observation, which tends to push the instruments to their limits.

This type of prototype study is paving the way for future 3D spectrographic deep fields that will be at the heart of the scientific goals of the future generation of 3D spectrographs like MUSE (Multi-Unit Spectroscopic Explorer) and KMOS (K-Band Multi-Object Spectrometer) for the VLT (very large telescope). Dedicated software will be needed for the co-addition of these tens of cubes (cumulative integration times that can reach 80 hours).

4.5.3 *Mosaicking*

In the mosaicking approach, a large number of (overlapping or not) sub-fields are obtained and are then merged to create the final field of view. In the following, we will outline the differences with the deep-field approach that has been described in detail in the previous section and we will give examples of mosaics of 3D spectrographic fields.

Differences with the deep-field approach

The observing procedure to obtain a mosaic of field is very similar to the one used for the deep-field approach, but for the offsets between the fields:

- one or two observations are obtained per field of view. If two are obtained, a small offset will be applied between them in order to facilitate the elimination of artifacts and to limit the introduction of systematic errors;
- known offsets are applied between fields of view. Depending on the scientific goals and on the observation constraints, these individual fields can be contiguous or not, and the degree of overlap can vary.

The relatively small number of exposures per field of view compared to the case of the deep field makes the problem of systematic errors much less critical. However, the registration (see Section 4.5.2, *Relative positioning*) and normalization (see Section 4.5.2, *Relative normalization*) steps are more troublesome in the case of a mosaic of fields.

Relative registration and normalization of the exposures

The problem comes from the lack of overlap between the fields of view. The methods for the relative registration and normalization of the exposures described for the deep-field approach rely heavily on the comparison of data obtained on the same patch of the sky. In the case of a mosaic of fields like the one shown in Figure 4.16, it is clear that:

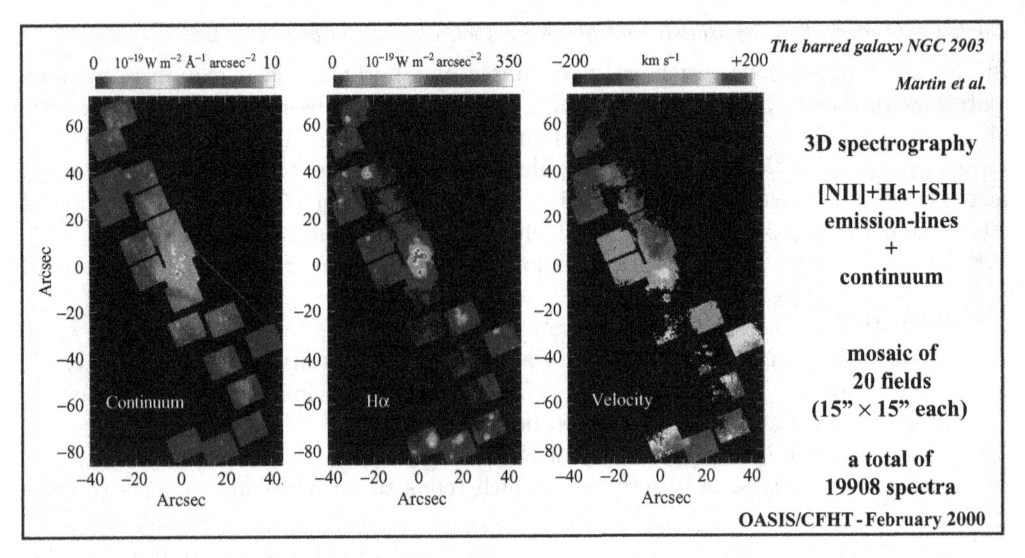

FIGURE 4.16. Reconstructed images of NGC 2903 for a mosaic of 20 sub-fields obtained with the OASIS/CFHT instrument.

- The number of spatial feature that could be used for registration purposes is very small and, quite often, they appear in a single sub-field (no overlap).
- Any normalization would have to be propagated from one datacube to the next, quite often using very small regions of overlap and yielding large errors.

A common solution is usually to make sure that a wide-field image in the same wavelength range is available. It is then possible to systematically compare reconstructed images to this wide-field image and derive the necessary registration and normalization information. This allows for a single common reference for all the sub-fields but does not solve the problems of the lack of reference sources for the registration in some sub-fields and the problems of changes of observing conditions from one sub-field to another (heterogeneous dataset).

Simple illustrated examples

In Figure 4.16 and Figure 4.17, we give two simple examples of mosaic of sub-fields observed in 3D spectrography. The first one on NGC 2903 is a prototypical example of a mosaic of a large number of non-overlapping sub-fields, yielding a sparse, incomplete coverage of a very extended field of view. As described in the previous section, a wide-field image has been used to register and normalize all the individual datacubes. The final, merged datacube contains more than 19,000 spectra.

The second example on NGC 1068 is intermediate between a mosaic and a deep-field. Offsets have been applied between the sub-fields, but a large overlap has usually been preserved to increase the cumulated integration time on some regions. This type of observation is extremely sensitive to changes in seeing conditions. Due to the complex structure of the object, these changes will introduce modifications in the observed morphology that cannot be corrected by a simple normalization. Prominent artifact – usually located at the boundaries of the sub-fields – can then appear when co-adding (merging) this set of heterogeneous datacubes.

Last, we outline in the right panel of Figure 4.17 that additional artifacts can be introduced by the merging software. It is important that this software does not interpolate between sub-fields if the distance between the points is too large. In this example, the

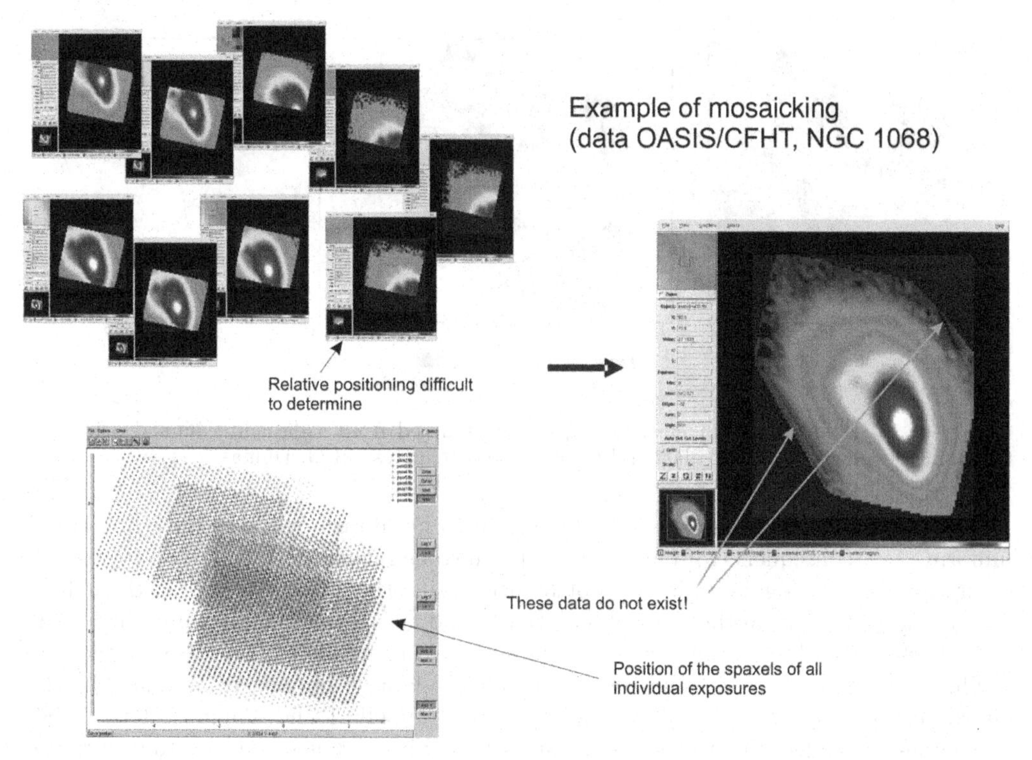

FIGURE 4.17. Example of a mosaic of fields obtained using the OASIS/CFHT 3D spectrograph on the active galaxy NGC 1068.

'no-fill' option of the XOasis 'merge datacubes' tool has been disabled, leading to the creation of non-physical interpolated spaxels in the final datacube (and they are not that easy to spot due to the quality of the interpolation).

4.6 Smoothing and binning

4.6.1 *Smoothing data*

Smoothing of data consists in convolving the data with a spatial or spectral filter. As for the interpolation, numerous different methods exist. The most straight-forward one is the 'boxcar' type of smoothing, where a pixel/spaxel is averaged with its closest neighbours. Most of the time smoothing is applied in order to increase the signal to noise in the data. In terms of noise coupling, it has effects equivalent to those of resampling data (pixels or spaxels are not independent anymore and their noises are correlated).

We will not develop the topic of smoothing further as it is usually better to use a good binning technique (see next section) instead. In fact, the binning method presented below can be seen as an advanced smoothing technique.

4.6.2 *Binning*

Binning has the same objectives as smoothing and consists of putting together different pixels/spaxels in order to increase the signal to noise in the data. The main difference with smoothing is that it actually reduces the number of pixels/spaxels (there is no data duplication). In the following, we will concentrate on spatial binning.

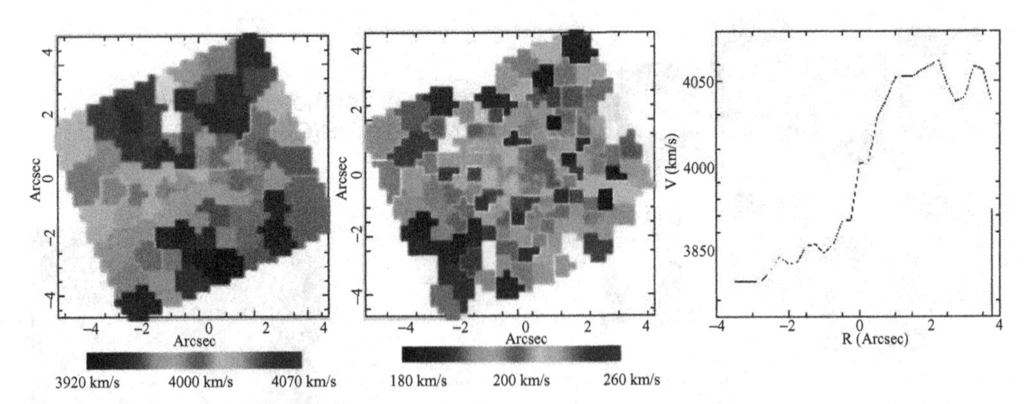

FIGURE 4.18. Example of kinematical maps reconstructed from a datacube binned using the technique developed by Cappellari and Copin (2003) (courtesy of G. Dumas).

The main difficulty is the fact that the signal to noise ratio within a datacube is not uniform. As a consequence, if binning is useful in some regions of a field of view (e.g. the external regions of a galaxy) it may not be needed in others (e.g. the core of the galaxy that would be present in the same observation). And, as spatial binning will decrease the spatial resolution, binning may not only be useless in some regions but also unwelcome.

This has led to the emergence of the so-called adaptive binning techniques like the one based on Voronoi tessellation developed by Cappellari and Copin (2003). These techniques are called adaptive since the size of the bins is tuned in order to ensure the signal to noise ratio in the binned datacube is as uniform as possible. This means that spaxels with a high enough signal to noise ratio will remain untouched while the binning will only be applied to areas where it is needed. The main constraints on the bins used by Cappellari and Copin (2003) are the following:

- The bin shall not overlap (this constraint is not satisfied by the smoothing techniques) and no holes shall be introduced.
- In order to try to avoid exotic bin shapes, the bins shall be as 'compact' (round) as possible.

In their paper, these two authors present a method satisfying these constraints and its practical implementation. It has been widely used in the SAURON project, and Figure 4.18 shows an example of maps reconstructed from an adaptively binned datacube.

4.7 Data mining and crowded-field spectrophotometry

4.7.1 *Data mining*

Data mining could be defined as the art of detecting small and weak objects within a huge datacube. It is becoming more and more important with the development of large 3D spectrographs and the increasing importance of spectrographic deep fields. Solutions have already been explored in various astronomical communities:

- by the X-ray community for the automated detection of X-ray sources in surveys (with a 3D touch as they can have energy-sensitive detectors).
- by the radio community for the detection of sources in the HI surveys as an example. A quite interesting example of data mining application of HI self absorption features in a survey can be found in Gibson *et al.* (2005).
- for the detection of sources and construction of catalogues in wide-field imaging surveys; see, for example, the SExtractor software from Bertin and Arnouts (1996).

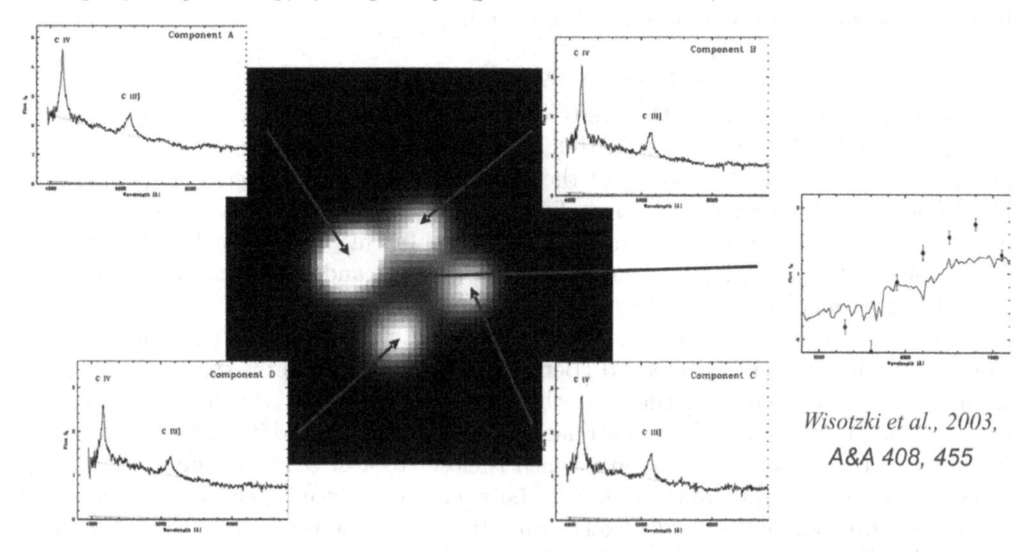

FIGURE 4.19. Example of application of the crowded field spectro-photometry with sources (the four quasar images) located on top of a non-uniform background (the lensing galaxy) (courtesy of M. Roth).

In the optical and infrared 3D community, these techniques are only just starting to emerge, and in the last few years work has been conducted by some members of the Euro3D network (e.g. S. Foucaud and G. Covone). Two prototype tools have been developed: one based on the systematic use of SExtractor on monochromatic slices of a cube, and the other based on the detection of correlations between spectra.

4.7.2 *Crowded-field spectrophotometry*

Another expanding domain is the field of crowded-field spectrophotometry, which aims at extracting the spectrophotometry of a collection of sources within a 'crowded' field of view (e.g. see example in Figure 4.19). A lot of work has been conducted in this field by the team at AIP (Astrophysikalisches Institut Potsdam), and the applications are in fact not limited to surveys of numerous sources but can also be efficiently applied to observations of single objects (planetary nebula, supernova, etc.). As in the case of imaging, the major issues are:

- How to efficiently and cleanly get rid of the backgrounds, which are usually not uniform spatially?
- How to efficiently collect the flux of an object that is spread over several spaxels and may be contaminated by the flux of neighbouring objects?

In particular, for the second point, an accurate knowledge of the PSF is key for an accurate spectrophotometry.

4.8 Model fitting

After a few chapters meandering at the frontier between processing and analysis, we really enter the domain of pure data analysis. We also enter the domain where things become extremely specialized (depending on the type of objects and the scientific goals). The model can in fact be extremely sophisticated both technically (e.g. when some models are used to produce synthetic datacubes that are compared to the observed ones) and scientifically. In the following, we will focus on two simple types of model fittings: the fitting of the stellar continuum and of emission lines.

4.8.1 *The example of stellar continuum fitting*

In extragalactic observations, the study of the stellar continuum of the objects allows extraction of information about their stellar contents (e.g. to derive the stellar population and also to trace back the history of the stellar formation) and their kinematics. The techniques used for stellar continuum fitting in 3D spectrography have been directly adapted from those used in regular spectrography. In this domain, a lot of work has been conducted by the SAURON team (amongst others), and we will present briefly the pixel-fitting technique developed by members of the this team.

These developments have been driven by the fact that usual methods like Fourier Correlation Quotient (FCQ) method (Bender, 1990) work in Fourier space and are very sensitive to the presence of artifacts in the spectra or simply of emission lines. In fact, this last point turned out to be an extremely strong limitation, and it led them to develop further a method, called pixel-fitting, which is closer to a direct fit of the spectra. It is based on the minimization of the difference between the observed spectra and a template spectrum shifted and broadened. Pixels with artifacts or regions where emission lines can be found are simply rejected.

The procedure used to determine both the stellar contents and the stellar kinematics is iterative and as follows:

- Using a single stellar template spectrum, a first guess of the stellar kinematics of the object is derived.
- Using this first guess as a constraint (i.e. the kinematics is kept fixed during the fit), the best linear combination of stellar template spectra is determined (determination of the 'optimal' template).
- The first step is then repeated, this time using the optimal template, yielding a new, refined estimate of the stellar kinematics.

Further iterations are possible but are usually not necessary.

Establishing which regions should be rejected during the fit due to emission lines is usually difficult, and they are usually oversized in order to be applicable to the complete field of view (it would be cumbersome to examine each spectrum individually to adapt the size of these exclusion windows). Given that, the SAURON group and in particular J. Falcón-Barroso went slightly further and they have coupled the fitting of the stellar continuum and the emission lines (single Gaussian profile, mainly the Hβ and [OIII] lines) in the SAURON data. They have therefore developed the tools to perform the optimal template and emission-line fitting simultaneously; an example of fit on a single spectrum is shown in Figure 4.20 and examples of maps derived using this pixel-fitting software can be seen in Falcón-Barroso *et al.* (2006).

Although this tool has proven very efficient and useful in the case of the galaxies observed with SAURON, it starts to reach its limits when the emission lines become strong compared with the continuum and when they start to display complex profiles that cannot be described by a single Gaussian profile. In these conditions, it is difficult to avoid fitting the stellar continuum and the emission lines independently, each with dedicated software.

4.8.2 *A step-by-step example of emission-line fitting*

As outlined above, it is sometimes unavoidable to fit the stellar continuum and the emission lines independently. As for establishing the stellar contents and kinematics, it is possible to proceed iteratively. The corresponding workflow is shown in Figure 4.21.

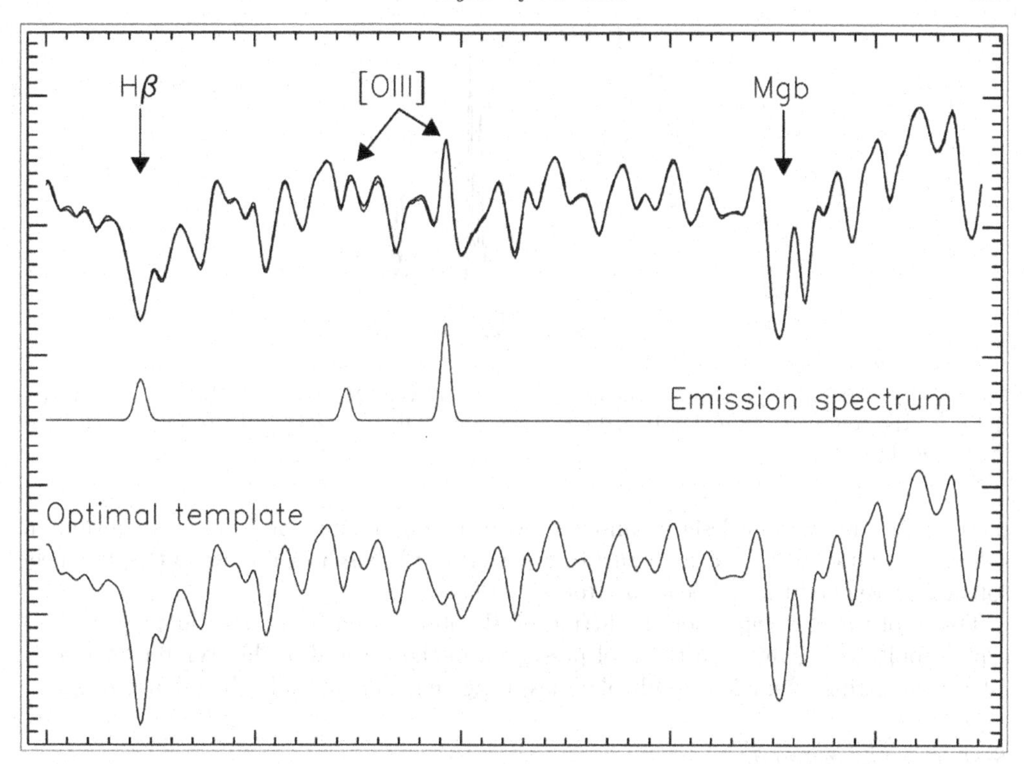

FIGURE 4.20. SAURON spectrum showing the superposition of the stellar continuum and of emission lines (top panel) and the corresponding optimal template spectrum (linear combination of stellar templates from a library) as determined using the pixel-fitting software developed by J. Falcón-Barroso (courtesy of J. Falcón-Barroso).

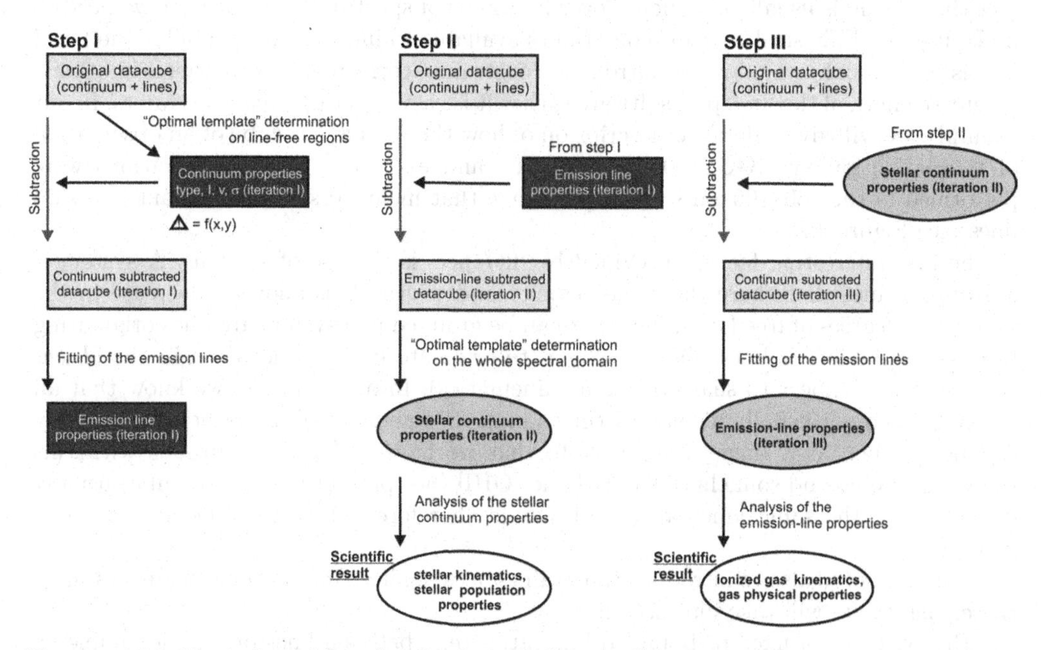

FIGURE 4.21. Typical workflow for the fitting of the stellar continuum spectrum and of the emission lines when they cannot be fitted simultaneously.

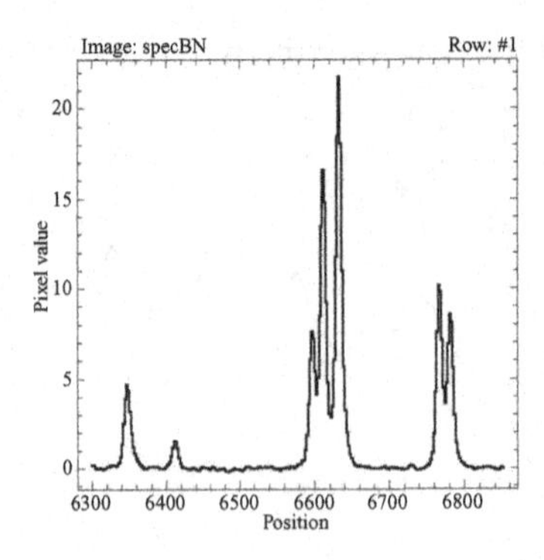

FIGURE 4.22. Typical spectrum of the central region of NGC 2110 that will be used to prepare the fit of the complete datacube. It includes lines from [OI] (doublet), [NII] (doublet), Hα and [SII] (doublet).

In the following we will describe an example of fitting of emission lines in a continuum subtracted datacube (i.e. a small part of this workflow), based on the use of the 'fit/spec' software developed by A. Pécontal (Rousset, 1992).

This tool was developed for the fitting of 3D observations of active galaxies and has a philosophy very similar to the tool presented in the tutorial of this Winter School. It allows the fitting of single or multiple lines using single or multiple Gaussian profiles.

Systems and constraints

In many cases, fitting the emission lines with a single Gaussian profile is not enough and it is necessary to use multiple ones. This increases the number of free parameters and the fitting process becomes more complex and also much less stable. If we add to that the fact that fitting is usually conducted over hundreds of spectra, it is clear that we need to make use of all physical *a priori* constraints available to improve and stabilize the fit. If this is not enough, we may also introduce constraints corresponding to approximations.

One strength of the 'fit/spec' software is that it is easy to set up such constraints. In our example, we will give a detailed description of how the fit of a datacube of an observation of the active galaxy NGC 2110 was prepared and executed. These observations were performed in the 630–700 nm wavelength range that included seven important emission lines (see Figure 4.22).

The first interesting feature provided by 'fit/spec' is the use of systems. A system is a group of lines that share the same velocity and width. This allows a decrease in the number of degrees of freedom. Lines that can be grouped in a system are lines originating from the same ion or lines that we expect to originate from the location in the object (and therefore expect to share the same kinematics). In our example, we know that all seven lines do not usually display distinct kinematics and we will therefore assume they can be grouped as a single system. Note that we have to be very cautious with this type of grouping and some lines like Hβ and [OIII] (not present in our datacube) usually do not share the same kinematics and should therefore not be included in the same system.

Second, we need also to introduce constraints between lines in a system. In our example, these constraints will take into account:

- the existence of fixed or bounded line ratios (e.g. between lines of a same doublet);
- the fact that the width (specified here as the full-width-at-half-maximum, FWHM) of a line shall be larger than the instrumental one;

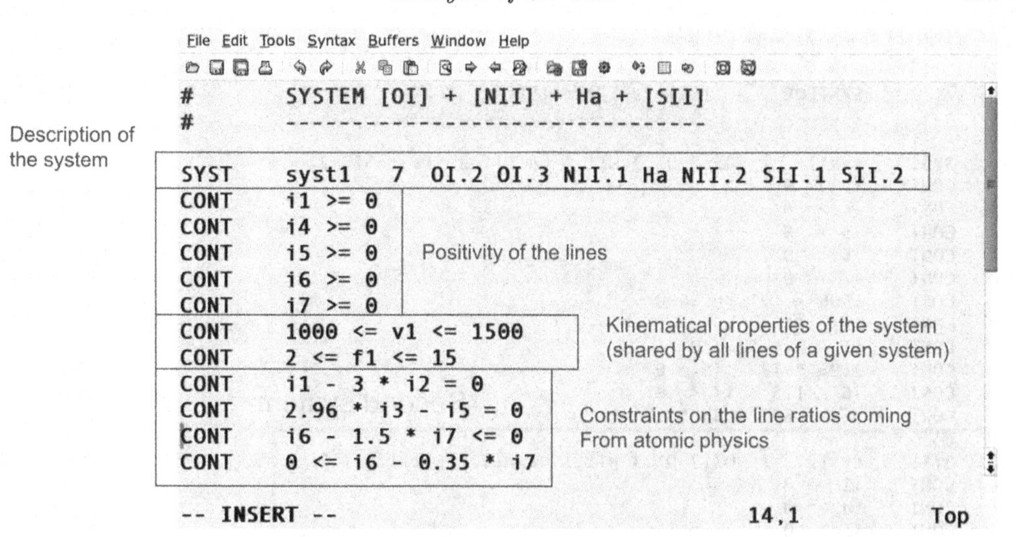

FIGURE 4.23. Configuration file describing the fit to be conducted and the constraints that will be used. Lines in the system are assigned numbers 1 to 7 (as they appear in the initial description of the system).

- the fact that the velocities shall be within a few hundreds of $\mathrm{km\,s^{-1}}$ around the systemic one;
- the fact that the intensity of the lines shall be positive.

The implementation of these constraints in the software is done through an ASCII configuration file called the 'stack' file, which is displayed in Figure 4.23. At this stage, the fit is performed with a single Gaussian profile.

Initial conditions

We also need to prepare a set of initial conditions for the fit. They will be stored in a second ASCII configuration file called the 'start' file that is a mirror of the 'stack' file and they can be obtained graphically in 'fit/spec'. Note that in complex objects, it is quite often necessary to prepare different initial conditions depending on the location within the object.

Fitting the data

So far, the preparation of the fit has been mainly done manually based on a few spectra extracted from the whole datacube. A first automated fit of all the spectra (>1000 in this case) is then performed and the resulting fits and maps inspected. In our case, despite the fact that these first maps look good, a check of the fit in the central regions of the object shows that an additional broad kinematical component is present in the nuclear region of the galaxy and that it is therefore necessary to describe the line profiles as the sum of one narrow and one broader Gaussian profile. The 'stack' file is modified accordingly (see Figure 4.24), a new set of initial conditions is prepared and the fit is run again, yielding much better results. This iterative process is performed until we are satisfied with the fit (in our case this was enough and the results of the fit can be seen in Ferruit *et al.*, 2004).

Determining the error bars

After all the emphasis in the first chapters on the tracking of the errors during data processing and analysis, we should not forget that we now need to determine the error

```
File Edit Tools Syntax Buffers Window Help
#        SYSTEME 2 x ([OI]+[NII]+Ha+[SII])
#        -------------------------------

SYST     syst1   7  OI.2 OI.3 NII.1 Ha NII.2 SII.1 SII.2
CONT     i1 >= 0
CONT     i4 >= 0
CONT     i5 >= 0
CONT     i6 >= 0
CONT     i7 >= 0
CONT     1500 <= v1 <= 3000
CONT     2 <= f1 <= 15
CONT     i1 - 3 * i2 = 0
CONT     2.96 * i3 - i5 = 0
CONT     i6 - 1.5 * i7 <= 0
CONT     0 <= i6 - 0.35 * i7

SYST     syst2   7  OI.2 OI.3 NII.1 Ha NII.2 SII.1 SII.2
CONT     i1 >= 0
CONT     i4 >= 0
CONT     i5 >= 0
CONT     i6 >= 0
CONT     i7 >= 0
CONT     1000 <= v1 <= 2500
CONT     30 <= f1 <= 50
CONT     i1 - 3 * i2 = 0
CONT     2.96 * i3 - i5 = 0
CONT     i6 - 1.5 * i7 <= 0
CONT     0 <= i6 - 0.35 * i7
                                                  1,1          All
```

Second system

Broader

FIGURE 4.24. Configuration file for a fit with two Gaussian profiles.

bars on the results from the fit. One possible way is to use the error estimate inferred from the minimization itself (from how 'sharp' the minimum toward which the program has converged is as a function of each fit parameter).

This functionality is not present in the 'fit/spec' software and we have therefore used a Monté-Carlo type of approach. The steps were the following:

- Create a synthetic, noise-free spectrum with spectral characteristics representative of those of the spectra in the object. Note that it may be necessary to prepare several of these template spectra if the spectral characteristics of the object vary significantly within the field of view.
- Create a set of noisy spectra with given signal to noise ratios by adding noise to the template. For each signal to noise ratio value, a large number of spectra (usually 500) are created in order to be able to perform a statistical analysis of the results.
- Automated fits are performed for all prepared spectra (i.e. 500 per signal to noise value).
- A statistical analysis is performed on the results to derive the accuracy of the results as a function of the signal to noise in the input spectrum.

The bottom panels of Figure 4.25 show the results for the velocity measurements. For this parameter the results are in fact quite straight-forward, but for parameters like the FWHM, they are usually much trickier to analyse. Note also that this rather simplistic Monté-Carlo approach does not 'see' some contributions to the actual error bars, such as the errors introduced by the finite accuracy of the wavelength calibration for the velocity measurements and by the subtraction of the instrumental FWHM for the FWHM measurements. These additional contributions will have to be determined and added. One

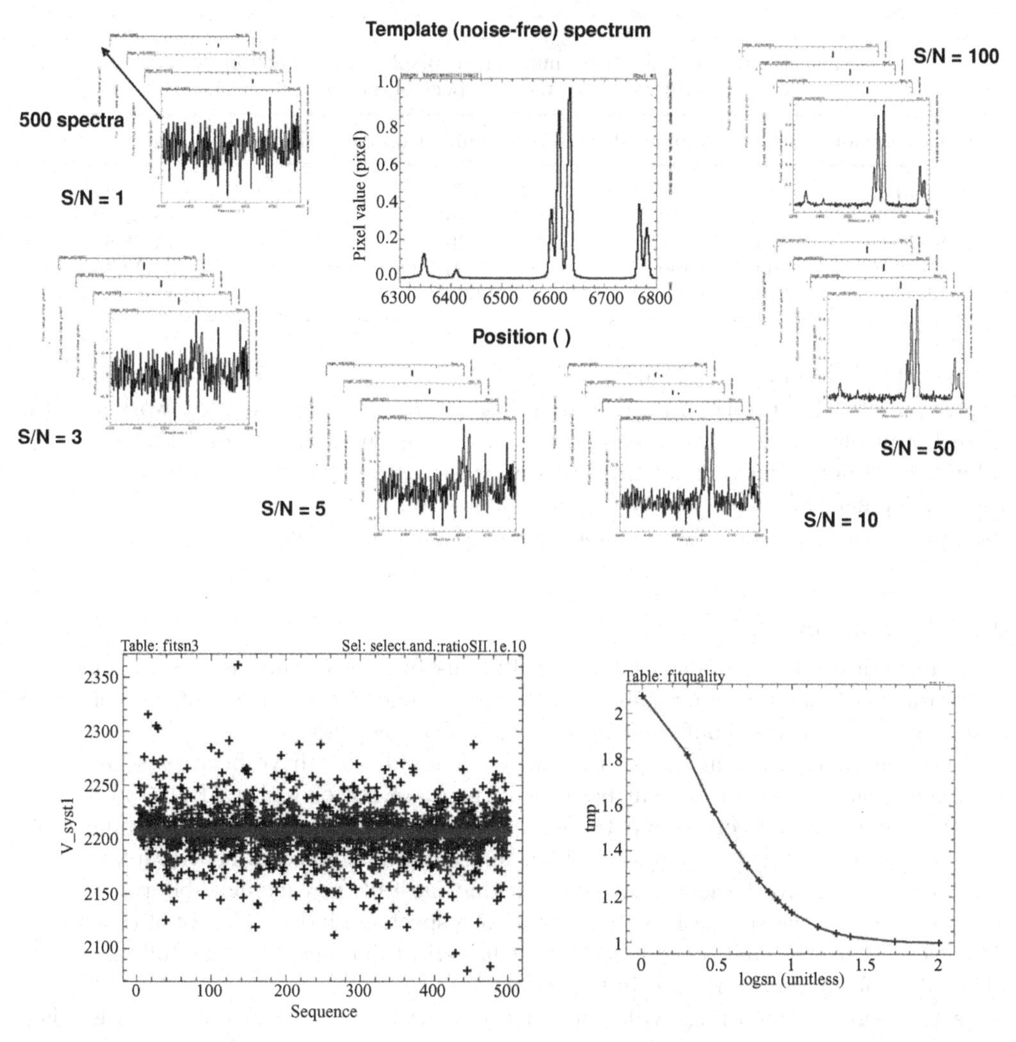

FIGURE 4.25. Top panel: sketch illustrating how a Monté-Carlo type of approach is used to determine the error bars. Bottom panels: fit of the results of Monté-Carlo approach to derive an analytical relation giving the 1σ error bar on the centroid velocity of a line as a function of the signal to noise in the input spectrum.

last contribution which is quite often not accounted for because it is extremely difficult to estimate is the contribution of errors on the stellar continuum subtraction (so-called template mismatch in particular).

Once all these steps have been performed the results are ready to be included in a paper.

The all-in-one and ultimate solutions

So far, we have described the processing/analysis of the data as a sequence of various tasks that are performed independently and with usually a heterogeneous set of tools. For instruments built around a given, well-defined scientific project, the trend is now to integrate the analysis tools within the data processing software (the all-in-one solution). This ensures in particular that all errors are carefully controlled throughout the complete process.

TABLE 4.1. Evolution of the total number of 'pixels' in a datacube for a set of arbitrary selected 3D spectrographs.

Instrument	Year	Number of spaxels	Number of spectral pixels	Total
TIGER	1987	572	270	154,440
OASIS	1997	1,200	360	432,000
SAURON	1999	1,577	540	851,580
VIMOS	2002	6,400	550	3,520,000
MUSE	2012	90,000	4,096	368,640,000

Last, there are also discussions for some instruments to use an extremely conservative approach with respect to any resampling of the data and to keep working with 'detector pixels' until the very end of the processing/analysis process (the 'ultimate' solution). Although very interesting, this approach is extremely complex and requires the development of specific analysis and visualization tools for every single step of the data processing/analysis. This has not been implemented to date.

4.9 Conclusions

This course has presented a short but hopefully useful tour of various analysis methods and strategies. The important message is to never forget to keep track of the noise and systematic errors throughout the complete data analysis process.

If we make a critical analysis of the various methods that have been presented, one relatively negative conclusion can be drawn: there are not enough real 3D tools. Most of the tools presented correspond to generalizations of classical, already-existing tools to a collection of spectra. They do not make use of the additional spatial information usually present in the datacube (e.g. the fact that most of the time neighbouring spaxels are correlated by the spatial PSF) and treat each spectrum independently of the others. This clearly outlines that a significant effort has still to be made to take full advantage of the 3D information contained in the datacubes.

Last, looking at how things will probably evolve within the next 5–10 years, there is a huge challenge in front of us. This is illustrated by the numbers in Table 4.1 that show the increase in size of the datasets over the years. Many of the analysis steps presented in this document still rely heavily on a 'manual' handling of the data. This will not be possible with data from, for example, the MUSE instrument. A huge effort will therefore have to be made to prepare the automated tools that will be needed for the next generation of 3D spectrographs.

REFERENCES

ARRIBAS, S., MEDIAVILLA, E., GARCÍA-LORENZO, B., DEL BURGO, C., FUENSALIDA, J.J. (1999), *A&AS*, **136**, 189

BENDER, R. (1990), *A&A*, **229**, 441

BERTIN, E., ARNOUTS, S. (1996), *A&AS*, **117**, 393

CAPPELLARI, M., COPIN, Y. (2003), *MNRAS*, **342**, 345

FALCÓN-BARROSO, J., SARZI, M., BACON, R., BUREAU, M., CAPPELLARI, M., DAVIES, R.L., EMSELLEM, E., FATHI, K., KRAJNOVIĆ, D., KUNTSCHNER, H., MCDERMID, R.M., PELETIER, R.F., DE ZEEUW, T. (2006), *New Astronomy Review*, **49**, 515

FERRUIT, P., MUNDELL, C.G., NAGAR, N.M., EMSELLEM, E., PÉCONTAL, E., WILSON, A.S., SCHINNERER, E. (2004), *MNRAS*, **352**, 1180

GIBSON, S.J., TAYLOR, A.R., HIGGS, L.A., BRUNT, C.M., DEWDNEY, P.E. (2005), *ApJ*, **626**, 214

HOOK, R., FRUCHTER, S. (2000), *Astronomical Data Analysis Software and Systems IX*, *ASPCS*, **Vol. 216**, edited by Nadine Manset, Christian Veillet, and Dennis Crabtree

KISSLER-PATIG, M., COPIN, Y., FERRUIT, P., PÉCONTAL-ROUSSET, A., ROTH, M.M. (2003), 'Euro3D data format, format definition', http://www.aip.de/Euro3D, *Euro3D data format: Format definition*, issue 1.2

ROUSSET (1992), *'Contribution des méthodes numériques au dépouillement des données du spectrographe intégral de champ TIGER'*, *PhD thesis*, Univ. J. Monnet de Saint-Etienne

SÁNCHEZ, S.F. (2004), *AN*, **325**, 167

WILMAN, R.J., GERSSEN, J., BOWER, R.G., MORRIS, S.L., BACON, R., DE ZEEUW, P.T., DAVIES, R.L. (2005), *Natur*, **436**, 227

5. Science motivation for integral field spectroscopy and Galactic studies

FRANK EISENHAUER

This lecture introduces the scientific motivation for integral field spectroscopy (IFS) and describes the results from this novel technique in Galactic studies as of 2005. The following chapter by Luis Colina then picks up on extragalactic studies and rounds out the picture giving an outlook to future integral field spectroscopic studies. Following the five one-hour lectures, this chapter is broken down into sections on:

- science motivation;
- the Galactic Centre black hole;
- the Galactic Centre stellar population;
- star formation; and
- the Solar System.

The publications discussed in this lecture are selected on the basis of their scientific impact as measured by the citation index and/or because they specifically qualify for a lecture from a didactic point of view.

5.1 Science motivation

Motivation for the development and application of integral field spectroscopy is manifold and is closely tied to personal interest and background. This is most obvious in the history of instrument development, outlined in the beginning of this section. The 'ideal' objects for integral field observations can be well characterized by their angular size, substructure and spectral characteristics. These qualifiers are used to identify the ideal objects on the distance ladder. The section ends comparing the apparent interest of the Galactic and extragalactic astronomical communities in integral field spectroscopy.

5.1.1 *Motivation for integral field spectroscopy*

The development of the first-generation integral field spectrographs was largely driven by an 'experimental' motivation. These instruments were built by small teams that wanted to invent a new technology. Typical examples for this early epoch are the ARGUS, MPE 3D, MPFS and Tiger instruments. This is well illustrated by a sentence from Bacon *et al.* (1995) on the Tiger instrument: 'Although it was primarily developed to validate the lenslet integral field spectroscopy (IFS) concept, it has provided a number of scientific results on a large variety of astrophysical objects.'

The second-generation integral field spectrographs were then strongly motivated by the astronomical applications, and these instruments were built by medium-size groups, partly in international collaborations: typical examples for this present generation of instruments are: OASIS, SAURON, VIMOS, NIFS, OSIRIS and SINFONI. Many of the instruments are permanently installed as facility instruments at large observatories. One finds both instruments for a wide variety of applications and instruments dedicated to very specific goals. The change from the experimental to the astronomical motivation is nicely illustrated by another citation from the above authors, now on SAURON, the successor of Tiger: 'Its design is optimized for studies of the stellar kinematics, gas kinematics, and line-strength distributions of nearby early-type galaxies' (Bacon *et al.*, 2001a).

The development of a third generation of integral field spectrographs has recently started. These instruments are built by large international consortia and are usually justified by a wide variety and mainstream science cases. Examples of this third generation of instruments are MUSE (approximately 8.4 million euro, 172 staff years, 7 institutes) and KMOS (approximately 6.5 million euro, 165 staff years, 6 institutes).

Presently, several integral field spectrographs are also being built or developed for upcoming missions, for example PACS on the Herschel satellite and Near-Infrared Spectrograph (NIRSpec) on the *James Webb Space Telescope* (JWST).

5.1.2 *Fundamental considerations based on technical constraints*

The parameter space of integral field spectrographs is spanned by the number of spatial and spectral elements. Although there is no fundamental limit on the number of these elements, most integral field spectrographs are in practice (still) limited by the number of available detector pixels, and the fact that array detectors are more or less squarish. Therefore it turns out that the various integral field spectrographs are quite comparable in terms of number of spatial and spectral elements. To estimate the typical numbers, we assume a detector with 2k $\times$ 2k pixels. The number of spectra is within a factor of a few equal to the number of rows, i.e. approximately 2000. This is obvious for mirror-slicer integral field units (see Bershady's chapter 3 on instrumentation), but also most lenslet integral field units turn out to have a comparable number of spectra because the gain by stacking several spectra in one column is compensated by the gaps between the spectra. Assuming a square field of view, the typical integral field spectrograph then covers of the order $\sqrt{4000} \times \sqrt{4000} \approx 50 \times 50$ pixels. The number of pixels for every spectrum is limited by the number of pixels per column. Mirror-slicer integral field units take full advantage of them; lenslet IFUs typically stack 2–4 spectra per column. Therefore the typical number of spectral elements is about 1000. For comparison, the typical number of elements (spatial $\times$ spatial $\times$ spectral) for a long-slit spectrograph are $1 \times 1000 \times 1000$, for an echelle spectrograph $1 \times 1 \times 1{,}000{,}000$ and for an Fabry–Perot spectral imager $1000 \times 1000 \times 1$.

The size of each spatial pixel on the sky, and therefore the total field of view, is also quite constrained: the argument for the lower limit is simply the diffraction limit of the telescope, which is approximately 20 mas for a 10 m telescope at a wavelength of 1 μm. Therefore, the total field of view of a diffraction-limited integral field spectrograph is of the order $50 \cdot 20\,\text{mas} \times 50 \cdot 20\,\text{mas} = 1'' \times 1''$. The maximum size of a pixel on the sky can be derived from the Lagrange invariant in optics, the small physical size of a detector pixels, and the practical limitation on building a camera with an f-number faster than 1. The Lagrange invariant implies that the étendue = 'area-solid angle product' is preserved in imaging optics (and in general can only increase, for example through the focal-ratio degradation of a fibre). Applied to an integral field spectrograph, this implies that *telescope diameter* $\times$ *pixel size on sky* [in rad] = *physical pixel size* $\times$ *beam opening angle at detector* [in rad]. If we assume a typical pixel size of 10 μm, and again a 10 m telescope, the maximum pixel size on sky for an $f/D = 1$ camera (i.e. beam opening angle at detector is approx. 1 rad) is approximately 10^{-6} rad or $0.2''$. Therefore, the maximum field of view of an integral field spectrograph is of the order $50 \cdot 0.2'' \times 50 \cdot 0.2'' = 10'' \times 10''$.

In summary, we find that a typical integral field spectrograph has of the order 50×50 spatial pixels with a field of view of $1'' \times 1''$ to $10'' \times 10''$, and of the order 1000 spectral pixels.

5.1.3 *Ideal and unsuitable objects for integral field spectroscopy*

The typical number of spatial and spectral pixels, and the typical field of view are criteria for 'ideal' and 'unsuitable' objects for integral field spectroscopy. Clearly unsuitable are objects that:

- cover a very large field of view (larger than several square arcminutes);
- are sparse and point-like (only few objects per several square arcminutes);
- require observations of a large spectral range at very high spectral resolution (larger than several 10,000, covering a full atmospheric band).

Typical examples for unsuited science cases are:

- (all-sky) surveys, for example the Wisconsin Hα Mapper Northern Sky Survey is done with a Fabry–Perot spectral imager;
- multi-object spectroscopy of loose stellar fields or extragalactic planetary nebulae, done with fibre-fed (e.g. 2dF at the Anglo-Australian Telescope, AAT) or slit-mask (e.g. Focal Reducer and Low Dispersion Spectrograph, FORS at the Very Large Telescope, VLT) multi-object spectrographs;
- high spectral resolution (echelle) spectroscopy for planet search from Doppler velocity measurement (e.g. High Accuracy Radial Velocity Planet Searcher, HARPS at the European Southern Observatory, ESO, 3.6 m telescope) or the Lyman-alpha forest observed in quasars (e.g. High Resolution Echelle Spectrometer, HIRES at the Keck Telescope).

Ideal objects fulfil one or more of the following criteria:

- resolved at the seeing- or diffraction-limited resolution of optical telescopes;
- position is not accurately known;
- no (or unknown) spatial symmetry;
- rich spectrum or unknown redshift.

Examples of ideal objects in our Solar System are the planets, their moons and asteroids. Ideal Galactic objects are, for example (roughly in the order of distance), brown dwarfs and low-mass stars, stellar accretion discs, young stellar objects (YSOs), jets, Herbig–Haro objects, asymptotic giant branch (AGB) stars, proto and young planetary nebulae, young star clusters, the Galactic Centre region with its black hole, and globular clusters. Examples of ideal extragalactic objects are (also ordered by distance) the nuclear and old stellar populations of nearby galaxies, nearby massive black holes, nearby active galactic nuclei (AGN), nearby starbursts, ultra-luminous infrared galaxies (ULIRGs), quasar host galaxies, high redshift star-forming galaxies and Lyman-break galaxies.

The science case for the above examples is described in detail in various instrument proposals and publications, for example in 'Science with NIFS, Australia's first Gemini Instrument' (McGregor *et al.*, 2001) and 'SINFONI a High Resolution Near-Infrared Imaging Spectrometer for the VLT' (Genzel *et al.*, 1997).

5.1.4 *Examples of integral field spectroscopic studies*

After the pioneering work with the 'first generation' principal investigator instruments, the race began with the public availability of facility instruments on large telescopes. This section highlights a number of refereed publications from recent years based on integral field spectroscopic observations.

As outlined above, the ideal objects are spatially resolved at the diffraction- to the seeing-limit of optical telescopes, i.e. those objects having structure on the 0.02–0.2″

scale and cover a field of view of approximately 1–10″. Accordingly, the physical size and substructure of ideal objects scale with distance. Therefore, the following list is ordered by the distance, climbing the distance ladder by typically a factor 10 with each rung.

The typical distance scale for the Solar System is one astronomical unit. An angular separation of 1″ corresponds to approximately 700 km, the ideal (1–10″) field of view (FoV) of an integral field spectrograph corresponds to 700–7000 km, e.g. the size of moons, continents of planets and asteroids. The angular resolution (0.02–0.2″) corresponds to 10–100 km, e.g. big volcanoes. Although many Solar System objects have already been visited by spacecraft, there is still need for ground-based observations, for example to obtain high-resolution spectra, and for monitoring long-term trends and seasons. Examples of Solar System studies taking advantage of integral field spectroscopy are 'Martian surface mineralogy from 0.8 to 1.05 μm TIGER spectra–imagery measurements in Terra Sirenum and Tharsis Montes formation' (Martin *et al.*, 1996), 'The three-dimensional distribution of Titan haze from near-infrared (NIR) integral field spectroscopy' (Ádámkovics *et al.*, 2006) and 'Near-Infrared Compositional Mapping of Vesta' (Dumas *et al.*, in preparation).

The next step on the distance ladder leads to the most nearby stars at about 1 pc (1″ = 1 AU), for example Proxima Centauri. At that distance the 'typical' FoV covers 1–10 AU, therefore matching the size of planetary systems. The resolution is about 0.02–0.2 AU, comparable to the separation of planets. However, present-day adaptive optics and instruments do not provide enough contrast and no publications have emerged from such observations yet.

The most nearby star-forming regions – for example the Taurus molecular cloud – are located at a distance of about 100 pc (1″ = 100 AU). The FoV of 100–1000 AU is comparable to the size of circumstellar discs, and the angular resolution is 2–20 AU, comparable to the orbital radius of Jupiter. So far, publications have concentrated on prototypical stars, jets and their interaction with the ambient interstellar medium: 'A near-infrared spectral imaging study of T Tau' (Herbst *et al.*, 1996; Kasper *et al.*, 2002), 'Sub-arcsecond morphology and kinematics of the DG Tauri jet in the [OI]λ6300 line' (Lavalley *et al.*, 1997) and 'The three-dimensional structure of HH 32 from GMOS IFU spectroscopy' (Beck *et al.*, 2004).

The closest high-mass star formation regions, for example the Orion Nebula, are at a distance of about 1 kpc (1″ = 1000 AU). The FoV of 1000–10,000 AU and the resolution of 20–200 AU is still appropriate to cover and resolve circumstellar discs and jets. The publications address the processes involved in high-mass star formation, for example 'Collimated molecular jets from high-mass young stars: IRAS 18151-1208' (Davis *et al.*, 2004), and the impact of winds and radiation from the high-mass stars on circumstellar discs of nearby stars, for example in 'GEMINI multi-object spectrograph integral field unit spectroscopy of the 167-317 (LV2) proplyd in Orion' (Vasconcelos *et al.*, 2005).

The Galactic Centre is located at a distance of about 10 kpc (10″ ≈ 0.5 pc). At that distance the FoV of 0.05–0.5 pc corresponds to the core size of a typical star cluster and the resolution of 200–2000 AU is close to a light-day. Being a unique laboratory for studying supermassive black holes and their environment, the Galactic Centre has been the subject of a larger number of publications from integral field spectroscopy, for example on 'The nuclear cluster of the Milky Way: star formation and velocity dispersion in the central 0.5 parsec' (Krabbe *et al.*, 1995) and 'SINFONI in the Galactic Centre: young stars and infrared flares in the central light month' (Eisenhauer *et al.*, 2005). A distance of 10 kpc is also the typical scale for our Galaxy. Consequently, integral field observations cover a wide variety of objects at this distance. Examples are 'Integral field spectroscopy of faint halos of planetary nebulae' (Monreal-Ibero *et al.*, 2005) and 'The X-ray binary X2127+119 in M15: evidence for a very low-mass, stripped giant companion' (Van Zyl *et al.*, 2004).

At a distance of 1 Mpc ($1'' \approx 5$ pc) we get to the closest disc galaxies, for example the Andromeda galaxy M31. The typical FoV covers 5–50 pc, e.g. nuclear starburst regions, and the spatial resolution is about 0.1–1 pc, the typical separation of stars in spiral arms. Publications have so far concentrated on the nuclear regions, for example in 'The M 31 double nucleus probed with OASIS. A natural vec m = 1 mode?' (Bacon *et al.*, 2001b).

The closest clusters of Galaxies – for example the Virgo cluster – are located at a distance of about 10 Mpc ($1'' \approx 50$ pc). The FoV corresponds to about 50–500 pc, and the resolution of 1–10 pc is still good enough to resolve individual starburst regions. Again, publications focus mainly on the dynamics and properties of nuclear regions, for example 'The SAURON project: III. Integral-field absorption-line kinematics of 48 elliptical and lenticular galaxies' (Emsellem *et al.*, 2004), and 'SINFONI adaptive optics integral field spectroscopy of the Circinus galaxy' (Müller-Sánchez *et al.*, 2006).

Yet another factor 10 more distant (about 100 Mpc, $1'' \approx 0.5$ kpc) are massive galaxy clusters – for example the Coma Cluster – which host 'extreme' galaxies like ultra-luminous infrared galaxies (ULIRGs). The field of view at that distance is approximately 0.5–5 kpc, about the size of galactic bulges. The resolution is about 10–100 pc. This is sufficient to spatially resolve the velocity field of merging galaxies, a published example is 'Stellar dynamics and the implications on the merger evolution in NGC 6240' (Tecza *et al.*, 2000).

The most distant objects observed so far with integral field spectrographs are at red-shifts of $z \approx$ few ($1'' \approx 5$ kpc at $z = 1$). This early epoch – several Gyr after the Big Bang – is particularly interesting because of its peak in cosmic star formation. The FoV of about 5–50 kpc (at $z = 1$) covers one or more galaxies. The resolution of 0.1–1 kpc allows us to resolve regions with sizes comparable to galactic bulges. The publications focus on characterizing the physical properties of various galaxy samples, for example 'Optical and near-infrared integral field spectroscopy of the SCUBA galaxy N2 850.4' (Swinbank *et al.*, 2005) and 'SINFONI integral field spectroscopy of $z \approx 2$ UV-selected galaxies: rotation curves and metallicity gradients' (Förster Schreiber *et al.*, 2006), but also cover various other topics like the feedback mechanism to the intergalactic medium, for example in 'Deep SAURON spectral imaging of the diffuse Lyman α halo LAB1 in SSA 22' (Bower *et al.*, 2004).

The above examples illustrate the start of the race, and many more exciting results are expected in the coming years. In parallel, several groups have started with the next-generation integral field spectrographs and telescopes. The outlook for this future science is given in the lecture by Luis Colina.

5.1.5 *Galactic versus extragalactic science*

The science cases outlined in the previous section cover Galactic and extragalactic objects equally. However, recent conferences indicate that the major science driver is extragalactic. This can be seen from the rough fraction of Galactic versus extragalactic proceedings articles in the last five big conferences on science with integral field spectrographs:

- 'Extragalactic science 3D optical spectroscopic Methods in Astrophysics', Marseilles 1994 (Comte and Marcelin, 1995): 40% Galactic versus 60% extragalactic;
- 'Imaging the universe in 3D', Walnut Creek 1999 (Van Breugel and Bland-Hawthorn, 2000): 10% Galactic versus 90% extragalactic;
- 'EURO 3D science conference', Cambridge 2003 (Astronomische Nachrichten, 325(2)): 10% Galactic versus 90% extragalactic;

- 'Workshop on AO assisted IFS', La Palma 2005 (Rutten *et al.*, 2006): 40% Galactic versus 60% Extragalactic;
- 'Science perspectives for 3D spectroscopy Conference', Garching 2005 (Kissler-Patig *et al.*, 2006): 20% Galactic versus 80% extragalactic.

In summary, about 25% of the proceedings articles of these conferences are on Galactic science, and 75% on extragalactic science. A similar trend can be seen at this IAC Winter School, for which only less than 20% of the astronomical poster contributions are on Galactic topics, and more than 80% on extragalactic topics.

5.1.6 *Overview of Galactic science*

The following four sections give an overview of Galactic science with integral field spectrographs as of 2005. All subsequent citation counts are up to late 2005. The most successful application of integral field spectroscopy to Galactic science up to 2005 is Galactic Centre research with about 600 citations, followed by star formation research with about 125 citations. Solar System studies and other topics have had limited impact so far and only a few citations. For the papers discussed in this lecture we find:

- On the Galactic Centre: Eisenhauer *et al.* (2003a), ≈ 71; Eisenhauer *et al.* (2005), ≈ 16; Genzel *et al.* (1996), ≈ 130; Genzel *et al.* (2000), ≈ 120; Genzel *et al.* (2003a), ≈ 89; Krabbe *et al.* (1995), ≈ 178;
- Related to star formation: Beck *et al.* (2004), ≈ 3; Davis *et al.* (2004), ≈ 7; Dougados *et al.* (2003), ≈ 1; Herbst *et al.* (1996), ≈ 38; Kasper *et al.* (2002), ≈ 6; Lavalley *et al.* (1997), ≈ 34; Lavalley-Fouquet *et al.* (2000), ≈ 35; Vasconcelos *et al.* (2005), ≈ 0;
- On Solar System objects: Ádámkovics *et al.* (2006); Martin *et al.* (1996), ≈ 5.

However, it should be noted that this list is to a large extent limited to the 'pioneering' phase of IFS, and that the real impact is just about to start with the installation of several facility instruments at large observatories.

5.2 The black hole in the Galactic Centre

This section is about the black hole at the Galactic Centre and how IFS contributes to prove its existence. The subsections introduce the subject of supermassive black holes, give the line of argument for proving the existence of the black hole, describe the measurement of its mass and distance, and discuss the paradox of its faintness. Many of the publications also address the stellar population close to the Galactic Centre. For didactic reasons, however, this topic is discussed in a separate section.

Excursion: history of supermassive black holes

Presently, we know two kinds of black holes: stellar black holes with a mass of ≈ 10–$30\,M_{\odot}$ and supermassive black holes with a mass of $\approx 10^5$–$10^9\,M_{\odot}$. While the formation of a stellar black hole from the collapse of a massive star in a Type II supernova is well understood, and pulsar binary systems provide a variety of good evidence for the black hole hypothesis, the case for supermassive black holes has been less clear for a long time.

The first strong evidence for such extreme mass concentrations was the discovery of the quasar 3C273 by Schmidt (1963), who even made it to the cover page of *Time* magazine. Based on the observed Doppler shift, this 13th magnitude 'star' moves away from us at 16% of the speed of light. If this redshift is correctly interpreted in terms of the Hubble flow, the object is located at a distance of approximately 1 Gpc, corresponding to a distance modulus of ≈ 40. Therefore the absolute magnitude of 3C273 must be about -27 mag, and its luminosity about $10^{13}\,L_{\odot}$. As the brightness of the object varies over a period of days, its size must be smaller than light-days, i.e. of the order less than 1000 AU.

FIGURE 5.1. Adaptive optics K-band (1.95–2.45 μm) image of the Galactic Centre region. The field of view is approximately $20'' \times 20''$. The arrows indicate the position of the black hole, which is not visible in this image. Only during sporadic flares (every couple of hours) is the emission strong enough to be detected. The angular resolution in the image is approximately 50 mas. For comparison, the Schwarzschild radius of the Galactic Centre black hole is approximately 10 μas.

In other words, 3C273 emits the light of the whole Galaxy from a region comparable to the size of the Solar System.

Subsequent observations of radio halos of quasars indicated that the energy source is gravitation and not nuclear fusion: the energy contained in such radio halos is up to about 10^{54} J. If nuclear fusion were the energy source – for which the efficiency of converting mass to radiation is of the order 1% – about 10^9 $M_\odot$ would be required to power the radio halos. However, the self-gravitational energy of this mass when squeezed into a couple of light-days would exceed 10^{55} J, more than 10 times the fusion energy. Therefore, most of the energy released in quasars must have a gravitational origin.

The evidence that the light source in quasars is as small as a black hole, however, was weaker and more indirect: it was based on observations of radio jets. These jets reach far out into intergalactic space and as such must 'remember' their ejection direction for quite a long time. The orientation of the emitting region must thus be very stable, like a 'gyroscope'. This indicates that the emitting object is probably spinning very fast. The other key observation was that radio jets move close to the speed of light. This implies indirectly that the speed of the spinning material is also close to that of light, i.e. it must originate close to the Schwarzschild horizon of a black hole.

To summarize the epoch of 1970–90, most astronomers believed the supermassive black hole picture. They found an enormous amount of indirect evidence, i.e. no other model could reproduce the measurements equally well. But there has been no strict proof that supermassive black holes exist, i.e. that the mass is enclosed within the Schwarzschild radius, and that there is an event horizon. So the search for supermassive black holes became a very hot topic, both in external galaxies (see lecture by Luis Colina) and in our Milky Way (this lecture).

Excursion: line of argument for proving the Galactic Centre black hole

The line of arguments to prove the existence of the supermassive black hole in the Galactic Centre is outlined in Figure 5.2. The first step is to measure its mass at ever-smaller radii. Down to a few thousand Schwarzschild radii (R_S) this is accomplished by measuring the orbits of stars with high angular resolution imaging (Figure 5.1) and integral field spectroscopy (Figure 5.5). The combined astrometric and radial velocity observations also provide the best measurement of the distance to the Galactic Centre, a prerequisite for

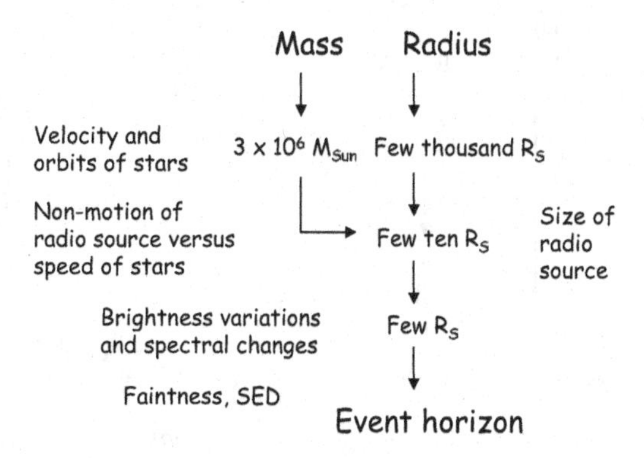

FIGURE 5.2. Line of arguments to prove the existence of the supermassive black hole in the Galactic Centre.

pinning down the absolute mass and radius. Already, these measurements exclude most alternatives (for example a very dense cusp of stars or stellar remnants) to the black hole hypothesis. Radio measurements of the size of Sgr A* (the name of the radio source which is supposed to harbour the black hole) and the fact that the radio source hardly moves when compared to the stars then make the case that the mass is indeed within just a few tens of R_S. The observed brightness variations and changes in the slope of the spectra are interpreted to arise partly from orbital motions as close as a few R_S. Finally, it is the faintness of the Galactic Centre and the detailed shape of its spectral energy distribution (Figure 5.6) that provide the theoretical argument that there is indeed a black hole with a Schwarzschild horizon.

5.2.1 *The nuclear cluster of the Milky Way: star formation and velocity dispersion in the central 0.5 parsec*

The first integral field spectroscopy observations of the Galactic Centre were published by Krabbe *et al.* (1995). Their observations show that the radial velocity dispersion of 35 early- and late-type stars with distances of 1–12″ from Sgr A* is 154 ± 19 km s^{-1}, which strongly favours the presence of a central dark mass of 3×10^6 $M_\odot$ within 0.14 pc of the dynamic centre.

The observations were carried out in 1994 with the world's first infrared integral field spectrograph MPE 3D (Weitzel *et al.*, 1996) visiting the ESO-MPI 2.2 m telescope in La Silla. The spectra cover the wavelength range 1.9–2.4 μm at a spectral resolution $\lambda/\delta\lambda \approx 1000$. In total, 12 early-type emission line stars and 9 late-type stars with CO band heads (see excursion) could be identified in the central 5″ radius.

Excursion: Stellar spectra of early- and late-type stars in the infrared

Near-infrared spectra of stars have characteristic emission and absorption lines that allow accurate spectral classification. In particular, it is very easy to distinguish between early-type emission line stars and late-type stars with CO band-head absorption. Figure 5.3 shows three stars from the Galactic Centre region with typical spectra. Early-type emission line stars are young massive (more than about 20 $M_\odot$) stars, for example Wolf–Rayet and Of stars (f indicating emission). They show strong emission from one or several lines of hydrogen, helium I and II, and ionized nitrogen and carbon, which originate mostly in stellar winds. Late-type stars – for example, K and M giants – do not show emission lines, but have deep CO band-head absorption from the outer parts of their cool (less than about 4500 K) atmospheres.

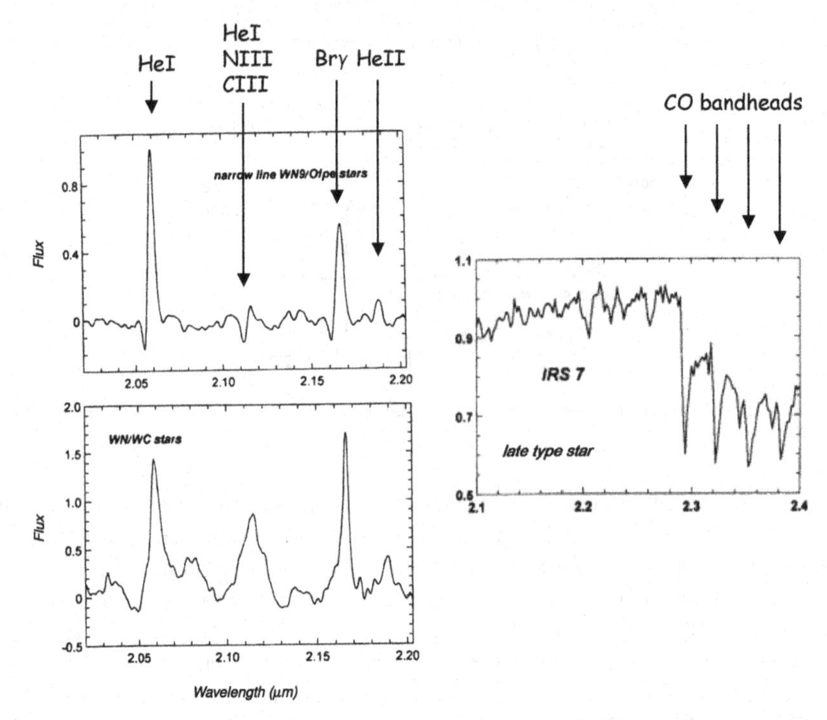

FIGURE 5.3. Typical spectra of early-type emission line stars (left) and late-type stars (right). The figure shows three stars from the Galactic Centre region (courtesy of Genzel *et al.*, 1996).

The emission lines from the early-type stars and the CO band-heads of late-type stars allow the measurement of the radial velocity of the stars from the Doppler shift. The measured radial velocities and projected distances from the Galactic Centre then give the enclosed mass. From the virial theorem (see excursion) the estimate for the enclosed mass within 0.14 pc is $4.0 \times 10^6\ M_\odot$. The Bahcall and Tremaine (1981) estimator (see excursion) for the enclosed mass within 0.19 pc is $6.3 \times 10^6\ M_\odot$. When adding the velocities from the more distant stars, and also considering the gas dynamics at larger radii, the gravitational potential in the Galactic Centre can be well fitted with the combination of a $3 \times 10^6\ M_\odot$ central (black hole) mass and an isothermal stellar cluster with $3 \times 10^5\ M_\odot$ within a core radius of 0.5 pc (Figure 5.4). However, the data are still consistent with a central mass dominated by a very compact cluster of stellar remnants (e.g. about 10^5 stellar black holes with a typical mass of $10\ M_\odot$).

Excursion: from velocity to mass I: the virial theorem

The measurement of radial velocities and projected distances allow the determination of the central mass from the virial theorem. The proof is well known, and is repeated here (following Bahcall and Tremaine, 1981) to give the basis for the various mass estimates used in the following papers. Assuming that the gravitational potential in the centre of the Galaxy is dominated by a point mass, the stars can be described as test particles, which follow the gravitational law:

$$\ddot{r} = -\frac{GMr}{r^3}, \tag{5.1}$$

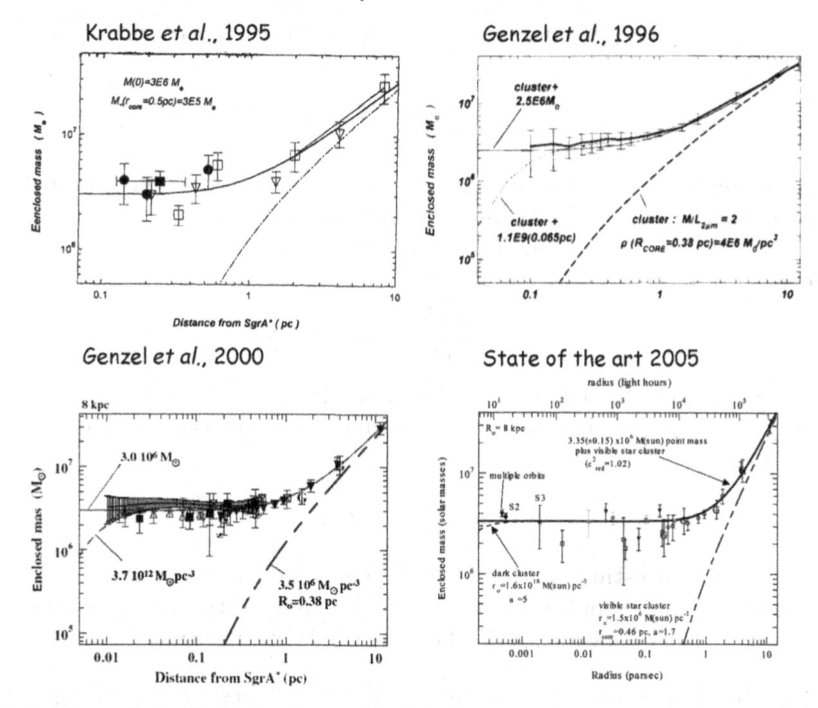

FIGURE 5.4. Mass of the Galactic Centre black hole: the plots show the various estimates for the encircled mass profile from the first integral field spectroscopic observations in 1995 to the state of the art in 2005 (courtesy of Genzel *et al.*, 1996; 2000; Krabbe *et al.*, 1995).

where r is the vector from the point mass to the star, and G the gravitational constant. If we then write

$$\frac{d^2}{dt^2}\left(\frac{1}{2}r \cdot r\right) = r \cdot \ddot{r} + \dot{r} \cdot \dot{r} \tag{5.2}$$

and use $r \cdot r = r^2$, $\dot{r} \cdot \dot{r} = v^2$, and average $<>$ over all stars, we get

$$\frac{1}{2}\frac{d^2}{dt^2} < r^2 > = -GM < \frac{1}{r} > + < v^2 >. \tag{5.3}$$

Assuming that the ensemble of stars is in a statistical steady state, $< r^2 >$ is independent of time and $\frac{d^2}{dt^2} < r^2 > = 0$. If we further assume that the system is spherically symmetric, then $< v^2 > = 3 < v_z^2 >$ and $< \frac{1}{r} > = \frac{2}{\pi} < \frac{1}{R} >$, where v_z is the radial velocity and R the projected distance to the centre. From that, we get the virial estimator for the central point mass:

$$M = \frac{3\pi}{2G}\frac{< v_z^2 >}{< \frac{1}{R} >}. \tag{5.4}$$

Excursion: from velocities to mass II: Bahcall–Tremaine estimator

As outlined by Bahcall and Tremaine (1981), the virial theorem estimator has some serious shortcomings: it is biased in the sense that the virial mass is not necessarily equal to the real mass when derived from only a finite number of stars. Second, the virial theorem estimator is inefficient, i.e. the variance in the estimate is larger than necessary, because it weights objects at small distances too strongly. The problem is inherent to the virial theorem estimator and can be avoided by using an estimator based on the projected mass, q:

$$q = \frac{v_z R}{G}. \tag{5.5}$$

This projected mass treats all test particles at all distances on an equal information basis. Doing some mathematics to derive the ensemble average $< q >$ for an arbitrary distribution of stars, Bahcall and Tremaine (1981) obtain the relation with the eccentricities of the stellar orbits:

$$< q > = \frac{\pi M}{32} (3 - 2 < e^2 >). \tag{5.6}$$

If all the stars are on circular or radial orbits, then $< e^2 > = 0$ or 1. For isotropic orbits $< e^2 > = \frac{1}{2}$. In the absence of any information on the eccentricities, the authors recommend using

$$M = \frac{24}{\pi G} < v_z^2 R >. \tag{5.7}$$

Other than the virial estimate, this Bahcall–Tremaine estimator is not biased when applied to only a small number of stars and also weights the test particles properly. The disadvantage is its dependence on the eccentricity distribution.

5.2.2 *The dark mass concentration in the central parsec of the Milky Way*

In this second paper, Genzel *et al.* (1996) present an extended dataset, and improve their mass estimate by considering a rotation component in the measured velocity dispersion and by simultaneously fitting the observed surface density of stars.

The field of view covered in the observations is approximately $20'' \times 20''$. Taking advantage of additional high angular resolution speckle imaging data, the authors could extract the spectra of 198 late-type stars, and 25 early-type stars. The projected stellar velocity dispersion increases statistically very significantly from about $55\,\mathrm{km\,s^{-1}}$ at a projected distance of 5 pc, to $180\,\mathrm{km\,s^{-1}}$ at 0.1 pc. When plotting the measured radial velocities against the declination, a significant rotation of the stellar population is obvious. The late-type stars follow general Galactic rotation, while the early-type stars show counter-rotation. One therefore has to subtract (in quadrature) the rotation component

$$M_{rot} = \frac{2v_{rot}^2(R)}{\pi G} < R > \tag{5.8}$$

from the mass as derived from the virial theorem or the Bahcall–Tremaine estimator (see previous section). $v_{rot}(R)$ is the best-fit rotation curve.

The shortcoming of the virial theorem and Bahcall–Tremaine estimator is obvious: their application assumes implicitly a homogeneous distribution of stars, but the measured distribution is by no means homogeneous.

Therefore, Genzel *et al.* (1996) have also derived a less-biased enclosed mass estimate from the first moment of the collisional Boltzmann equation, the 'Jeans equation' (for details see Binney and Tremaine, 1987). The problem is that the formula contains the three-dimensional distance and the space density, and not the measured projected distance and surface density. The transformation from volume to surface quantities is mathematically described by Abel integrals. Since the deconvolution of the projected quantities to the corresponding volume quantities is very sensitive to individual data points and their errors, the authors parameterize the velocity dispersion and density distribution and find the best-fitting model to the data. The analysis gives a central dark mass of 2.5–3.2 $\times 10^6$ $M_\odot$ at 6–8 σ significance (Figure 5.4). The data also show that the mass-to-light ratio significantly increases towards the Galactic Centre up to at least 100 $\frac{M_\odot}{L_{\odot,2\,\mu\mathrm{m}}}$, indicating the 'dark' nature of the central mass. It is therefore either a compact cluster of 10^5 stellar black holes with a typical mass of 10–20 $M_\odot$, or a single massive black hole.

5.2.3 *Stellar dynamics in the Galactic Centre: proper motions and anisotropy*

This paper by Genzel *et al.* (2000) combines proper motion data from speckle interferometric imaging with radial velocity measurements from integral field spectroscopy in a new analysis of the stellar dynamics in the Galactic Centre. They find that the sky-projected velocity components of the young, early-type stars indicate significant deviations from isotropy, with a strong radial dependence. Most of the bright emission line stars at separations from 1″ to 10″ from Sgr A* are on tangential orbits. This tangential anisotropy of the He I stars and most of the brighter members of the IRS16 complex is largely caused by a clockwise (on the sky) and counter-rotating (line of sight, compared to the Galaxy), coherent rotation pattern. The assumption of velocity isotropy in the above mass estimates is clearly violated. The authors thus explicitly include the velocity anisotropy in estimating the central mass distribution. This Leonard and Merritt (1989) mass estimator (see excursion) confirms previous conclusions that a compact central mass concentration (central density $\geq 10^{12.6}$ $M_\odot$ / pc^3) is present and dominates the potential at 0.01–1 pc. Depending on the modelling method used, the derived central mass is 2.6–3.3 $\times 10^6$ $M_\odot$.

Excursion: from velocity to mass III: the Leonard–Merrit estimator

The virial theorem and Bahcall and Tremaine mass estimator assume an isotropic distribution of stellar orbits. To overcome this limitation, Leonard and Merrit (1989) introduced – assuming a spherically symmetric, non-rotating system – an isotropy-independent mass estimator, which, however, requires data on all three space velocities:

$$M = < r(2\sigma_r^2 + 2\sigma_t^2) >, \tag{5.9}$$

where σ_r and σ_t denote the velocity dispersion parallel and perpendicular to the radius vector, r. When transformed to the projected quantities – as observed – the Leonard–Merrit mass estimator can be written as:

$$M = \frac{16}{3\pi}\left(2 < R\sigma_R^2 > + < R\sigma_T^2 >\right), \tag{5.10}$$

where σ_R is the line-of-sight velocity dispersion and σ_T is the dispersion measured in the velocity perpendicular to the projected radius vector, R.

Excursion: from accelerations and orbits to mass

All mass estimates for the Galactic Centre black hole presented above rely to some extent on assumptions concerning the distribution function of the stellar orbits, for example spherical symmetry or isotropy. The combination of radial velocity measurements with proper motions and the use of advanced mass estimators removed only part of this bias. The next step forward in the mass determination was the measurement of accelerations and stellar orbits of close stars (in particular 'S2'; see Figure 5.5) from multi-epoch speckle and adaptive optics imaging (Schödel *et al.*, 2002; Ghez *et al.*, 2003). With the exception of the ambiguity in the inclination, these astrometric observations constrain all orbital parameters. However, the conversion of the angular semi-major axis – in units of arcseconds – to a physical distance – in units of metres – and therefore the determination

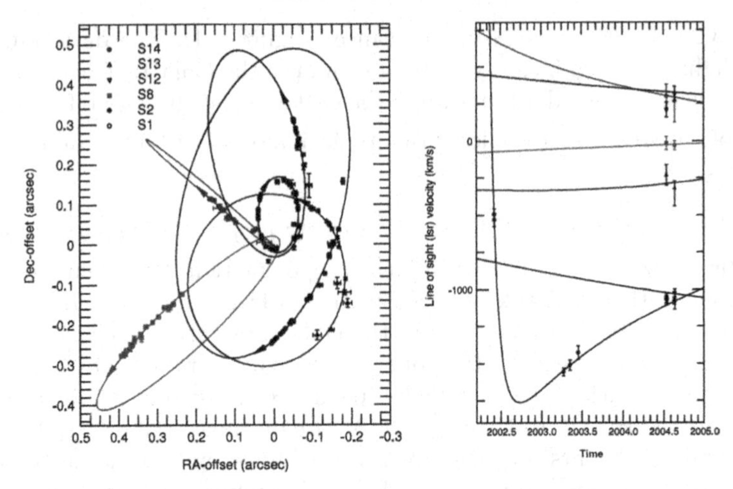

FIGURE 5.5. Three-dimensional orbits and distance determination for the Galactic Centre: fitting simultaneously the measured radial velocities and motion on the sky gives directly the distance to the Galactic Centre. The radial velocity also removes the ambiguity in the inclination when fitting the stellar orbits solely from their motion on the sky (courtesy of Eisenhauer *et al.*, 2005).

of the central point mass requires the distance to the Galactic Centre. This distance is known from various other measurements to be 8.0 kpc, but only to an accuracy of ± 0.5 kpc. Since the mass estimate scales with the third power of distance, the distance is also the major uncertainty (about 20%) in the mass determination.

5.2.4 *A geometric determination of the distance to the Galactic Centre*

This paper by Eisenhauer *et al.* (2003a) was the first to report a geometric determination of the distance to the Galactic Centre. The distance of 7.94 ± 0.42 kpc is derived from combining the astrometric data of the 'S2' star with radial velocity measurements from adaptive optics long-slit spectroscopy and new integral field spectroscopy.

The idea behind this distance estimate is quite simple and has been known and applied for decades to binary stars: from Kepler's law one can calculate all three space velocities. While the measured radial velocities are independent of distance, the orbit as observed on the sky (in angular units) is inversely proportional to the distance. Fitting the observed radial velocities and the orbits on the sky simultaneously thus directly gives the distance to the Galactic Centre (see also Figure 5.5). The derived distance of 7.94 ± 0.42 kpc was not only the most accurate primary distance measurement in 2003 but also has minimal systematic uncertainties of astrophysical origin.

5.2.5 *SINFONI in the Galactic Centre: young stars and infrared flares in the central light month*

This paper by Eisenhauer *et al.* (2005) focuses on the stellar population within the central light-month and flares from the black hole (see below), but also gives an update on the mass and distance to the Galactic Centre. The paper is based on the first adaptive optics-assisted integral field spectroscopy observations of the Galactic Centre with SINFONI (Eisenhauer *et al.*, 2003b; Bonnet *et al.*, 2004), which allowed measurement of the radial velocities of about a dozen stars in the central arcsecond, and to add another epoch to the S2 orbit (Figure 5.5). The updated estimate of the distance to the Galactic Centre from the S2 orbit fit is 7.62 ± 0.32 kpc, resulting in a central mass value of $3.61 \pm 0.32 \times 10^6\ M_\odot$ (Figure 5.4).

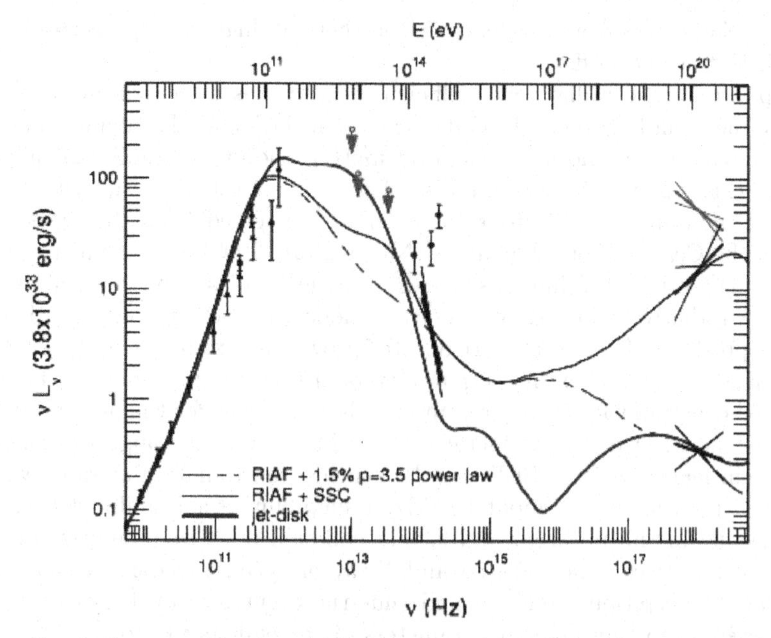

FIGURE 5.6. Spectral energy distribution (SED) of the Galactic Centre. The measurements of the slope at near-infrared frequencies strengthen the radiative accretion models and as such provide further evidence for an event horizon (courtesy of Eisenhauer *et al.*, 2005).

Excursion: Srg A does not move*

As outlined in the line of arguments in Figure 5.2, the stellar orbits measure the enclosed mass within a few thousand Schwarzschild radii (R_S). The evidence that this mass is indeed enclosed within a few ten R_S comes from radio observations: the argument is based on two findings; first, that the size of the radio source measured at mm radio wavelengths is less than about $10\,R_S$; and, second, that the radio source is indeed the massive central object. The evidence for the latter is that the radio source does not move. Very large baseline interferometry gives an upper limit of $0.5\,\mathrm{km\,s^{-1}}$ perpendicular to Galactic plane. Such low velocities can be expected only if the radio source cannot be kicked around by encounters with stars, i.e. is much, much heavier than stars and does indeed carry the mass as derived from the stellar orbits (Reid and Brunnthaler, 2004).

Excursion: hot spots on the last stable orbit

The evidence that the compact supermassive source is indeed as small as a few R_S comes from observations of infrared flares, which show a modulation with a period of approximately 17 min (Genzel *et al.*, 2003b). If this modulation reflects the orbital motion of the emitting region, the short period can only be explained if the 'hot spot' moves close to the last stable orbit, which is at $3\,R_S$ for a non-rotating black hole.

5.2.6 *Again: SINFONI in the Galactic Centre: young stars and infrared flares in the central light month*

This paper by Eisenhauer *et al.* (2005) was already mentioned for the mass and distance determination. The authors also report the first measurements of the slope of the spectral energy distribution (SED) in the infrared during a flare. They find that the SED at 1.7–2.45 μm is fitted by a featureless 'red' power law $S_\upsilon \propto \upsilon^{-\alpha'}$ of spectral index -4 ± 1 (Figure 5.6). The observed spectral slope strengthens synchrotron models in which the infrared emission comes from accelerated, non-thermal, high-energy electrons in a radiatively inefficient accretion flow.

This support for the radiatively inefficient accretion is a step towards 'proving' the existence of the event horizon of the Galactic Centre black hole. The main evidence for the presence of such an event horizon – i.e. that matter, energy and radiation cannot escape from inside this region – is the faintness of the Galactic Centre supermassive object. It is several thousand times fainter than what is expected from the mass–luminosity relation for the Centre of other galaxies (Nagar *et al.*, 2005) and about a million times smaller than the Eddington luminosity (see excursion). This very low luminosity is best explained by radiatively inefficient accretion models (see excursion), which inherently rely on the presence of an event horizon. To 'prove' the existence of the event horizon the observations must constrain the parameters of the accretion models. The strongest observational constraint for the accretion models – and thus for the presence of an event horizon – is the SED of the Galactic Centre. In the energy range from radio to X-rays, the main features of the SED are the synchrotron radiation at radio wavelengths rising with a power law up to about 10^{12} Hz, a maximum at sub-millimetre wavelengths, a strong decline in the infrared, and strong emission at X-ray energies during flares. The connection between the infrared and X-ray emission is particularly important in constraining the accretion models – and thus the event horizon hypothesis – because it is very sensitive to the physical parameters of the emitting region, e.g. size, density, temperature and magnetic field strength. Coordinated multi-wavelength observations and infrared integral field spectroscopic observations are targeting this topic.

Excursion: the Eddington luminosity

The Eddington limit is a simple relation for the maximum luminosity of a source with respect to its mass. It is the limit where the radiation force on the particles in the emitting region is equal to the gravitational force. While it does not directly apply to emission from accretion on to a black hole, it is nevertheless often used to characterize the emission from active galactic nuclei and black holes. In a plasma of electrons and hydrogen atoms, the radiation force, F_R, is dominated by the Thompson scattering of the photons with the electrons:

$$F_R = \frac{\sigma_T F}{c} \tag{5.11}$$

$$\sigma_T = \frac{8\pi}{3}\left(\frac{e^2}{m_e c^2}\right)^2 \tag{5.12}$$

$$F = \frac{L}{4\pi r^2} \tag{5.13}$$

where σ_T is the Thompson cross-section for electrons (mass m_e, charge e, speed of light c) and F is the radiation flux at the distance r for given luminosity L.

The gravitational force on the plasma from the central black hole is dominated by the much more massive protons:

$$F_G = \frac{GMm_P}{r^2}, \tag{5.14}$$

where G is the gravitational constant, m_P the mass of a proton and M the mass of the black hole. To maintain a stationary accretion flow in the plasma, gravitation must dominate over the radiation pressure ($F_G > F_R$). This leads to the Eddington limit:

$$L < L_{Eddington} = \frac{4\pi G m_P c}{\sigma_T} = 1.3 \times 10^{38} \frac{\mathrm{M}}{\mathrm{M}_\odot} \,\mathrm{erg\,s^{-1}}. \tag{5.15}$$

Excursion: radiatively inefficient accretion

There are several models to explain the faintness of the Galactic Centre black hole. A recent summary of the theoretical concepts can be found, for example, in Narayan (2002). In essence, two ingredients are necessary to reduce the flux by a factor of a thousand below what is expected from the mass–luminosity relation of active galactic nuclei. First, one needs a low-density accretion flow that can develop a two-temperature plasma. Here, the protons are at a low enough temperature ($\approx 10^{10}$ K) in order not to contribute significantly to the synchrotron radiation, which is dominated by the higher temperature ($\approx 10^{12}$ K) electrons. As the protons are much heavier than the electrons, this two-temperature, low-density plasma can effectively transfer most of the mass down to the black hole at low luminosity, and by passing the event horizon permanently trap the gravitational energy.

5.3 The stellar population in the Galactic Centre

This section is about the stellar population in the Galactic Centre. The subject is closely related to the black hole, not only because the stars are the 'test particles' for deriving its mass, but also because their formation and evolution is strongly linked to black hole feeding and growth. Indeed, many papers have already been introduced in the previous section. This section now focuses on the stellar population in the central parsec, the 'paradox of youth' and the stellar dynamics.

Because of the high extinction, deep imaging is limited to near-infrared wavelengths, and not much colour information can be retrieved for the stars. In addition, the highly variable extinction makes it very difficult to interpret the colour–magnitude diagram. Photometric studies cannot even distinguish early- and late-type stars. The need for spectroscopy was obvious. However, the region is so crowded that classical long-slit spectroscopy turned out to be impractical in most cases. In fact the Galactic Centre was a major scientific driver to build the first near-infrared integral field spectrograph MPE 3D (Weitzel *et al.*, 1996) and the first adaptive optics-assisted integral field spectrograph SINFONI (Bonnet *et al.*, 2004; Eisenhauer *et al.*, 2003b) at a 10 m-class telescope.

5.3.1 *The nuclear cluster of the Milky Way: star formation and velocity dispersion in the central 0.5 parsec*

This paper, by Krabbe *et al.* (1995), showed that the central parsec of the Galaxy is powered by a cluster of about two dozen luminous hot stars.

The authors could identify two main groups of emission line stars and model their spectra with non-LTE (local thermal equilibrium) stellar atmospheres with winds: the first group are the so-called 'He-stars', which show strong emission in He I and Brγ (see also Figure 5.3). These stars are classified as WN9 (late nitrogen-rich Wolf–Rayet stars)/ Ofpe (O stars with peculiar emission lines) stars with temperatures of 20,000–30,000 K. The emission lines show P Cygni profiles, indicating winds with a mass loss rate of $> 10^{-5}\,\mathrm{M}_\odot\,\mathrm{yr}^{-1}$. Several stars are found to have a very large He I/Brγ ratio and seem to be pure He-stars. The second group of emission lines stars are of type WC (carbon rich Wolf–Rayet stars) with temperatures $\gtrsim$ 30,000 K. These stars show C III and N III emission lines with a width of several 1000 km s^{-1}, indicating very strong winds. The progenitor stars must have had zero age main sequence (ZAMS) masses up to about $100\,M_\odot$. The hot star cluster can fully account for bolometric and ionizing luminosity of central parsec of the Galaxy.

The authors could also identify more than a dozen late-type stars by the strong CO band-head absorption in their spectra. They found four supergiants with spectral type K/M. These stars have K-band luminosities 10,000 times larger than the Sun. There are also more than 10 late-type giants within the central 8″, maybe so-called asymptotic giant branch (AGB) stars.

Putting together the findings from their integral field spectroscopy, the authors can model the stellar population and star-forming history (see excursion on stellar population modelling). The following observations have to be reproduced by the model: $L_{\rm bol} \approx 10^7\,{\rm L}_\odot$, $\frac{L_{\rm bol}}{L_{\rm Lyc}} \approx 10$, $\frac{L_K}{L_{\rm Lyc}} < 0.1$, $N({\rm OB} + {\rm WR/HeI}$ with $L_{\rm bol} > 5.5) \approx 3$, $\frac{N({\rm KMI})}{N({\rm OB+WR/HeI})} \approx 0.1$ and $\frac{N({\rm OB})}{N({\rm WR/HeI})} < 0.1$.

Even without detailed modelling, some scenarios are obviously excluded; for example, continuous star formation – i.e. constant star formation over more than Gyrs – cannot account for the high number of Wolf–Rayet stars. If one uses our Galaxy as a template for continuous star formation – where we observe only of the order 100 Wolf–Rayet stars for a total stellar mass of $2 \times 10^{11}\ {\rm M}_\odot$ – one would expect no (< 0.001) WR star in the central 10″ with its $< 10^7\,{\rm M}_\odot$, but several such stars are observed. Also, constant star formation – i.e. star formation that started some time ago and has continued at a constant rate since then – can be excluded: the large $\frac{L_{\rm bol}}{L_{\rm Lyc}} > 10$ would require that the star formation has started more than 10^7 years ago. This would imply that the first late-type supergiants had formed, so that $\frac{N({\rm KMI})}{N({\rm OB+WR/HeI})} > 1$. However, the observed ratio is less than 0.1.

The best-fitting model is a small starburst approximately 7 ± 1 Myr ago, with a decay time of 3–4 Myr. About $10^{3.5}$ stars have formed in this small starburst. In this scenario the Galactic Centre stars are presently in a short-lived, post main sequence wind phase.

The presence of the intermediate mass late-type stars cannot be explained by that scenario and indicates another star formation event about 10^8 years ago. Another problem with the small starburst model is that it predicts about 15 early-type (OB) main sequence stars, which have not been found in the data of Krabbe *et al.* (1995). However, this non-detection is solely based on the sensitivity limits at that time, and these stars have been found with the next generation integral field spectrograph SINFONI at the VLT.

Excursion: stellar population modelling

Several computer programs are available to simulate the integrated properties of stellar populations, for example 'Starburst99' (Leitherer *et al.*, 1999). In these programs the stellar population is parameterized by the stellar initial mass function (IMF, typically described by a broken power law with upper and lower mass cut-off), star formation history (e.g. the age of the burst and its decay time) and stellar evolution and atmospheres (e.g. depending on metallicity). As a result the programs return, for example, luminosities, number counts and line ratios, which can then be compared to observations.

5.3.2 *The dark mass concentration in the central parsec of the Milky Way*

In this second paper, Genzel *et al.* (1996) also analyse in detail the spatial distribution and dynamics of the central star cluster. In particular, they find that the early-type stars (He stars, WR/Of stars) are strongly concentrated within a radius of 12″, and that these stars show a large and coherent counter-rotation. This strongly supports their origin in a well-defined epoch of star formation (see above), presumably when a dense cloud fell into the central parsec less than 10 Myr ago. Apart from the early-type stars, the late-type stars (supergiants and bright AGB stars) form a ring at 4–10″. The authors find that the most likely explanation for the absence of the largest red stars (with a stellar radius of ≈ 50–$900\,R_\odot$) in the $\approx 5''$ central 'hole' is their destruction by collisions with main sequence, solar-type stars.

Excursion: The paradox of youth

The large number of young (7 Myr), massive ($> 10\,M_\odot$) stars identified by Krabbe *et al.* (1995) and Genzel *et al.* (1996) so close (< 1 pc) to the central black hole is puzzling, because all 'simple' explanations seem to fail. This puzzle got even more severe when Ghez *et al.* (2003) could spectroscopically show that the star S2 orbiting the central black hole with a semi-major axis of only ≈ 4 mpc is an early-type star, which led them to introduce the term 'paradox of youth'. The reason why 'these young, massive stars have no right to exist' is that the strong tidal field from the central black hole prevents in situ formation, and that the life-time of the massive stars is too short to bring them from further out at a high enough rate. In somewhat more detail:

The in situ formation of stars from molecular clouds at a distance of 10″ would require gas densities of $> 10^{9.5}$ hydrogen atoms cm^{-3}. However, the densest clouds within 30–60″ have a factor 1000 smaller densities ($\approx 10^6$ hydrogen atoms cm^{-3}).

In the migration scenario, where the stars form further out, dynamical relaxation can slowly bring the stars to the centre. The relaxation timescale for the Galactic Centre depends on the mass, M, of the stars and is of the order $\frac{1500\,\mathrm{Myr}}{M[\mathrm{M}_\odot]}$. For comparison, the main sequence lifetime of a star is of the order $\frac{20{,}000\,\mathrm{Myr}}{M[\mathrm{M}_\odot]^2}$. In order to get the stars down to the Galactic Centre, the life time has to be larger than the relaxation time, which is fulfilled only for stars $< 1.3\,M_\odot$. However, the observed stars are much more massive.

Another scenario is the infall of a young cluster, which undergoes core collapse and loses its outer stars by dynamical friction close to the black hole. The timescale for dynamical friction is inversely proportional to the cluster mass, and can therefore be as short as few Myrs. The required cluster mass is of the order $> 10^5\,M_\odot$ at a distance of 5 pc. To avoid tidal disruption before the cluster can core-collapse, the density must be of the order $10^7\,M_\odot/\mathrm{pc}^3$ and its diameter < 0.4 pc. But there are at present no such clusters nearby. The most massive clusters Arches and Quintuplet are at about 30 pc and have only few $10^4\,M_\odot$.

Another scenario proposed is the collision of lower-mass stars, which would rejuvenate the stars' atmospheres and produce 'super-blue stragglers'. The collision timescale is of the order $1000\,\mathrm{Myr} \times \frac{R['']^2}{M[\mathrm{M}_\odot]}$. For the 10 Myr young stars this scenario could thus only be plausible for the central 0.1″, but these stars are observed up to 10″.

The rotation of the early-type stars discovered by Genzel *et al.* (1996) gives a hint on how to resolve the 'paradox of youth', supporting star formation from dense molecular clouds falling into the Galactic Centre. This scenario is supported by the fact that the stars are counter-rotating the Galaxy, because counter-rotating gas streams have a higher probability of colliding with 'normal' rotating clouds, so that they can lose their angular momentum and sink to the centre.

5.3.3 *Stellar dynamics in the Galactic Centre: proper motions and anisotropy*

As described in the discussion on the black hole mass, Genzel *et al.* (2000) find that many of the emission line stars at 1–10″ radius not only show rotation in the radial velocity but also rotate clockwise on the sky. This supports even more strongly the notion that the rotation of the young star cluster may be a remnant of the original angular momentum

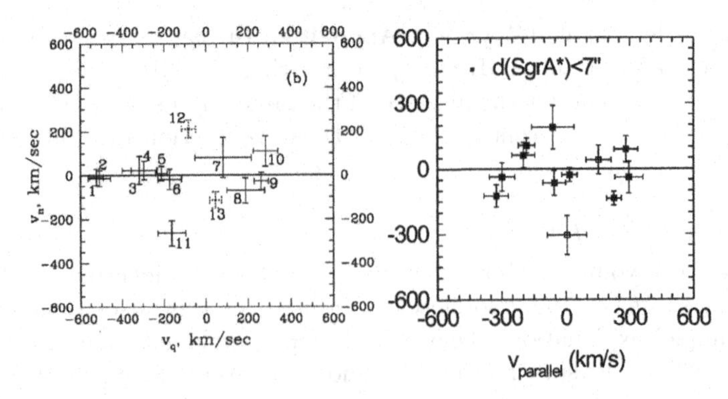

FIGURE 5.7. Two stellar discs in the Galactic Centre. Left: measured velocities in the clockwise disc and perpendicular to it (courtesy of Levin and Beloborodov, 2003). Right: the same for the counter-rotating disc (courtesy of Genzel *et al.*, 2003a).

pattern in the interstellar cloud from which these stars were formed. The authors also report the first proper motions of the 'S-stars' within the central 1″. The absence of CO emission in the integrated spectrum of that region indicates that these S-stars are also early-type stars. If they are on the main sequence, they would be of type B0 to B2. As most of the S-star cluster members are also on clockwise orbits, the authors propose that the S-stars are those members of the early-type cluster that happen to have low angular momentum and can thus plunge into the immediate vicinity of Sgr A*.

5.3.4 *Stellar disk in the Galactic Centre: a remnant of a dense accretion disk? and the stellar cusp around the supermassive black hole in the Galactic Centre*

Levin and Beloborodov (2003) and Genzel *et al.* (2003a) were the first to interpret the observed rotation of the early-type stars in the central 10″ as a thin stellar disc.

Levin and Beloborodov (2003) find that out of 13 stars whose three-dimensional velocities have been measured by Genzel *et al.* (2000), 10 lie in a thin disc. If the stars are rotating in a common disc, the velocity vectors v_i of all stars must be perpendicular to the normal vector, n, of the disc. The authors thus search for the vector n that minimizes

$$\chi^2 = \frac{1}{N-1} \sum_{i=1}^{N} \frac{(n \cdot v_i)^2}{(n_x \sigma_{xi})^2 + (n_y \sigma_{yi})^2 + (n_z \sigma_{zi})^2}, \tag{5.16}$$

where $n_{x,y,z}$ are the components of the normal vector and $\sigma_{xi,yi,zi}$ the errors in the measured velocity components of the stars. N is the number of stars. Figure 5.7 shows the velocities of the stars parallel and perpendicular to the disc for the best-fitting orientation. Simulating an artificial sample of flared discs, the authors deduce that the half-opening angle of the disc is consistent with zero within the measurement errors, and does not exceed 10°.

Levin and Beloborodov (2003) propose that a recent burst of star formation has occurred in a dense gaseous disc around Sgr A*. In such a self-gravitating disc the particle density is at least four orders of magnitude larger than that of the densest molecular clouds in the Galaxy, enough to overcome the tidal forces to start fragmentation. The initial characteristic mass (Jeans mass) is accordingly high, about 1.8 $M_\odot$. Therefore, star formation in a disc could naturally explain the observed dominance of high-mass stars.

Taking advantage of the adaptive optics at the VLT, Genzel *et al.* (2003a) could significantly extend the sample of stars with proper motions. Following Levin and Beloborodov (2003), Genzel *et al.* (2003a) find a second disc of stars (Figure 5.7). The two discs are

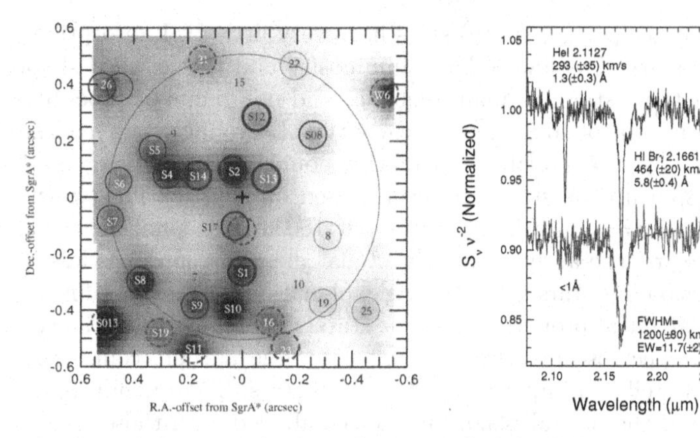

FIGURE 5.8. The 'S-star' cluster in the central arcsecond of the Galactic Centre. Left: integrated SINFONI K-band (1.95–2.45 μm datacube with spectral classification of the S-stars: solid circles indicate early type stars, dashed circles indicate late-type stars). Right: average spectrum of the bright ($m_K < 15$, top) and faint ($m_K = 15 - 16$, bottom) early-type stars (courtesy of Eisenhauer *et al.*, 2005).

inclined at large angles and counter-rotate with respect to each other. The authors also find that the proper motions of the many more stars without radial velocities are not randomly orientated but group at maximum absolute projected angular moment along the line of sight:

$$\frac{J_z}{J_{z,max}} = \frac{xv_y - yv_x}{\sqrt{(x^2 + y^2)(v_x^2 + v_y^2)}} \tag{5.17}$$

where x, y and v_x, v_y are the sky coordinates and velocity components. $J_z/J_{z,max} = +1, 0, -1$ for clockwise tangential, radial, counter-clockwise tangential orbits. The grouping of the stars in absolute projected angular moment fits the two discs well.

Based on the spectra from integral field spectroscopy (see above), the stellar content of the two discs is essentially the same, indicating that they formed at the same time. Although other scenarios cannot yet be fully excluded, the 'paradox of youth' on a parsec scale may be solved: the authors conclude that of the possible formation scenarios for these massive stars the most probable one is that 5–8 Myr ago two clouds fell into the centre, collided, were shock-compressed and then formed two rotating (accretion) discs, which then fragmented and formed the stars.

5.3.5 *SINFONI in the Galactic Centre: young stars and infrared flares in the central light month*

This paper by Eisenhauer *et al.* (2005) reports on 75 mas resolution, NIR imaging spectroscopy within the central arcsecond of the Galactic Centre. The authors find that to a limiting magnitude of $K \approx 16$, 9 out of 10 stars in the central 0.4″, and 13 out of 17 stars out to 0.7″ from the central black hole have spectral properties of B0–B9 main sequence stars (Figure 5.8). The spectral classification is based on the detection of the Brγ and He I lines. Different from the emission line stars in the stellar discs at 1–10″, the S-stars show Brγ in absorption, which is characteristic of B-type stars. Another indication is the lack of the typical CO band-head absorption of late-type stars (see Figure 5.3). Averaging the spectra of the brighter and fainter stars, the signal-to-noise ratio is large enough even to distinguish the spectral sub-type from the equivalent width (EW) of the He I line. The bright stars ($m_K < 15$) show strong (EW > 1 Å) absorption, typical for B0–B3 stars, while the fainter ($m_K = 15$–16) stars have very little He I absorption (EW < 1 Å), typical of B5–B9 stars. The low K-band luminosity and the detailed shape of the Brγ line indicate that the S-stars are main sequence and not giants or supergiants.

The fact that nine out of 10 S-stars are early-type (= young) stars makes the 'paradox of youth' even stronger. A possible explanation could be stellar mergers. Such mergers would probably spin up the stars. However, based on the He I line width, all brighter S-stars have normal rotation velocities of about $150\,\mathrm{km\,s^{-1}}$, similar to solar neighbourhood stars. Therefore, the authors rate the stellar merger scenario as less likely.

As for the emission line stars at 1–10″, the explanation for the 'paradox of youth' could come from the dynamical properties of these stars. Combining the radial velocities from the integral field spectroscopy with imaging data, the authors could derive improved three-dimensional stellar orbits for six of these S-stars. The orientations of the orbits appear random and the orbital planes are not co-aligned with those of the two discs of massive young stars 1–10″ from Sgr A*. The authors thus exclude the hypothesis that the S-stars as a group inhabit the inner regions of these discs.

But have the S-stars ever been in a disc? One mechanism to change the direction of the angular momentum is Lens–Thirring precession from the rotating black hole. The rotating space time breaks the spherical symmetry, which would normally preserve the angular momentum. The period of this Lens–Thirring precession is (e.g. Levin and Beloborodov, 2003):

$$P_{LT} = 5.8 \times 10^6 \; a^{-1} \left(\frac{3 \times 10^6 \mathrm{M}_\odot}{M}\right)^2 \left(\frac{1-e^2}{1-0.87^2}\right)^{3/2} \left(\frac{r_0}{4.6 \times 10^{-3}\ \mathrm{pc}}\right)^3 \mathrm{yr}, \qquad (5.18)$$

where a is the dimensionless spin parameter ($a = 1$ for maximum spin), M the mass of the central black hole, and r_0 and e the semi-major axis and eccentricity of the orbit. For a spin parameter of $a = 0.52$ (Genzel *et al.*, 2003b), however, the lifetime (10 Myr) of two out of the six B-stars is too short for Lens–Thirring precession to account for the change in angular momentum. Therefore, other mechanisms are necessary to explain the S-stars (see also the excursion on the 'paradox of youth').

One potential explanation is a dissolving star cluster with an intermediate mass black hole (IMBH), possibly formed by runaway merger in the cluster core. But it is not clear if an IMBH can have such a tightly bound cluster, and the IMBH would have to be two orders of magnitude more massive than can be explained by formation from a runaway merger. Another explanation might be a massive binary exchange, where the S-stars were originally in a massive ($\approx 100\,\mathrm{M}_\odot$) binary system and switched partner in a close interaction with the black hole. However, there is no evidence of such massive WR/O stars on highly eccentric orbits, and more massive stars should also have been captured, which is not observed. The last scenario discussed here is exchange capture with stellar black holes, in which the B-stars play billiards with equal mass stellar black holes in a central cusp. The advantage of this scenario is that it would favour the observed eccentric orbits and would prefer B-stars because they best match the mass of the about $10\,\mathrm{M}_\odot$ stellar black holes. This scenario is also supported by X-ray observations indicating the presence of about 10^4 neutron stars and stellar black holes in the Galactic Centre region. However, this scenario requires a large reservoir of B-stars at larger radii, which are not found.

In summary, the 'paradox of youth' for the S-stars has not yet been solved, although the observations indicate that these stars are probably brought into the central light-month by strong individual scattering events.

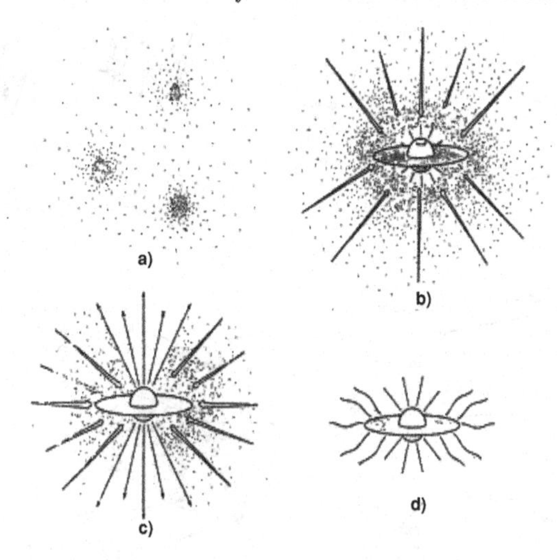

FIGURE 5.9. The four stages of star formation: a) cores form within molecular clouds; b) protostar with surrounding nebular discs; c) stellar wind creating bipolar outflow; d) newly formed star with circumstellar disc (courtesy of Shu *et al.*, 1987).

5.4 Star formation

This section is about integral field spectroscopic studies related to star formation. A recent review of this topic can be found in García (2006). Compared to Galactic Centre observations, the impact of integral field spectroscopy on star formation research has been much smaller. A possible reason might be the wide variety of objects and questions so that the early integral field spectroscopic studies are less coherent and less straightforward. Therefore, the papers presented in this section are not ordered by publication date but are grouped by topic. Some papers (for example on planetary nebulae) have been omitted to present the subject in a more coherent way. After a brief introduction to theory of star formation, this section discusses integral field spectroscopic observations of T Tauri stars and their immediate vicinity, their jets and the emerging Herbig–Haro objects. This is followed by a discussion of studies related to the formation of high mass stars, and finally on the interaction between the winds and radiation of high mass stars with nearby protoplanetary discs.

Excursion: star formation in a nutshell

This section introduces the basic concepts of star formation theory. Many good lectures are available on the internet for download, for example from Cornelis Dullemond.[1] There are also a number of good review articles, for example Shu *et al.* (1987). Specifically recommended for further reading are the conference proceedings 'Protostars and Planets I–IV' published by The University of Arizona Press. The different phases of star formation are illustrated in Figures 5.9 and 5.10.

The first phase is the collapse of a molecular cloud (Figure 5.9a). Collapse occurs if the mass and density are large enough to overcome the thermal and magnetic support of the cloud. The so-called Jeans mass, M_J, defines the limit for a thermally supported cloud:

$$M_J \propto T^{\frac{3}{2}} R^{-\frac{1}{2}}. \tag{5.19}$$

[1] http://www.mpia-hd.mpg.de/homes/dullemon/lectures/starplanet/

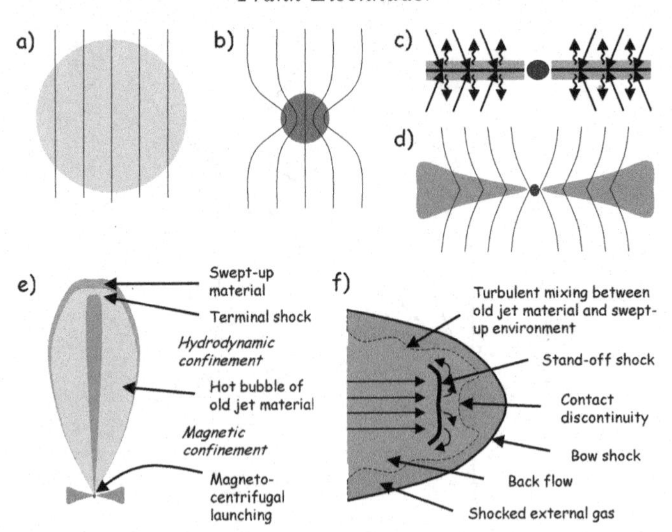

FIGURE 5.10. The formation of discs and outflows: a,b) gravitational collapse of a magnetically supported molecular cloud; c) formation of an accretion disc; d) magnetically threaded disc; e) bipolar outflow; f) head of the jet (courtesy of Cornelis Dullemond).

The critical mass for magnetic support is:

$$M_B \propto BR^2, \tag{5.20}$$

where T is the temperature of the cloud, R the radius and B the magnetic field. The major barrier to continued collapse is that the magnetic field is frozen in the molecular cloud, even if the fraction of ionized gas is very small (Figures 5.10a and 5.10b). The contraction also increases the magnetic field, and the collapse would stop as soon as the critical mass, M_B, is reached. This barrier is overcome by ambipolar diffusion, the slow decoupling of the neutral molecules and dust particles from the electrons and ions frozen in the magnetic field.

In the next phase, a circumstellar disc is formed. As the cloud rotates, infalling matter collides with matter from the other side, forms shocks and converts kinetic energy into heat (Figure 5.10c). This heat is radiated away, so that the matter cools and settles on the mid-plane.

Because of friction, the matter in the disc slowly loses its angular momentum and moves inwards. The plasma in the disc pulls the magnetic field with it, forming a magnetically threaded disc (Figure 5.10d). The magnetic field creates an effective gravitational potential along the field line, and as soon as the angle between the field lines and the disc gets below 60°, matter will start to outflow along the magnetic field, like in a slingshot. These are the so-called 'disc winds'. Further in – but still at a distance significantly larger than the stellar radius – the interaction of accreting mass with the magnetosphere of the protostar truncates the gas disc. At that radius, R_x, the disc accretion divides into a funnel inflow on to the star, and an outflow called the 'x-wind'.

As the gas wants to preserve its angular momentum, it will bend and wind the field lines, thus confining the outflow. This bipolar outflow (Figure 5.10e) is surrounded by a bubble of hot gas. A terminal shock forms at the head of the jet (Figure 5.10f). Most of the energy is dissipated in a stand-off shock at the boundary between the jet and the external medium. This external matter is pushed away and forms a bow shock. The shocks in the ambient medium are observed as Herbig–Haro objects at distances up to thousands of astronomical units. The jet is flanked by a region with turbulent mixing between old jet material and the entrained environment. As the jet material is much more tenuous than the external medium, the propagation of the bow shock is much slower than the jet.

The protostar in the centre is a self-gravitating ball of hot gas continuously being fed by accreting matter. Initially, there is no nuclear energy production, but the star is heated from the gravitational energy released in the accretion and contraction (Figure 5.9b). The heating is countered by cooling from radiation, and the star contracts on the Kelvin–Helmholz timescale until the central density and temperature are high enough to start deuterium burning at $T \gtrsim 10^6$ K. At this stage, the star enters it pre-main sequence phase (Figure 5.9d). The star becomes fully convective, and inflates again up to 3–5 $R_{\odot}$. In a second, slow contraction the central temperature and density then increase until hydrogen burning starts, and the star becomes a main sequence star.

Excursion: T Tauri and RW Auriga stars

The so-called T Tauri and RW Auriga stars are examples of low mass pre-main sequence (PMS) stars. They are named after their prototypes, and are classified from the following observational characteristics. The brightness is irregularly variable, typically with variations of 1 mag within days. The stars also show strong emission lines from ionized H, Ca and other elements. Their spectral types are in the range F–K (indicating low-mass stars), and they are located above the main sequence in the Hertzsprung–Russell diagram. These stars are found in star-forming regions, indicative of their young age.

A major problem for the interpretation of the characteristic signatures is that these signatures arise from multiple regions, including the stellar photosphere, circumstellar discs, jets and shocks. In addition, many of these stars live in multiple systems so that it is difficult to disentangle the components. This is particularly true for the 'prototype', T Tau.

5.4.1 *A near-infrared spectral imaging study of T Tau*

This paper by Herbst *et al.* (1996) presents K (2.01–2.42 μm) and H (1.5–1.8 μm) band imaging spectroscopy of the complex environment of the young stellar object T Tau. T Tau is a pre-main-sequence binary system, with an optically visible primary star, and a highly obscured infrared component, separated by approximately 0.6″ (see Figure 5.12). The observations by Herbst *et al.* (1996) reveal at least five distinct sources of molecular hydrogen emission in the central 4″ × 4″ of the system: diffuse knots on primary (N) and infrared (S) components, a bright NW knot (T Tau NW) with a SE counterpart, and an east–west jet. The ratio of various H_2 lines (see excursion on H_2 line diagnostics) indicates that all components are shock-heated.

The strongest source in molecular hydrogen emission is T Tau NW. It is detected in 11 different lines. The spectrum is typical for a Herbig–Haro Object (Figure 5.11a). The H_2 spectral-line ratios point to shock heating with temperatures in the range 1000–2500 K (Figure 5.11b). However, T Tau NW is also detected in two H-band forbidden transitions of [Fe II] (square brackets indicate forbidden lines). Normally, a shock that produces [Fe II] quickly destroys H_2. This implies that we see a range of excitation conditions. The authors also find a slight offset between H_2 and [Fe II], indicating that [Fe II] is coming from the tip of the shock where the energy is dissipated, and that the H_2 emission is flanking the shock. An apparent SE counterpart supports the Herbig–Haro hypothesis. The authors further detect an H_2 filament towards the west, which appears to be a signature of a second outflow and its counter-jet to the east. The near perpendicularity

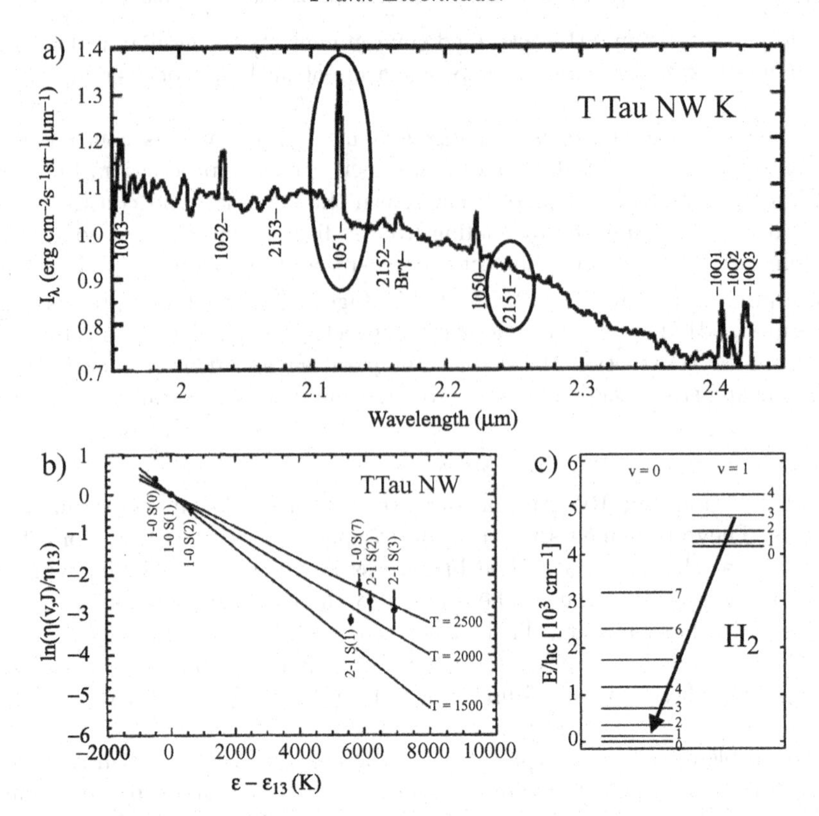

FIGURE 5.11. Emission from molecular hydrogen in T Tau: a) K-band spectrum of T Tau NW. The flux ratio of $H_2 \nu = 1-0$ S(1) and $H_2 \nu = 2-1$ S(1) (highlighted) give a simple first estimate for the excitation mechanism; b) excitation diagram (both figures courtesy of Herbst *et al.*, 1996); c) level diagram with H_2 $\nu = 1-0$ S(1) transition.

of T Tau NW and the west filament suggests that one star of the binary system produces each outflow. Indirect evidence suggests that T Tau NW is associated with the primary star, and the west filament with the infrared companion. The thermal, diffuse H_2 emission observed at the location of the stars could be caused either by collimation of the NW and W outflows or by accretion shocks, which would also explain the large extinction and infrared luminosity of the companion.

Excursion: molecular hydrogen lines in the infrared

The vibrational and rotational excitations of a molecule determine its spectrum in the infrared. The electronic excitations have much higher energies and are only observed in the visible or at even shorter wavelengths. The energy levels for vibrational and rotational excitations are:

$$E_{vib} = h\nu_0 \left(\nu + \frac{1}{2} \right) \tag{5.21}$$

$$E_{rot} = \frac{h^2}{8\pi^2 I} J\,(J+1)\,, \tag{5.22}$$

where ν and J are the vibrational and rotational quantum numbers, respectively, h the Planck constant, ν_0 the vibrational frequency, and I the moment of inertia. Figure 5.11c illustrates the energy levels for the hydrogen molecule.

The line emission observed in the infrared results from vibrational transitions with or without rotational changes. The nomenclature for the spectral lines gives the initial and final vibrational quantum number, ν, the change in the rotation quantum number – corresponding to a spectral branch – and the final rotation state, $H_2\,\nu = from - to$ branch (final J). The change in rotation – the branch – is coded by capital letters: for the Q-branch we have $J \rightarrow J$, for the S-branch we have $J \rightarrow J - 2$. Because H_2 is lacking a dipole moment, the R-branch ($J \rightarrow J - 1$) is forbidden in this molecule.

Excursion: H_2 line diagnostics

The Boltzmann equation gives the fractional number of particles per degenerate sub-level $\eta_{\nu,J}$ occupying a set of ro–vibrational states ν, J for thermal equilibrium:

$$\eta_{\nu,J} \equiv \frac{N(\nu,J)}{g_{\nu,J}} = \frac{N_{\rm tot}}{Z(T)}\, e^{-\epsilon\ \frac{\nu,J}{T}} \tag{5.23}$$

where $N(\nu, J)$ and $N_{\rm tot}$ are the sub-level and total number of particles, respectively, $g_{\nu,J}$ the degeneracy, and $\epsilon_{\nu,J}$ the energy of the state expressed in Kelvin.

Forming the ratio of $\eta_{\nu,J}$ with that of a reference state (η_{ν_0,J_0}, ϵ_{ν_0,J_0}) eliminates several of these quantities, giving

$$\log \frac{\eta_{\nu,J}}{\eta_{\nu_0,J_0}} = -\frac{1}{T}\left(\epsilon_{\nu,J} - \epsilon_{\nu_0,J_0}\right). \tag{5.24}$$

The corresponding log–normal plot is called an excitation diagram (see Figure 5.11b), and the slope in this diagram is inversely proportional to the temperature. It can be used to distinguish the two major excitation mechanisms: UV radiation with high excitation temperature (>2500 K), and shock heating with comparable low excitation temperature (<2500 K). A simple first estimate for the excitation mechanism can be derived from the flux ratio of $H_2\nu = 1 - 0\,S(1)$ – usually the strongest H_2 line in the K-band – and its $H_2\nu = 2 - 1\,S(1)$ counterpart for the next higher vibrational transition: a flux ratio $\frac{H_2\nu=1-0S(1)}{H_2\nu=2-1S(1)} > 5$ indicates shock heating, a flux ratio $\approx$2 indicates ultraviolet (UV) excitation.

5.4.2 *Spatially resolved imaging spectroscopy of T Tauri*

The small separation of the stars in the T Tau system has prevented spectroscopy of its individual components in the past. This paper by Kasper *et al.* (2002) presents near-infrared data (H- and K-band) on this object taken at the Calar Alto 3.5 m telescope using the ALFA adaptive optics system in combination with the MPE 3D integral field spectrograph. The diffraction-limited observations allow us for the first time to clearly resolve the spectra of the primary (N) star and the infrared (S) companion (Figure 5.12).

The spectrum of the primary T Tau N (Figure 5.12 top right) shows photospheric absorption lines from Mg, Al and Ca. The measured line ratios are consistent with spectral type K. However, the equivalent width of the lines is significantly smaller, requiring a large amount of continuum veiling, for example from the hot inner part of a circum stellar disc. This K band excess is of the order 80%.

The spectrum of the infrared component T Tau S (Figure 5.12 bottom right) is very red and does not show photospheric lines. The stellar photosphere must be extremely veiled, if not completely invisible. The Bracket line emission from T Tau S and T Tau N could not be resolved with the 3.5 m telescope, implying that it arises within 6 AU from the stars. The line ratios between Brγ (2.166 µm) and Br10 (1.736 µm) towards T Tau N and T Tau S are similar, suggesting comparable selective extinctions of $A_V \approx 6$ toward the

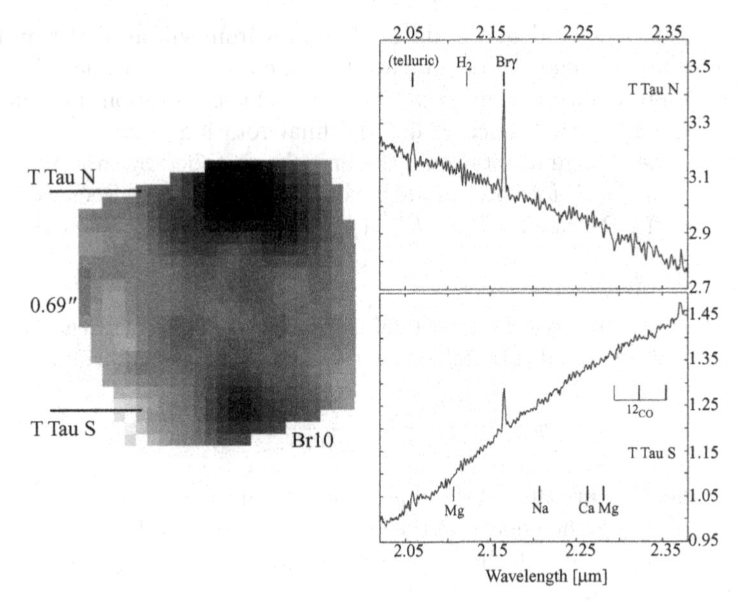

FIGURE 5.12. Integral field spectroscopy of T Tau. Left: Br10 (1.736 μm) image; right: K-band spectrum of T Tau N (top) and T Tau S (bottom) (courtesy of Kasper *et al.*, 2002).

respective emitting regions. All this indicates that T Tau S is not stellar light reddened by strong absorption from cold dust, but that the emission arises from hot dust. Therefore a protostar with an infalling envelope seems an unlikely explanation for T Tau S. On the other hand, the results are consistent with a model that describes T Tau S as a pre-main sequence star surrounded by a small edge-on disc. The disc is observed at an inclination such that it obscures the star, at the same time leaving the line of sight towards the Bracket emission from the polar regions relatively unobscured.

Excursion: microjets in pre-main sequence stars

Outflows and jets are common phenomena in pre-main sequence stars. In particular, we find so-called microjets, which are only a few hundred AU long, and which are launched at 0.1–10 AU radius. The microjets show forbidden line emission and high velocities. There are two long-standing questions on microjets:

- What is driving the jets? There are a number of competing explanations (see excursion on star formation): disc winds, x-winds and stellar winds. Unfortunately, the available magnetohydrodynamical models do not predict line ratios, and simple spectroscopy cannot answer the question. Instead, the models predict velocities, densities and magnetic field strengths. We thus have to measure quantities like diameter, density, terminal velocity and rotation speed to identify the driving mechanism of the jets.
- What is the heating mechanism? Given the small launching radius (0.1–10 AU) of the jets, dramatic adiabatic cooling should occur in the expansion region, and extended forbidden line emission should not be detectable. This is in disagreement with the observations. Several heating processes are discussed for the microjets: shocks against the environment or within the jet, turbulence in a viscous mixing layer, ambipolar diffusion heating and compression by jet instabilities.

Several integral field spectroscopic studies have been aiming at resolving these questions.

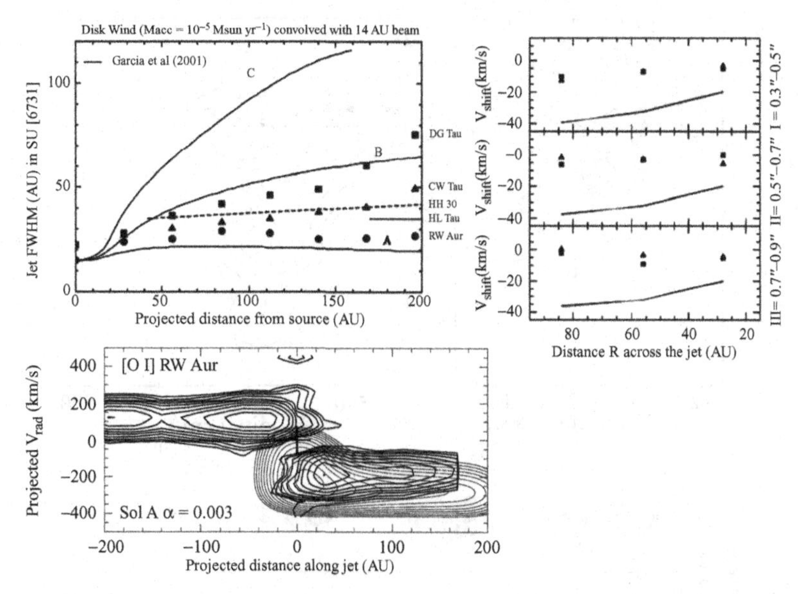

FIGURE 5.13. Comparison of the RW Aur microjet with cold disc wind models. Top left: jet full-width-at-half-maximum (FWHM) measured from the [SII]λ6731 line; right: transverse velocity shifts between symmetric positions on either side of the jet axis (both figures courtesy of Dougados *et al.*, 2003); bottom: position–velocity diagram measuring the terminal velocity (courtesy of García, 2006).

5.4.3 *The RW Aur microjet: testing MHD disc wind models*

This paper by Dougados *et al.* (2003) aims at answering the question of what is driving the jets (see excursion). Based on observations with the adaptive optics bonnette PUE'O and the OASIS (McDermid *et al.*, 2004) spectro-imager at the CFHT, the authors conduct a detailed comparison of the RW Aur microjet on scales < 200 AU with predictions from cold disc wind models.

The observed collimation and width (≈ 30 AU) of the microjet of RW Aur can well be modelled with cold disc winds with moderate to high ejection efficiencies (Figure 5.13 top left). However, the observed rotation signature – transverse velocity shifts between symmetric positions on either side of the jet axis – of $< 10\,\mathrm{km\,s^{-1}}$ is significantly below what is expected from disc wind models (Figure 5.13 right): cold disc winds, extending radially out to > 1 AU, would have transverse shifts of 30–40 $\mathrm{km\,s^{-1}}$. Cold disc wind models are also in conflict with the observed terminal velocities of the RW Aur jet (Figure 5.13 bottom left). While the observed terminal velocities are ≈ 150–$350\,\mathrm{km\,s^{-1}}$ (García, 2006), the disc wind models predict much higher velocities of $\approx 500\,\mathrm{km\,s^{-1}}$. The x-wind models, on the other hand, would predict velocities that were too low.

So the answer to the question of what is driving the jets is still pending. There is evidence in favour of and against all considered mechanisms, and more observations and modelling are required to close the topic.

5.4.4 *Sub-arcsecond morphology and kinematics of the DG Tauri jet in the [OI]λ6300 line and DG Tau: a shocking jet?*

These two papers by Lavalley *et al.* (1997) and Lavalley-Fouquet *et al.* (2000) present the first integral field observations of the DG Tauri jet. The goal is to answer the second question of what is heating the jet (see excursion).

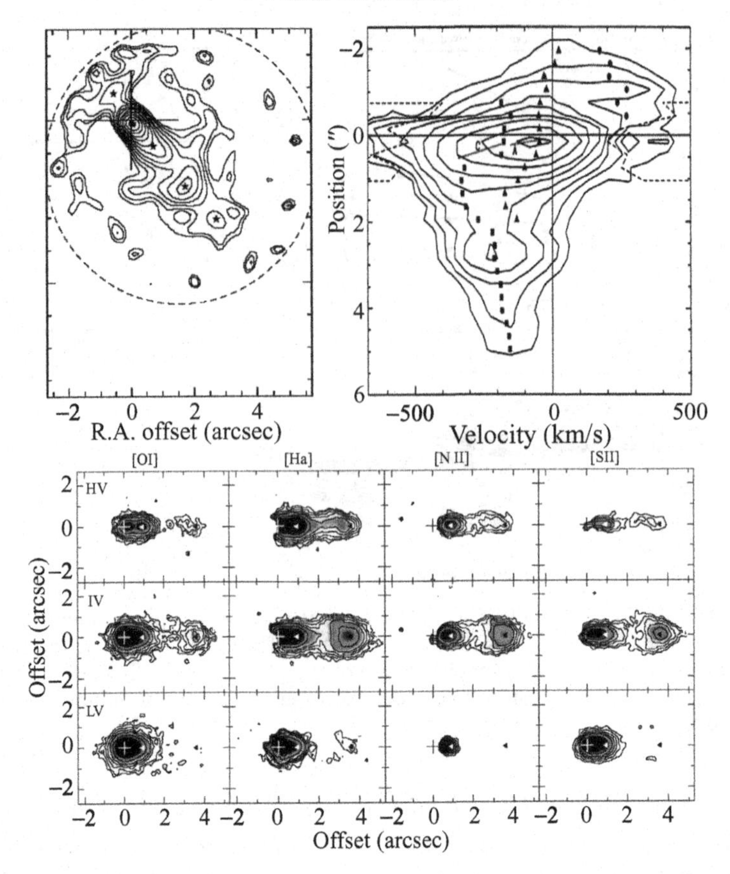

FIGURE 5.14. DG Tauri jet. Top left: deconvolved [O I]λ6300 map with an effective resolution of 0.35″; top right: position–velocity diagram along the jet axis, symbols indicating the centroid velocities of the multi-component line fit (both figures courtesy of Lavalley *et al.*, 1997); bottom: channel maps for [O I]λ6300, Hα, [N II]λ6583 and [S II]$\lambda\lambda$6716,6731 lines at [−400, −250] km s^{-1} (top), [−250, −100] km s^{-1} (middle), [−100, +10] km s^{-1} (bottom) (courtesy of Lavalley-Fouquet *et al.*, 2000).

The observations were obtained with the Tiger and OASIS instruments at the CFHT in 1994 and 1998 and cover the [O I]λ6300, Hα, [N II]λ6583 and [S II]$\lambda\lambda$6716,6731 lines (the numbers following λ indicate the wavelength in Å). The morphological structure of the mass outflow is revealed with unprecedented spatial resolution of approximately 0.35–0.5″. The deconvolved [O I] map (Figure 5.14 top left) shows an unresolved inner peak containing two thirds of the total line flux, followed by a collimated jet-like body extending out to ≈1.5″ from the star, two resolved knots at distances of 2.7″ and 4″, and a counter-jet. The unresolved inner peak is slightly displaced (≈20 AU) from the star. It is apparently stationary and possibly linked to the initial jet collimation. The size limit and absolute photometry for this peak allows Lavalley *et al.* (1997) to set new constraints on the jet mass loss rate of 1.3×10^{-8} – 6.5×10^{-6} M$_\odot$ yr^{-1}, lowering previous estimates by a factor 5 to 25. From the [O I], Hα, [N II] and [S II] channel maps at different velocities (Figure 5.14 bottom), the fast jet core appears surrounded by a slower-moving flow or cavity. The two velocity components are analysed in more detail from the position–velocity diagram (Figure 5.14 top right). A multiple component line fit for each position clearly shows the two distinct velocity components at distances less than 1.5″: the high-velocity component quickly accelerates up to 350 km s^{-1}; the slow component follows the fast component at proportionally slower velocities up to about $100\,\mathrm{km\,s^{-1}}$. Accordingly, the slow component is interpreted as circumstellar material entrained in the jet. Beyond 1.5″ (200 AU) only one velocity component is detected, and the jet starts wiggling around (Figure 5.14 top left and bottom). The knot at 2.6″ shows a curved morphology (Figure 5.14 top left). This and the transverse velocity gradient are strongly suggestive of a resolved bow shock. Radial velocity jumps of $\approx 50\,\mathrm{km\,s^{-1}}$ are observed between the bright knots at 1″ and 3.6″. In order to determine the heating process of the jet, Lavalley-Fouquet *et al.* (2000) also compare the measured [N II]λ6583/[O I]λ6300 and [S II]λ6716/[S II]λ6731 versus [S II]λ6731/[O I]λ6300 line ratios with available model predictions for shocks, ambipolar diffusion and turbulent mixing: the measurements are best reproduced by shocks with speeds of $\approx$ 50–100 $\mathrm{km\,s^{-1}}$ and pre-shock densities of $\approx 10^5$–$10^3\,\mathrm{cm^{-3}}$.

Combining the evidence – the bow shape structure of the knot, the velocity jumps at 1–4″, the wiggling beyond 1.5″, the line ratios and the variability in the ejection velocity reported elsewhere – indicates that 'shocks' is the answer to the question of what is heating the microjets.

5.4.5 *The three-dimensional structure of HH 32 from GMOS IFU spectroscopy*

In this paper Beck *et al.* (2004) present high-resolution spectroscopic observations of the Herbig–Haro object HH 32 made with the GMOS IFU (Allington-Smith *et al.*, 2002) at Gemini North Observatory.

HH 32 is one of the brightest sources in the original catalogue of Herbig and Haro. Previous studies have shown that HH 32 has a high excitation spectrum with strong emission from H, He and a wide range of metals. It also exhibits H_2 emission from shocks as the flow encounters ambient cloud material. The outflow is clearly associated with a T Tauri star. Several condensations up to several 10″ have been detected. The three-dimensional spectral data from Beck *et al.* (2004) cover a field of $\approx 9'' \times 6''$ centred on the HH 32 A knot complex. The spectral range is 4820–7040 Å. The position-dependent line profiles and radial velocity channel maps of the Hα line show a variety of spatial and spectral structures. The most prominent features are double-peaked radial velocity profiles, two arclike structures at low and high velocities and two compact components at intermediate velocity. These qualitative findings already indicate that HH 32 A is a combination of at least two bow shocks.

The authors also present various line ratio velocity channel maps from the [O III]λ5007, Hα, [O I]λ6300, [N II]λ6583 and [S II]$\lambda\lambda$6716,6731 lines. The line emission and the line ratios vary significantly on spatial scales of $\approx 1''$ and over velocities of $\approx 50\,\mathrm{km\,s^{-1}}$. At low velocities, the density tracer (see excursion on line ratios) [S II]λ6716/[S II]λ6731 is diffuse and extended. The ratio of $\approx$ 0.7–1.0 corresponds to an electron density of $\approx$ 230–$150\,\mathrm{cm^{-3}}$. The line ratio is getting more compact and is dropping at higher velocities. At velocities above $\approx 110\,\mathrm{km\,s^{-1}}$, the observed line ratio in the compact component is ≈ 0.55, corresponding to an electron density of $\approx 650\,\mathrm{cm^{-3}}$. The line ratio [N II]/Hα – which traces the excitation temperature (see excursion) – peaks in the diffuse, low-density regions. We thus see high-density knots surrounded by a highly excited, low-density plasma. As expected from a shock scenario, the line ratio [O III]/Hα – which traces the shock velocity – increases with the observed velocity in the knot.

To interpret the observations, Beck *et al.* (2004) compare the measured channel maps and line ratios with a '3/2-dimensional' bow shock model (see excursion). The model is qualitatively successful at reproducing the general features of the radial velocity channel maps, for example the transition between a single arclike feature at low radial velocities and a more concentrated and centrally peaked condensation at high radial velocities. But

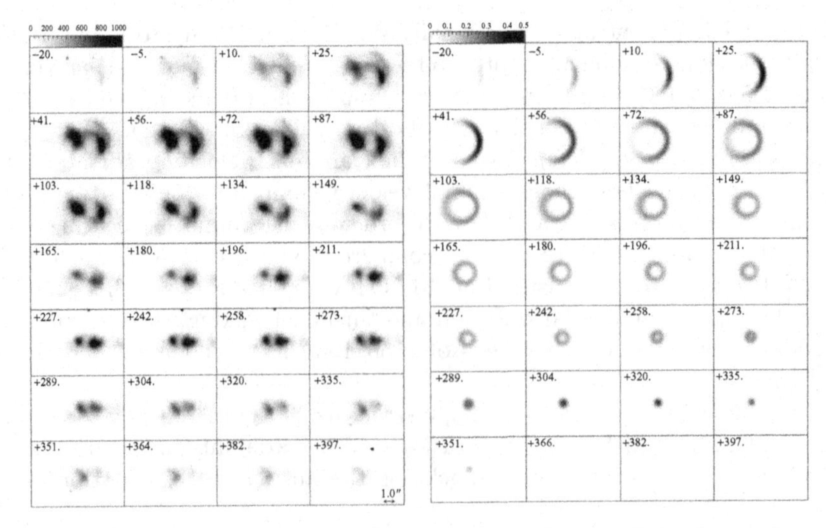

FIGURE 5.15. Herbig–Haro object HH 32 A. Left: observed Hα radial velocity channel maps; right: channel maps computed from a 3/2-dimensional bow shock model (courtesy of Beck *et al.*, 2004).

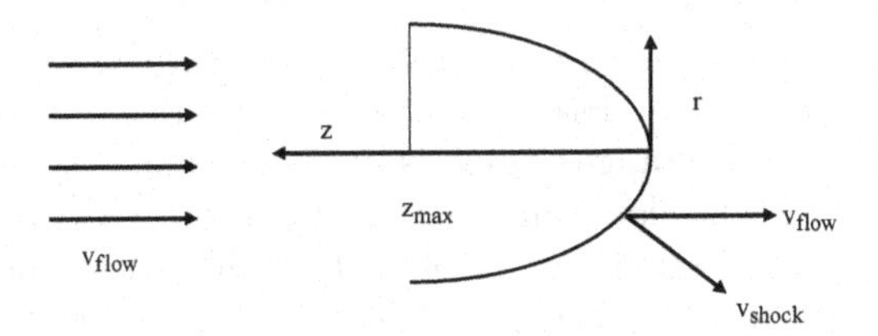

FIGURE 5.16. Geometry of a 3/2-dimensional shock model.

the model does not show the same complexity as the data, and in particular it fails to reproduce the line ratios. The reason for the failure is that the parameterization is much too simple for the spatially complex region. Another shortcoming of the 3/2-dimensional model is that it treats the shock as locally planar, i.e. by assuming that post-bow shock cooling distance is small compared to the bow shock. In HH 32 A, however, the cooling distance is comparable to the size of knots.

In summary, the integral field spectroscopic observations of HH 32 A reveal two or three bow shocks along the redshifted body of the HH 32 outflow.

Excursion: 3/2-dimensional bow shock model

The so-called '3/2-dimensional' models simplify the complex three-dimensional structure of a bow shock by assuming:

- a simple axially symmetric surface, $z \propto r^p$;
- that the local shock velocity v_{shock} is the projected pre-shock flow velocity v_{flow} perpendicular to the shock surface;
- a finite shock size described by $z_{\max}$;
- that the shock is locally described by a planar shock.

Figure 5.16 illustrates the geometry of the 3/2-dimensional model. The result from such a 3/2-dimensional model for the Herbig–Haro object HH 32 A is shown in Figure 5.15.

Excursion: line ratio diagnostics in the optical

The optical wavelength range is rich in atomic lines caused by transitions between electronically excited states. The strength of an emission line is sensitive to the physical conditions in the emitting region. Therefore line ratios provide a diagnostic tool for deriving temperature, densities and other properties. A good introduction to this topic can be found in Osterbrock (1989). Here are a few examples of diagnostic line ratios used in the analysis of stellar jets and Herbig–Haro objects:

- $\frac{[OIII]\lambda\lambda(4959+5007)}{[OIII]\lambda4363}$ and $\frac{[NII]\lambda\lambda(6548+6583)}{[NII]\lambda5754}$: [O III] and [N II] are amongst the best examples of ions that have emission lines from two different upper levels with significant different excitation energies and for which both lines are easily observable in a single spectrum. For low densities $< 10^5$ cm^{-3}, where collisional excitation is negligible, the level population follows the Boltzmann formula, and the line ratio traces the excitation temperature (larger ratio indicating lower temperature). At high densities $> 10^5$ cm^{-3} collisional de-excitation begins to play a role. Because the lower excitation state has a considerably longer lifetime – a lower radiative transition probability – it can be more quickly collisionally de-excited, thus weakening its line emission. In this case the line ratio traces the density (higher ratio indicting lower density);
- $\frac{[OII]\lambda3729}{[OII]\lambda3726}$ and $\frac{[SII]\lambda6716}{[SII]\lambda6731}$: these line ratios are examples of density tracers. Because the two lines are emitted from levels with nearly the same excitation energy, the excitation rate is almost independent of temperature. The relative strength of the optical transitions is set by the competing collisional de-excitation, which de-populates the state more efficiently with the longer lifetime, i.e. a smaller radiative transition probability (see above). Because this collisional de-excitation is strongly enhanced at higher densities, the above line ratios are tracing directly the density of the plasma. A high line ratio indicates a low electron density;
- $\frac{[NII]\lambda6583}{H\alpha}$ and $\frac{[OI]\lambda6300}{H\alpha}$: if the emission lines of the above indicators are too weak to be observed, line ratios from different species and ionization states can be used to determine the physical parameters of the plasma; for example, the ratio of lines with high excitation energies (e.g. [N II]λ6583 and [O I]λ6300) and the low excitation Hα line trace the excitation temperature of the plasma. A high line ratio indicates a high excitation temperature.

Excursion: high-mass star formation

Other than for low-mass stars, the formation of high-mass stars ($\gtrsim 20\,M_\odot$) is poorly understood and little explored observationally. The problem is that high-mass stars evolve very quickly – on a Myr timescale – so that nuclear fusion starts before the end of accretion. The observational problem is that at the time when the high-mass stars clear their circumstellar clouds and get visible the formation process has already finished. The theoretical problem is that the radiation from the star will stop its growth, and it is not clear how the stars can acquire enough material before this happens. Several scenarios have been proposed for the formation of high-mass stars: through disc accretion (similar to low-mass stars), via collision of stars, and from Bondi accretion (collecting material from travelling through the interstellar medium). A possible route to test the scenarios is to search for accretion discs and associated outflows and jets.

5.4.6 *Collimated molecular jets from high-mass young stars: IRAS 18151−1208*

This paper by Davis *et al.* (2004) presents integral field spectroscopy of the high-mass protostar IRAS 18151−1208, its jet and the associated shocks. IRAS 18151−1208 is a class I young stellar object with a double-peaked spectral energy distribution dominated

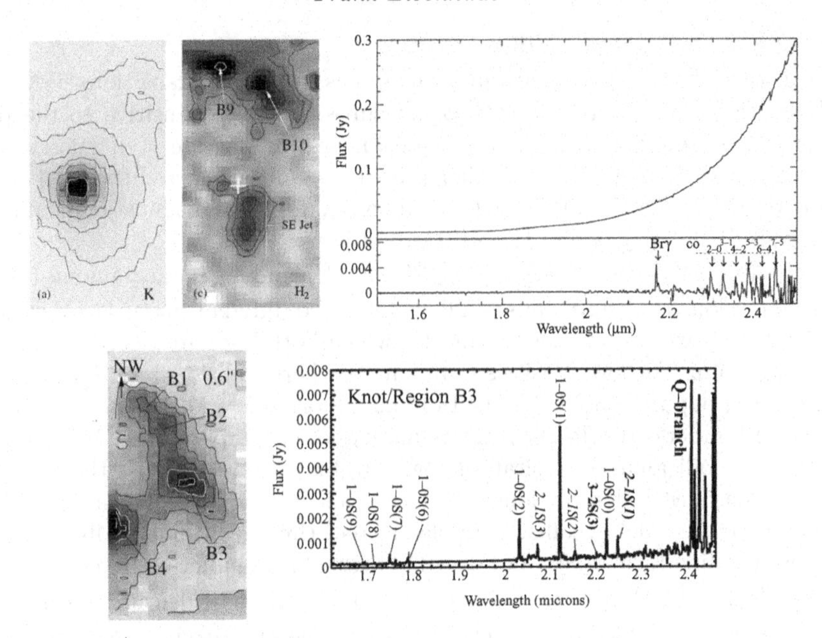

FIGURE 5.17. IRAS 18151−1208. Top left: K-band continuum image of the IRS1 high mass protostar; top middle: south-eastern structure pointing in the direction of the large scale jet; top right: spectrum of IRS1; bottom: $H_2\nu 1-0S(1)$ line maps and spectrum of a typical knot in the jet (courtesy of Davis *et al.*, 2004).

by cold (peak emission at 100 μm) and warm (peak emission at 20 μm) dust. Its bolometric luminosity is $\approx 20{,}000\,L_\odot$, indicating that IRAS 18151–1208 is a high-mass star in formation. Near-infrared images in the emission lines of molecular hydrogen H_2 have revealed two collimated jets in the IRAS 18151−1208 region, one of which is almost a parsec in length. If these jets originate from the young stellar object, this would imply that high-mass stars form via disc accretion. The K-band image also shows a bright central peak (Figure 5.17 top left), denoted IRS1, roughly at the position of the mid-infrared source and aligned with one of the jets. But is IRS1 really the high-mass star? And is the jet really coming from that star? Unfortunately the error ellipse for the position of the mid-infrared source is about $25'' \times 7''$. The position of IRS1 in the near-infrared colour-magnitude diagram supports the view that this star is indeed the high-mass star. But this photometric identification is not conclusive, because emission from circumstellar discs and dust extinction prevents the straightforward conversion from near-infrared colours and magnitudes to temperature and luminosity.

The integral field spectroscopy from Davis *et al.* (2004) shows the H- and K-band spectrum of IRS1 steeply rising at longer wavelengths (Figure 5.17 top right), indicating a deeply embedded object. IRS1 also shows CO band-heads in emission, tracing dense molecular gas ($n_H \approx 10^{12}$–$10^{13}\,\mathrm{cm}^{-3}$ and $T \approx 2000$–$5000\,\mathrm{K}$) probably heated by UV photons impinging on the inner accretion disc surface. Alternatively, the CO emission could come from the outer regions of accreting funnel flows, where the gas is heated by adiabatic compression. The line maps from the [Fe II]$\lambda 1.644$ μm and H_2 emission (Figure 5.17 top middle) show an $\approx 1''$-long south-eastern structure extending from IRS1 in the direction of one of the large-scale jets. Additional echelle spectroscopy supports the hypothesis that this feature is indeed the base of the jet. The distant jet components show only H_2 (Figure 5.17 bottom right), but not [Fe II], indicating low excitation. The line ratio $\frac{H_2\nu=1-0S(1)}{H_2\nu=2-1S(1)}$

is approximately 7–11, typical for shock heating (see excursion on H_2 line diagnostics). The analysis of the excitation diagram gives a temperature of $\approx$2000–2600 K. The knots have a bow-like morphology (Figure 5.17 bottom left), and the kinematics are comparable to what is seen in Herbig–Haro objects. In summary, Davis *et al.* (2004) find that the knots in the IRAS 18151−1208 jets are similar to their counterparts from low-mass protostars. The close association between the H_2 features and the high-velocity CO emission observed at radio wavelengths suggests that the CO represents gas entrained by the collimated jets. From the mass and momentum of the molecular gas and the luminosity of the H_2 features, it is clear that the flow must be powered by a massive source. To all intents and purposes, the molecular jet appears to be a scaled-up version of jets from low-mass young stellar objects.

Collectively, the observations add further support to the idea that massive stars are formed through vigorous disc accretion, and that massive protostars drive collimated jets in their earliest stages of evolution.

5.4.7 *GEMINI multi-object spectrograph integral field unit spectroscopy of the 167–317 (LV2) proplyd in Orion*

This paper by Vasconcelos *et al.* (2005) presents high spatial resolution spectroscopic observations of the proplyd 167–317 (LV2) in the Orion Nebula obtained with the GMOS integral field unit at the Gemini South Observatory.

Proplyds are protoplanetary discs that are being exposed to an intense ultraviolet radiation field which renders them visible. The characteristics of a proplyd are a bow-shaped head that faces the ionizing star, a tail that is directed away from the source, often a young star and a disc, sometimes seen in silhouette against a background H II region. There are more than 100 proplyds in the Orion star-forming region, photoionized by the Trapezium cluster. 167–317 (LV2) is a particularly bright proplyd. Previous studies have found spectral signature of a collimated outflow, and its mass loss rate has been estimated to approximately $8.2 \times 10^{-7}\,M_\odot\,yr^{-1}$.

Using integral field spectroscopy, Vasconcelos *et al.* (2005) have detected 38 forbidden and permitted emission lines associated with the 167–317 (LV2) proplyd. The [N II]λ5755, [N II]λ6548, Hα, [N II]λ6583, [He I]λ7065 and [Ar III]λ7135 are strong enough to construct intensity and line profile maps. Figure 5.18 top left shows the intensity and line profile map for Hα. In order to identify the jet of 167–317 (LV2) and to analyse its properties, the authors extract velocity moment maps (see excursion). These velocity moments provide an unbiased and robust simplification of the velocity substructure. Both Hα velocity dispersion and skewness (Figure 5.18 top middle and right) peak towards the south-east, indicating a redshifted wing in the profile. Indeed, the Hα line profile (Figure 5.18 bottom left) shows a significant asymmetry, and subtracting a simple Gaussian fit reveals the red and blue components (Figure 5.18 bottom right). The redshifted component is associated with the previously reported jet. It has a receding velocity of about 80–120 km s^{-1} and it is spatially distributed to the south-east of the proplyd. The approaching structure may possibly be associated with a faint counter-jet with a velocity of -75 ± 15 km s^{-1}. There is evidence that the redshifted jet has a variable velocity, with slow fluctuations as a function of the distance from the proplyd. The line ratio $\frac{[NII]\lambda\lambda(6548+6583)}{[NII]\lambda5754}$ in the jet is approximately 10, corresponding to a particle density of $\approx 2 \times 10^6$ cm^{-3} (see excursion on line ratio diagnostics in the optical). Vasconcelos *et al.* (2005) also provide an updated mass loss rate for the 167–317 (LV2) proplyd from a simple model for the observed Hα fluxes (see excursion): the mass loss for the photo-evaporated flow and the redshifted microjet are $\dot{M}_{\rm proplyd} = (6.2 \pm 0.6) \times 10^{-7}\,M_\odot\,yr^{-1}$ and $\dot{M}_{\rm jet} = (2.0 \pm 0.7) \times$

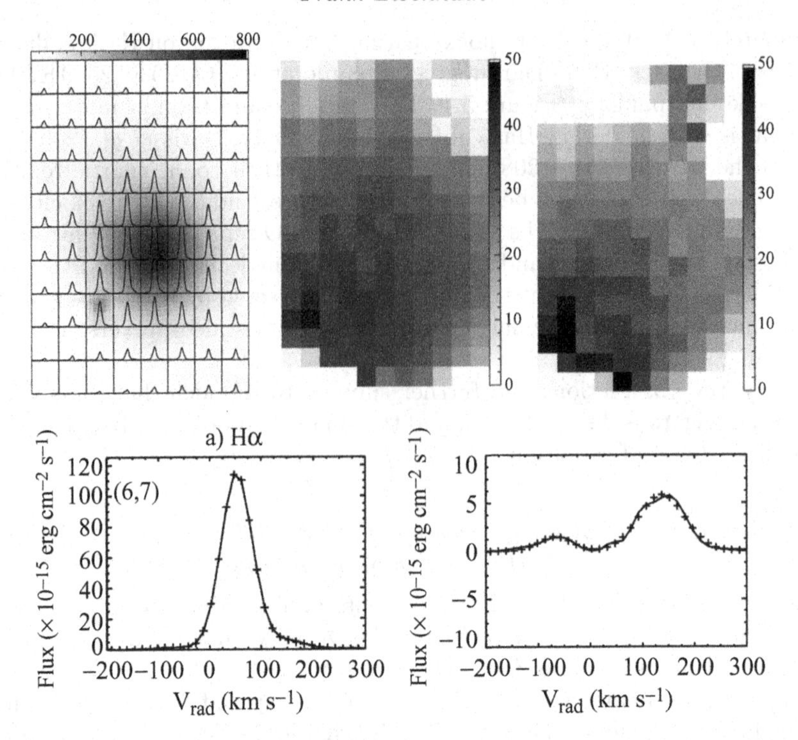

FIGURE 5.18. 167–317 (LV2) proplyd in Orion. Top: Hα intensity and line profile map (left), flux weighted rms (middle) and skewness (right); bottom: typical Hα line profile (left); data minus simple Gaussian fit (right), revealing a blue- and a redshifted component (courtesy of Vasconcelos *et al.*, 2005).

$10^{-8}\,M_\odot\,yr^{-1}$, respectively. For a typical proplyd mass of $\approx 0.1\,M_\odot$, the mass loss rate implies that the lifetime is less than $\approx 10^5$ yr.

Excursion: velocity moments

A common problem of integral field spectroscopy is to extract the 'important' information from the vast amount of data. Line profile maps (Figure 5.18 top left), which show the spectrum for every pixel, contain all the information, but are difficult to read and interpret. One possible way to condense the information is to fit the line profile for every pixel and to interpret the fit parameters. However, line fit has the disadvantage of being biased from the assumed number of lines and their profile (for example, Gaussian or Lorentzian), and is not always numerically stable (returning the wrong optimum). An alternative is velocity (v) moments. The first three moments are the mean $< v >$, variance Δv^2 and skewness Δv^3:

$$< v > = \frac{1}{I} \int I_v v dv \tag{5.25}$$

$$\Delta v^2 = \frac{1}{I} \int I_v \left(v - < v >\right)^2 dv \tag{5.26}$$

$$\Delta v^3 = \frac{1}{I} \int I_v \left(v - < v >\right)^3 dv \tag{5.27}$$

These quantities have the advantage of being unbiased and numerically stable. The variance indicates the width of the line and the skewness describes the asymmetry of the profile (positive skewness indicates a redshifted wing). Figure 5.18 top illustrates the technique for the velocity field of a protoplanetary disc. The result may be used as an initial guess for subsequent line profile fitting.

Excursion: simple model for mass loss rate

The mass loss rate $\dot{M}$ from a proplyd can be estimated from a simple model that assumes that the flow arises from a hemispherical wind that originates at the proplyd ionization front with the sound speed, v_S. Under these assumptions we can write:

$$\dot{M} = 4\pi R^2 \rho v_S \tag{5.28}$$

where R is the radius, $4\pi R^2$ the surface and ρ the density of the hemispherical ionized region. The density, ρ, can be estimated from various line ratio diagnostics (see the relevant excursion), or from the Hα luminosity $L_{H\alpha}$. The assumption for the latter estimate is that the probability for the electron–proton recombination – a two-particle process – is proportional to the square of the particle density, n:

$$\dot{n} = \alpha n^2 \tag{5.29}$$

$$L_{H\alpha} = \frac{4}{3}\pi R^3 \dot{n} h\nu \tag{5.30}$$

$$\rho = \mu n m_H, \tag{5.31}$$

where μ is the mean molecular weight of the ionized gas, m_H the mass of a hydrogen atom, $\frac{4}{3}\pi R^3$ the volume of the ionized region, $\dot{n}$ the recombination rate (in particles per cm^3 and second), $h\nu$ the energy of a Hα photon, and $\alpha \approx 5.83 \times 10^{14}$ cm^{-3} the effective recombination constant for the Hα line. Combining the formulae we get the estimate for the mass loss rate:

$$\dot{M} = 4\pi R^2 n \mu m_H v_S \tag{5.32}$$

$$n = \sqrt{\frac{3 L_{H\alpha}}{4\pi \alpha h\nu R^3}} \tag{5.33}$$

Typical numbers for a proplyd are $R \approx 10$ AU, $n \approx 10^6$ cm^{-3}, and $v_S \approx 10$ km s^{-1}, implying a mass loss rate $\dot{M} \approx 10^{-6}$ M$_\odot$ yr^{-1}.

5.5 The Solar System

This section is about integral field spectroscopic studies of Solar System objects. Very few publications have emerged from this field, probably because the Solar System is the domain of space missions. Also, the scientific impact has been minimal. Nevertheless, this section describes the early work on integral field spectroscopic observations of Mars, because this study uses the very powerful principal component analysis to condense the vast amount of information in the datacubes. The new adaptive optics-assisted instruments are about to infuse new life into the field, and several papers are in the press or in preparation. Therefore, the section also includes the discussion of a paper on Titan, which does not fulfil the criterion of being published before the IAC Winter School in November 2005.

5.5.1 *Martian surface mineralogy from 0.8 to 1.05 μm TIGER spectra-imagery measurements in Terra Sirenum and Tharsis Montes formation*

This early paper by Martin *et al.* (1996) presents 0.8–1.05 μm spectra of Mars from TIGER spectra-imaging. The primary goal of this investigation is to identify the physico-chemical origins of the observed spectral variations, to map the surface based on the

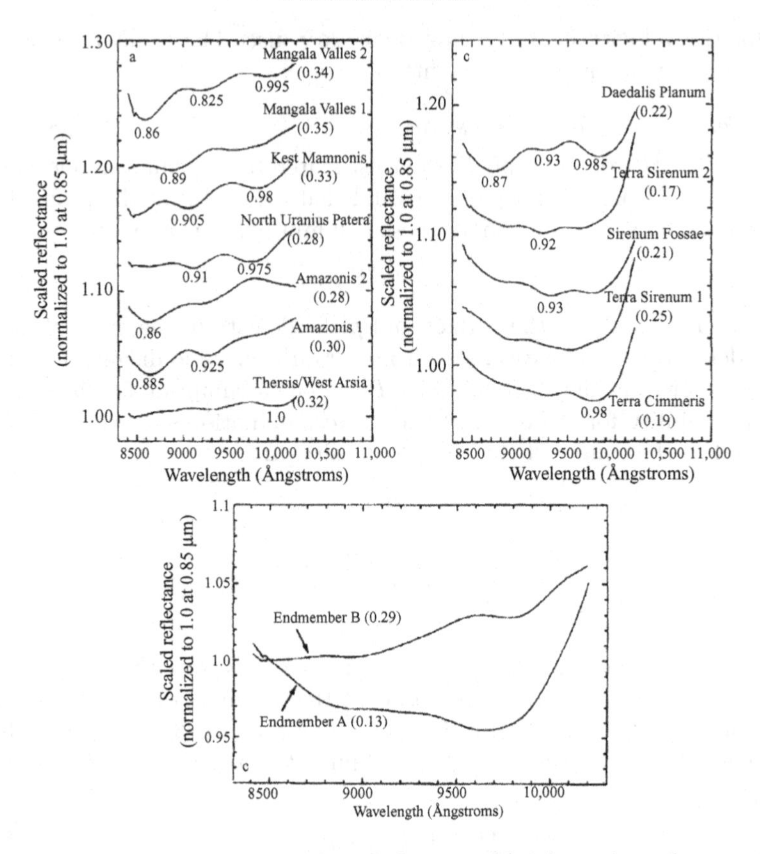

FIGURE 5.19. Spectra of the Marian surface. Left: typical bright regions characterized by the absorption band from ferric iron oxides at ≈ 0.86 μm; right: typical faint regions characterized by the absorption band from proxine at ≈ 0.93 μm and ≈ 0.96 μm; bottom: end-members used for the linear decomposition of the spectra (courtesy of Martin *et al.*, 1996).

spectroscopic properties and to draw conclusions on the compositional heterogeneities of bright and dark materials.

The measurements were obtained at Mauna Kea CFH Observatory during the 1990 opposition, and covered a systematic survey of the Tharsis volcanic region with a spatial resolution of 200–250 km. To first order, the Martian surface can be characterized by low and high albedo terrains. Figure 5.19 shows typical spectra of these regions. The spectrum of the bright – high albedo – terrains is characterized by the absorption band of ferric iron oxides ($Fe_2^{+3}O_3$) at 0.85–0.89 μm. Faint – low albedo – terrains show absorption from pyroxine ($XYSi_2O_6$ where X is a metal ion, and Y a smaller-sized ion) at 0.90–0.94 μm (from ferrous pyroxine, $X = Fe^{+2}$) and at 0.9–1.0 μm (from wollastonite, $X = Ca^{+2}$). To classify and characterize the ≈1500 spectra from the survey, the authors apply a principal component analysis (see excursion), which decomposes the spectra into the 'most important' components: the first principal component turns out to be indicative of the slope in the spectrum, which can be interpreted as the relative absorption from ferric (at shorter wavelengths) and ferrous (at longer wavelengths) minerals. The second principal component indicates the bending of the spectrum and therefore measures the heterogeneity of the material. Next, the authors identify in the principal component diagram two groups with typical spectra (called end-members), and look for the corresponding terrains: the first spectrum is representative of a cratered rocky region with low reflectance,

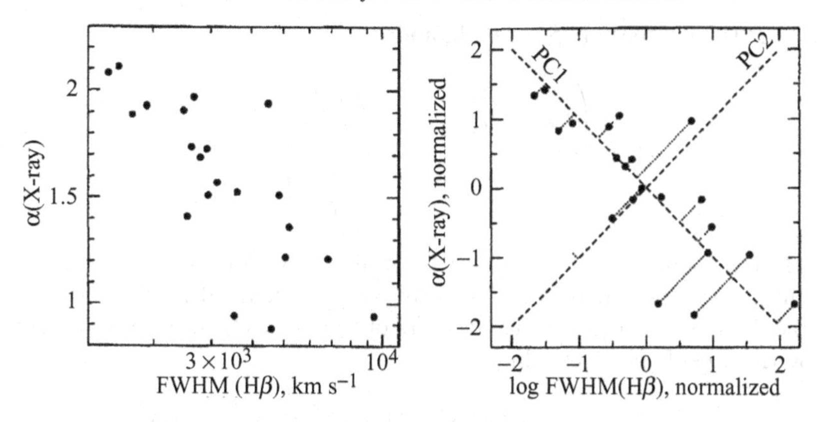

FIGURE 5.20. Geometric picture of principal component analysis (here illustrated for the X-ray spectral index–Hβ line width relation for quasars). Left: measurements in natural units; right: normalized units (mean subtracted, divided by standard deviation) and direction of principal components (courtesy of Francis and Wills, 1999).

and – by comparing with laboratory spectra – corresponds to a mixture of $\approx 60\%$ bulk haematite (from the Greek for 'bloodlike' because of the red colour of its powder) and $\approx 40\%$ of small (45–75 μm) particles of clinopyroxene (from Latin 'ignis' meaning 'fire', common in igneous rocks from volcanoes). The second spectrum maps a highly fractured terrain originating from lava flows covered with high albedo soils composed of active dust deposits. This spectrum can be represented by a mixture of $\approx 90\%$ haematite with a large fraction of nanoparticles and $\approx 10\%$ clinopyroxene. The two end-member spectra are then used for the linear decomposition of the dataset. This mixture modelling reveals that $\approx 90\%$ of the observed terrains can be explained by a combination of clinopyroxene-dominated and haematite-dominated material. In contrast, a linear mixture of low and high albedo soils mentioned in previous Martian observations cannot explain the spectrum of the intermediate albedo soils. Both ferric and ferrous signatures are found in the spectra of most dark and bright regions, demonstrating the strong heterogeneity of the soils.

Excursion: principal component analysis

Principal component analysis (PCA) is a technique that can be used to simplify a dataset. It is a linear transformation to a new coordinate system such that the greatest variance in the dataset is along the first axis ('first principal component'), the second greatest variance along the second axis, and so on. In many cases the low-order components contain the 'most important' aspects of the data and by showing which original variables correlate with one another, principal component analysis often leads to new physical insights. An introduction to the principal component analysis applied to astronomy can be found in Francis and Wills (1999).

Geometric picture of principal components: The principal components can be interpreted as a series of linear least squares fits to a dataset, all mutually orthogonal: the first principal component is an unconstrained minimum distance fit, the second principal component a fit in a direction plane perpendicular to the first principal component, and so on. Figure 5.20 illustrates the geometric picture of principal component analysis.

Algebraic picture of principal components: Here, we describe the data as a $(N \times P)$ matrix X, where the element $x_{i,j}$ contains the ith measurement $(i = 1, \ldots, N)$ of the jth parameter $(j = 1, \ldots, P)$, i.e. each row contains one spectrum, N is the number of spectra and P the number of spectral channels:

$$X = \begin{pmatrix} x_{1,1} & x_{1,2} & \cdots & x_{1,P} \\ x_{2,1} & x_{2,2} & \cdots & x_{2,P} \\ \vdots & \vdots & \vdots & \vdots \\ x_{N,1} & x_{N,2} & \cdots & x_{N,P} \end{pmatrix} \begin{matrix} \rightarrow \lambda \\ \\ \downarrow \\ ID \end{matrix} \tag{5.34}$$

Accordingly, the columns x^i ($i = 1, \ldots, N$) contain the various measurements for a given wavelength. The average value for a given wavelength is denoted $\bar{x^i}$. The principal components of the dataset are then the eigenvectors of the $(P \times P)$ covariance matrix $Cov(X)$ ordered by the size of the eigenvalues:

$$Cov(X) = \begin{pmatrix} cov(x^1, x^1) & \cdots & cov(x^1, x^P) \\ cov(x^2, x^1) & \cdots & cov(x^2, x^P) \\ \vdots & \vdots & \vdots \\ cov(x^P, x^1) & \cdots & cov(x^P, x^P) \end{pmatrix} \tag{5.35}$$

$$cov(x^j, x^k) = \frac{1}{N} \sum_{i=1}^{N} \left(x_{i,j} - \bar{x^j}\right)\left(x_{i,k} - \bar{x^k}\right) \tag{5.36}$$

where $j, k = 1, \ldots, P$. For our application, the principal components have the format of a spectrum.

Excursion: seasons on Titan

Titan is the largest moon of Saturn. It has drawn public attention from the recent visit and landing of the Cassini and Huygens mission in 2005. Titan's atmosphere is strikingly similar to the atmosphere of the pre-biotic Earth, dense in nitrogen with a pressure of $\approx$1.5 bar, and a surface temperature of $\approx$90 K. The second most abundant gas, methane ($\approx$5%), is close to its triple point, and takes part in a methane-based meteorological cycle that is similar to the Earth's hydrological cycle. A series of chemical reactions result in the formation of aerosols in the upper stratosphere at $\approx$600 km. When the aerosols settle into the troposphere (0–40 km), they serve as condensation nuclei for Titan's methane and ethane clouds. This is the starting point of an aerosol-climate feedback cycle: the aerosols and clouds absorb and scatter the sunlight and therefore determine the thermal structure of Titan's atmosphere; in turn, the temperature gradients drive the atmospheric circulation, which redistributes the aerosols and haze, closing the cycle.

5.5.2 *Titan's bright spots: multi-band spectroscopic measurement of surface diversity and hazes*

Ádámkovics *et al.* (2006) have obtained spatially resolved near-infrared spectra of Titan that allowed them to derive the three-dimensional haze distribution in the atmosphere and map the albedo of its surface.

The authors used the SINFONI integral field spectrometer with adaptive optics at the VLT in February 2005 to obtain spatially resolved 1.45–2.45 μm spectra of Titan. Figure 5.21 shows a sample of spectral images (top) and a typical K-band spectrum (bottom left) extracted from the datacube. All of Titan's observed near-infrared intensity is sunlight scattered by aerosols or reflected from the surface. The dominant source of opacity is the methane bands at wavelengths around 1.669 μm and 2.190 μm. Therefore, the H- and K-band centres show the strongest absorption, and we observe sunlight

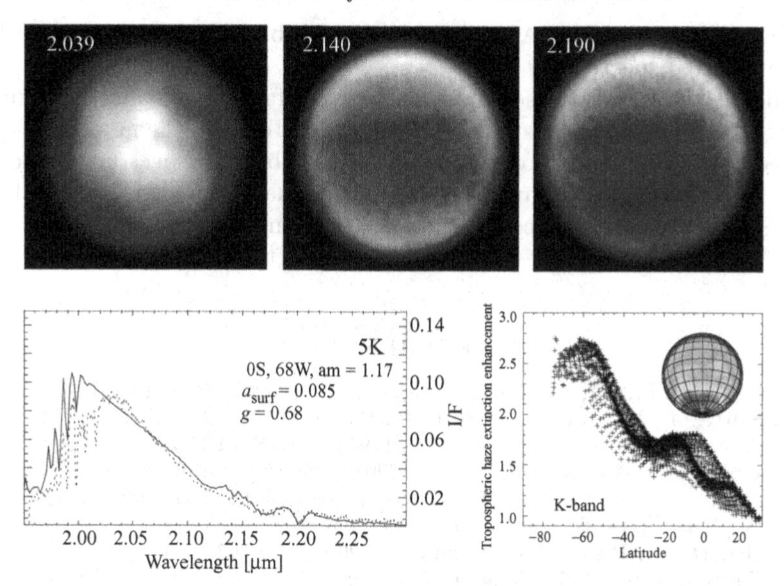

FIGURE 5.21. Integral field spectroscopy of Titan. Top: spectral images in the K-band probing the surface (left), troposphere (middle) and stratosphere (right); bottom left: typical K-band spectra (dashed line shows the best fitting radiative transfer model); bottom right: tropospheric haze enrichment (courtesy of Ádámkovics *et al.*, 2006).

scattered as aerosol haze high in the lower stratosphere (40–200 km). At shorter wavelengths (e.g. 1.588 μm and 2.039 μm) – towards the limits of the H- and K-bands – we see only weak absorption, and the observations trace the albedo of the surface. Intermediate wavelengths (e.g. 1.611 μm and 2.140 μm) probe accordingly intermediate altitudes, and the spectral images show the tropospheric (0–40 km) haze distribution. The stratospheric haze has a higher concentration near the north (winter) pole. In contrast, the tropospheric haze is enhanced near the south pole (Figure 5.21 bottom right). Surface albedos are recovered at 60 mas (375 km) resolution in both the 1.5 μm and the 2.0 μm windows, nicely tracing the high-resolution surface maps from the Cassini spacecraft. For the detailed analysis the authors apply a two-stream, plane–parallel, radiative transfer model to fit the observed spectra (see excursion). This modelling provides a tool to determine the vertical haze profile and retrieve latitudinal trends in both stratospheric and tropospheric aerosol extinction: the stratospheric aerosol extinction is measured to increase at a rate of $\approx 0.65 \pm 0.05\%$ per degree latitude from 40S to 60N. In contrast, the tropospheric haze is confined near the south pole at latitudes above 40S.

Excursion: radiative transfer modelling

The analysis of the three-dimensional haze distribution in Titan's atmosphere is based on a radiative transfer model. In such a model, the software solves the two-stream radiative transfer equation for multiple plane–parallel layers.

The model for the analysis of the Titan spectra is based on the atmospheric temperature and pressure profile measured in situ by the Huygens lander. The model takes into account the scattering at the atmospheric haze, the absorption from methane, and the reflection from the surface. A Henyey–Greenstein phase function is used to describe the angular distribution of scattered light by the aerosol and is parameterized by the average cosine of the scattering angle, g. As an intermediate result the model returns the so-called contribution function, which is the fraction of photons at the top of the atmosphere that arise from a particular altitude. After an additional limb correction, the model ultimately returns the reflected spectrum.

The model fit has only two free parameters, the surface albedo and the characteristic parameter, g, of the aerosol scattering phase function. To begin the fit, a series of spectra is calculated by varying g and compared against the observed spectrum of the methane absorption in the H- and K-band centres. Once g has been determined, the surface albedos are varied to fit the spectra below 1.60 μm and 2.05 μm. Figure 5.21 (bottom left) shows a typical K-band spectrum and the best-fitting radiative transfer model.

REFERENCES

Ádámkovics, M., de Pater, I., Hartung, M. *et al.* (2006), *JGR*, **111**, 7

Allington-Smith, J., Murray, G., Content, R. *et al.* (2002), *PASP*, **114**, 892

Bacon, R., Adam, G., Baranne, A. *et al.* (1995), *A&AS*, **113**, 347

Bacon, R., Copin, Y., Monnet, G. *et al.* (2001a), *MNRAS*, **326**, 23

Bacon, R., Emsellem, E., Combes, F. *et al.* (2001b), *A&A*, **371**, 409

Bahcall, J.N., Tremaine, S. (1981), *ApJ*, **244**, 805

Beck, T.L., Riera, A., Raga, A.C., Aspin, C. (2004), *AJ*, **127**, 408

Binney, J., Tremaine S. (1987), *Galactic dynamics*, Princeton University Press

Bonnet, H., Abuter, R., Baker, A. *et al.* (2004), *Msngr*, **117**, 17

Bower, R.G., Morris, S.L., Bacon, R. *et al.* (2004), *MNRAS*, **351**, 63

Comte, G., Marcelin, M. (eds. 1995), *ASPCS*, 71

Davis, C.J., Varricatt, W.P., Todd, S.P., Ramsay Howat, S.K. (2004), *A&A*, **425**, 981

Dougados, C., Cabrit, S., Lopez-Martin, L., Garcia, P., O'Brien, D. (2003), *Ap&SS*, **287**, 135

Dumas, C. *et al.* in preparation

Eisenhauer, F., Schödel, R., Genzel, R. *et al.* (2003a), *ApJ*, **597**, L121

Eisenhauer, F., Abuter, R., Bickert, K. *et al.* (2003b), *Proc. SPIE*, **4841**, 1548

Eisenhauer, F., Genzel, R., Alexander, T. *et al.* (2005), *ApJ*, **628**, 246

Emsellem, E., Cappellari, M., Peletier, R.F. *et al.* (2004), *MNRAS*, **352**, 721

Förster Schreiber, N.M., Genzel, R., Lehnert, M.D. *et al.* (2006), *ApJ*, **645**, 1062

Francis, P.J., Wills, B.J. (1999), *ASPCS*, **162**, 363

García, P.J.V. (2006), *New Astronomy Review*, **49**, 590

Genzel, R., Thatte, N., Krabbe, A., Kroker, H., Tacconi-Garman, L.E. *et al.* (1996), *ApJ*, **472**, 153

Genzel, R. *et al.* (1997), in *SINFONI a High Resolution Near-Infrared Imaging Spectrometer for the VLT*, www.mpe.mpg.de/SPIFFI/preprints/STC.210.ps.gz

Genzel. R., Pichon, C., Eckart, A., Gerhard, O.E., Ott, T. (2000), *MNRAS*, **317**, 348

Genzel, R., Schödel, R., Ott, T. *et al.* (2003a), *ApJ*, **594**, 812

Genzel, R., Schödel, R., Ott, T. *et al.* (2003b), *Natur*, **425**, 934

Ghez, A.M., Duchêne, G., Matthews, K. *et al.* (2003), *ApJ*, **586**, L127

Herbst, T.M., Beckwith, S.V.W., Glindemann, A. *et al.* (1996), *AJ*, **111**, 2403

Kasper, M.E., Feldt, M., Herbst, T.M., Hippler, S., Ott, T., Tacconi-Garman, L.E. (2002), *ApJ*, **568**, 267

Kissler-Patig, M., Roth, M.M., Walsh, J.R. (2007), in Science Perspectives for 3D Spectroscopy, *ESO Astrophysics Symposia*, eds. Kissler-Patig, M., Wals, J.R., & Roth, M.M. Springer

Krabbe, A., Genzel, R., Eckart, A., *et al.* (1995), *ApJ*, **447**, L95

Lavalley, C., Cabrit, S., Dougados, C., Ferruit, P., Bacon, R. (1997), *A&A*, **327**, 671

Lavalley-Fouquet, C., Cabrit, S., Dougados, C. *et al.* (2000), *A&A*, **356**, L41

Leitherer, C., Schaerer, D., Goldader, J.D. *et al.* (1999), *ApJS*, **123**, 3

Leonard, P.J.T, Merrit, D. (1989), *ApJ*, **339**, 195

Levin, Y., Beloborodov, A.M. (2003), *ApJ*, **590**, L33

Martin, P.D., Pinet, P.C., Bacon, R. *et al.* (1996), *P&SS*, **44**, 859

McDermid, R., Bacon, R., Adam, G., Benn, C., Cappellari, M. (2004), *Proc. SPIE*, **5492**, 822

McGregor, P.J., Dopita, M., Wood, P., Burton, M.G. (2001), *PASA*, **18**, 41

MONREAL-IBERO, A., ROTH, M.M., SCHÖNBERNER, D., STEFFEN, M., BÖHM, P. (2005), *ApJ*, **628**, L139

MÜLLER-SÁNCHEZ, F. ET AL. (2006), *A&A*, accepted, astro-ph/0604317

NAGAR, N.M., FALCKE, H., WILSON, A.S. (2005), *A&A*, **435**, 521

NARAYAN (2002), in *Lighthouses of the Universe: The Most Luminous Celestial Objects and Their Use for Cosmology*, Proceedings of the MPA/ESO Conference, 405

OSTERBROCK, D.E. (1989), *Astrophysics of gaseous nebulae and active galactic nuclei*, University Science Books

REID, M.J., BRUNTHALER, A. (2004), *ApJ*, **616**, 872

RUTTEN R.G.M., BENN, C.R., MÉNDEZ, J. (eds. 2006), *New Astronomy Reviews*, **49**, 487

SCHMIDT, M. (1963), *Natur*, **197**, 1040

SCHÖDEL, R., OTT, T., GENZEL, R. ET AL. (2002), *Natur*, **419**, 694

SHU, F.H., ADAMS, F.C., LIZANO, S. (1987), *ARA&A*, **25**, 23

SWINBANK, A.M., SMAIL, I., BOWER, R.G. ET AL. (2005), *MNRAS*, **359**, 401

TECZA, M., GENZEL, R., TACCONI, L.J. ET AL. (2000), *ApJ*, **537**, 178

VAN BREUGEL, W., BLAND-HAWTHORN, J. (2000), *ASPCS*, 195

VAN ZYL, L., CHARLES, P.A., ARRIBAS, S., NAYLOR, T., MEDIAVILLA, E., HELLIER, C. (2004), *MNRAS*, **350**, 649

VASCONCELOS, M.J., CERQUEIRA, A.H., PLANA, H., RAGA, A.C., MORISSET, C. (2005), *AJ*, **130**, 1707

WEITZEL, L., KRABBE, A., KROKER, H. ET AL. (1996), *A&AS*, **119**, 531

6. Extragalactic studies and future integral field spectroscopy science

LUIS COLINA

In this set of lectures, I review recent observational progress on extragalactic studies using integral field spectroscopy (IFS) techniques, highlighting the importance of IFS for the study of the nuclear regions of nearby galaxies, of low-z active galactic nuclei (AGN) and massive star-forming galaxies, and of high-z galaxies, including lensed quasars, lensing galaxies and bright submillimetre galaxies. Emphasis is given to the study of (ultra)luminous infrared galaxies as examples of low-z systems where the physical processes relevant to the formation and evolution of galaxies can be investigated in more detail. Research projects involving future ground-based facilities and satellites are also briefly presented.

6.1 Introduction

The use of IFS for extragalactic studies has burgeoned over the past 10 years and is already becoming a standard observational technique used by several groups in many different areas. Most IFS systems (INTEGRAL, GMOS, PMAS, SAURON, SINFONI, VIMOS, etc.) allow us to simultaneously obtain spectra covering a wide spectral range over a wide field of view (up to 1 arcmin square for VIMOS). These instruments in their standard configurations provide low–intermediate spectral resolution (R of 1000 to 4000) with a relatively low angular resolution ($0.5''$ to $3.0''$). In addition, a few IFS systems, such as OASIS on the William Herschel Telescope and SINFONI on the Very Large Telescope (VLT), can provide very high angular resolution (i.e. $0.1''$) in the optical (OASIS) and near-infrared (SINFONI) when combined with adaptive optics (AO) systems. Finally, multi-object integral field spectrographs (GIRAFFE on the VLT) are already available that are leading the way to future IFS systems currently in design (see Section 6.3). A full account of the different available IFS systems and their characteristics can be found in M. Bershady's contribution to this Winter School.

Studies of galaxies at low and high redshift benefit greatly from the flexibility of IFS systems. Detailed studies on different angular scales – from the nuclear, to circumnuclear and external regions – can only be executed efficiently with IFS. Moreover, these systems allow us to investigate simultaneously the two-dimensional structure of many astrophysically important properties such as chemical composition, stellar populations, ionization conditions, kinematics, etc. that are directly related to the formation and evolution of galaxies.

Among the several areas of extragalactic research covered by IFS-based studies, this contribution will focus on the study of (ultra)luminous infrared galaxies – (U)LIRGs – explaining in detail the impact and use of IFS in the investigation of several topics, such as the structure of the internal dust extinction, the properties of young stellar clusters, the presence of different ionizing sources such as hidden AGN, (extra)nuclear star-forming regions and shocks, and the kinematic properties of the nuclear and extended ionized gas. All these aspects are common to many other research areas, such as the study of the nuclear regions in nearby ellipticals and spirals, of the ionization structure in low-z AGN, of extended nebulae in low- and high-z galaxies, and of studies of high-z galaxies, including lensed objects. Due to space limitations, these studies will not be covered in

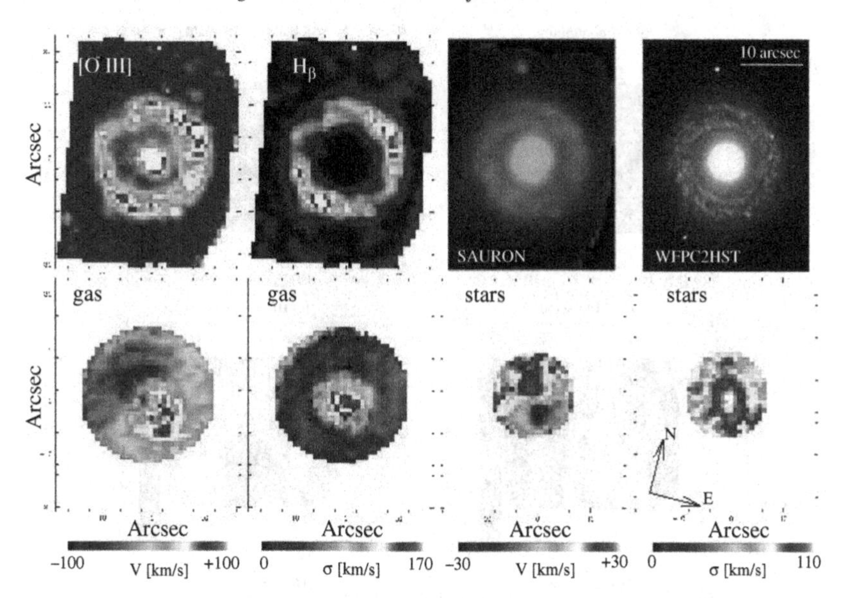

FIGURE 6.1. Maps of NGC 7742 taken with SAURON. The top panels show the emission line intensity distributions of [O III] and Hβ, followed by a SAURON reconstructed image composed of [O III], blue continuum and red continuum, and a similar image composed from HST/WFPC2 filters. The bottom row shows (from left to right) the derived gas velocity and velocity dispersion fields, and the stellar velocity and velocity dispersion fields (figure reproduced from de Zeeuw *et al.*, 2002).

detail here and the reader is referred to the relevant references given in the following as part of this introduction.

6.1.1 *Stellar populations and kinematics in early-type galaxies*

The formation and evolution of early-type galaxies is being investigated by studying the fossil record (i.e. the two-dimensional kinematics and ages of the stellar population) for a complete sample of nearby ellipticals and spirals using mostly SAURON, a dedicated IFS system on the William Herschel Telescope (Bacon *et al.*, 2001; de Zeeuw *et al.*, 2002; see Figure 6.1 for an example). On kiloparsec scales, many objects display circumnuclear stellar kinematic twists, kinematically decoupled components, central stellar discs and other peculiarities with almost no age differences with their host galaxy (Emsellem *et al.*, 2004). However, on scales of hundreds of parsecs, additional high angular resolution IFS studies with OASIS reveal the presence of previously undetected kinematically distinct nuclear components that also show a large fraction of young stars (McDermid *et al.*, 2006). On the other hand, the kinematics of the ionized gas is often decoupled from that of the stars, is rarely consistent with coplanar circular motions, but shows large-scale coherent motions and smooth variations (Sarzi *et al.*, 2006). The recent extension of these studies to late-type (Sb to Sd) spirals indicate different, more complicated features in the velocity fields of these galaxies with respect to the early types. These features include wiggles in the zero-velocity line, irregular distributions and ring-like structures (Ganda *et al.*, 2006).

6.1.2 *Ionization structure and kinematics in AGN*

Research on low-z AGN has been well represented in IFS science over the last 10 years. Interest in learning about the kinematics and ionization in the nuclear regions close to the AGN engine, and in the associated starburst–AGN connection has been the main topic

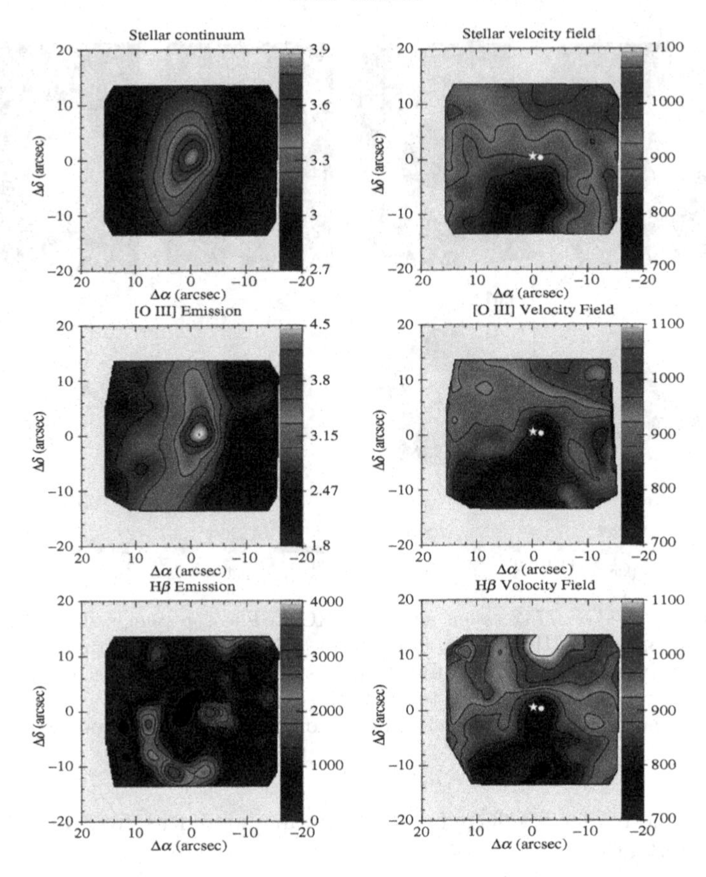

FIGURE 6.2. Intensity (arbitrary units) and velocity (in km s^{-1}) maps for NGC 5033 taken with INTEGRAL. The intensity maps correspond to the stellar continuum (4700–5840 Å wavelength interval), and the ionized gas traced by the [O III] and Hβ emission. The star and the circle in the velocity maps represent the location of the intensity peak in the stellar continuum and [O III] intensity maps, respectively (figure reproduced from Mediavilla *et al.*, 2005a).

in this area. Detailed studies of the galaxy NGC 1068, considered as the prototype of Seyfert 2 galaxies, include early work on the biconical ionization field (Arribas *et al.*, 1996) and on the kinematics of the central nuclear and circumnuclear region (García-Lorenzo *et al.*, 1997, 1999). Similar individual studies of other well-known nearby galaxies on scales of hundreds of parsecs and covering a wide range of activity from LINERs to Seyfert 1 galaxies have been undertaken from the early days of IFS to the present (see Figure 6.2 for an example; Arribas and Mediavilla, 1994; Arribas *et al.*, 1997, 1999; Mediavilla *et al.*, 1997, 2005a; García-Lorenzo *et al.*, 2001; Ferruit *et al.*, 2004). With the advent of near-IR AO-based IFS systems such as SINFONI, the star formation in the inner regions close to the AGN, the properties of the molecular gas in the central 100 pc region and its relation to the torus is starting to be investigated on spatial scales of a few to several parsecs (Davies *et al.*, 2006; Mueller-Sánchez *et al.*, 2006).

Not only has work on AGN covered radio-quiet AGN, but IFS has also been used to investigate the kinematics and ionization of the interaction of the radio jet with the surrounding interstellar medium in the radio galaxy 3C 120 (Sánchez *et al.*, 2004; García-Lorenzo *et al.*, 2005). VIMOS on the VLT is now also being used to obtain deep IFS of the giant nebulae with sizes in excess of 100 kpc that surround high-z radio galaxies, and

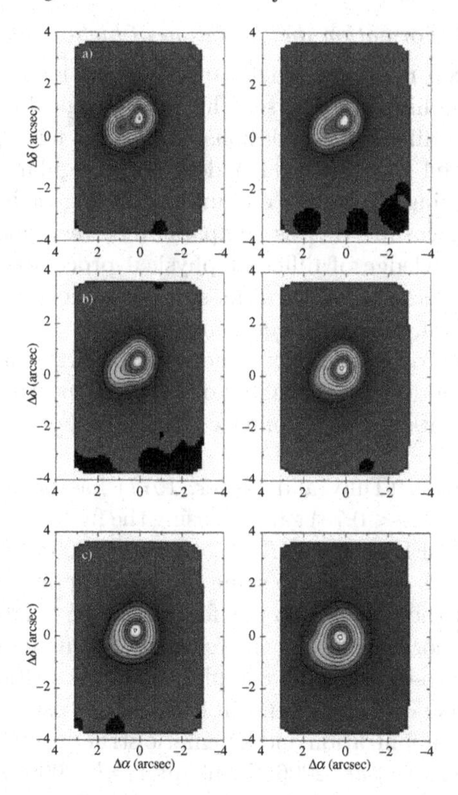

FIGURE 6.3. Image representing six of the twenty continuum maps obtained for the lensed system SBS 0909+532 using INTEGRAL. a) to c), left to right: Continuum central wavelengths are 8727 Å, 7233 Å, 5739 Å, 4447 Å, 4006 Å and 3487 Å. Changes in the relative brightness of the two components as a function of wavelength are due to internal extinction within the intermediate redshift lens galaxy (figure reproduced from Motta *et al.*, 2002).

to establish their origin. The kinematics of the nebulae around the galaxy MRC2104-242 at a redshift of 2.49 is consistent with rotation, therefore implying a dynamical mass of at least 3×10^{11} $M_{\odot}$ in this system (Villar-Martín *et al.*, 2006).

6.1.3 *Interstellar medium in high-z gravitational lens systems*

In the past few years IFS has been used to investigate the internal properties of both intermediate-z lens galaxies and high-z QSOs (quasi-stellar objects) and lensed galaxies. A precise determination of the internal extinction law in the lens galaxy of the system SBS 0909+532, at a redshift of 0.83, has been obtained combining IFS in the optical (Figure 6.3; Motta *et al.*, 2002) and HST spectroscopy (Mediavilla *et al.*, 2005b). The IFS study of the Einstein Cross (i.e. lensed QSO 2237+0305) has allowed the detection of the extended narrow line region (NLR) in this QSO, establishing a size of 0.7 to 0.9 kpc (Mediavilla *et al.*, 1998; Motta *et al.*, 2004). Finally, the gravitational potential of the cluster A2218 has been used as a lens magnifier to derive the resolved kinematics of the ionized gas in a lensed spiral galaxy at a redshift of 1.034, demonstrating that this galaxy lies close to the mean Tully–Fisher relation of present-day spirals (Swinbank *et al.*, 2003). Further analysis to investigate the emission line properties of several galaxies at intermediate redshifts ($z \sim 1$) have recently been reported (Swinbank *et al.*, 2006a) applying the same technique, and exploiting the gravitational lens effect due to low-z massive clusters.

6.1.4 *Formation and evolution of high-z galaxies*

Over the last decade deep multiwavelength surveys have identified a large diversity of galaxy populations in the high-z Universe. The morphological evolution of galaxies and the contribution of these different populations to the star formation history of the universe is being investigated mostly through deep high-angular resolution imaging and stellar population modelling. However, our understanding of how galaxies of different mass are assembled over cosmic time is comparatively very poor. In order to progress in this area a better knowledge of different physical processes – such as gravitational interactions due to mergers, feedback from star formation and AGN, exchange and loss of angular momentum, and metal enrichment of the interstellar and intergalactic medium – is needed. Due to the complex structure in a large fraction of high-z galaxies, IFS has a clear advantage over pure imaging or long-slit spectroscopy to investigate the formation and evolution of galaxies (Colina *et al.*, 2005; Förster Schreiber *et al.*, 2006; Swinbank *et al.*, 2006b).

The Tully–Fisher relation (Tully and Fisher, 1977) has been studied in a sample of intermediate redshift ($0.4 < z < 0.75$) galaxies using the first operational multi-object IFS system, FLAMES/GIRAFFE on the VLT (Flores *et al.*, 2006). Of the galaxies surveyed only 35% are rotating discs showing the same slope, zero point and scatter as the low-z TF relation. The rest of the galaxies do not follow the TF relation, indicating a strong dynamical evolution (minor or major mergers, merger remnant, or radial gas motions) traced by the observed two-dimensional complex kinematics (Flores *et al.*, 2006).

The new optical and near-IR IFS systems in 10 m-class telescopes are also being used to obtain the two-dimensional ionization and kinematic structure of galaxies at redshifts of 2 and above (Förster Schreiber *et al.*, 2006; Swinbank *et al.*, 2006b). Massive submillimetre galaxies, such as SMM J14011+0252 at a redshift of 2.565 (Tezca *et al.*, 2004) and SMM J16365+5047 (N2 850.4) at a redshift of 2.385 (Figure 6.4; Swinbank *et al.*, 2005), are already been studied. The presence of large peak-to-peak velocity differences, decoupled stellar and gas structures and evidence for AGN have already been detected (Tezca *et al.*, 2004; Swinbank *et al.*, 2005). Many of these systems comprise two or more dynamical subcomponents with velocity offsets of 180 km s^{-1} across scales of several kpc (Swinbank *et al.*, 2006b), suggesting that these systems are analogous to low-z Ultra Luminous Infrared Galaxies (ULIRGs; Colina *et al.*, 2005). A more extensive investigation of the Hα rest-frame emission in a sample of UV-selected galaxies at an average redshift of 2.2 has already been done using SINFONI on the VLT (Förster Schreiber *et al.*, 2006). Spatially resolved velocity fields with peak-to-peak velocities of 40 to 410 km s^{-1} over scales of 10 kpc have been identified in 60% of the galaxies. These velocity fields are generally consistent with orbital motions, although in only four galaxies (i.e. 35% of the sample) are they well described by rotating disks (Förster Schreiber *et al.*, 2006).

6.2 IFS studies of low-z ULIRGS

6.2.1 *Astrophysical relevance of (Ultra) Luminous Infrared Galaxies*

The importance of luminous and ultraluminous (LIRGs, $L_{\rm IR} = L[8\text{–}1000\,\mu\text{m}] = 10^{11}$–$10^{12}\,L_\odot$, and ULIRGs $L_{\rm IR} \geq 10^{12}\,L_\odot$, respectively) has been recognized since their discovery (Rieke and Low, 1972) and the detection of large numbers by the *IRAS* satellite (Soifer *et al.*, 1987). The process of interaction/merging (Farrah *et al.*, 2001; Bushouse *et al.*, 2002, and references therein) with accompanying starbursts (Lutz *et al.*, 1998) that produces (U)LIRGs, appears to be an important stage in the evolution of galaxies. The transformation of gas-rich (Solomon *et al.*, 1997) moderate-mass ($< m_*$) disk galaxies

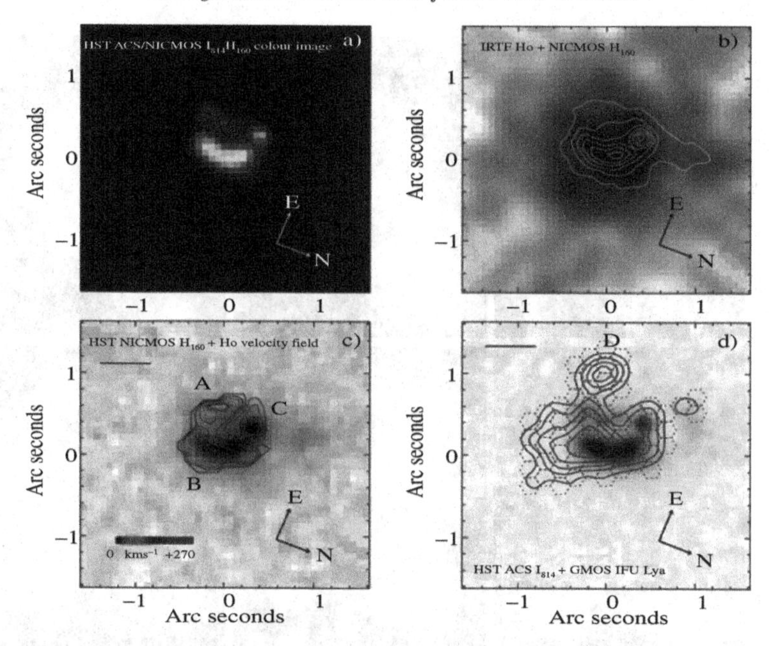

FIGURE 6.4. a) Image of N2 850.4 combining HST ACS and NICMOS imaging. The image shows a complex morphology, with at least three distinct components separated by 1 arcsec (8 kpc) in projection. b) Infra-Red Telescope Facility (IRTF) Hα narrow-band image with the contours from the NICMOS H-band image overlaid. This Hα narrow-band image shows a diffuse halo of material distributed asymmetrically around the galaxy. c) The velocity field derived from UIST IFS observations of the Hα emission line overlaid on the NICMOS H-band image. Components A and B are separated by $50 \pm 50\,\mathrm{km\,s^{-1}}$ whilst there is a velocity difference of $270 \pm 50\,\mathrm{km\,s^{-1}}$ between components A and C. d) HST ACS I-band image with the Lyα intensity from the GMOS IFS overlaid as contours (figures reproduced from Swinbank *et al.*, 2005).

(Colina *et al.*, 2001, 2005) into moderate-mass ellipticals through the process of merging (Genzel *et al.*, 2001; Tacconi *et al.*, 2002; Dasyra *et al.*, 2006) suggests a physical process for forming early-type galaxies in the early Universe, such as those recently detected at redshifts of 2–3 (Franx *et al.*, 2003; Cimatti *et al.*, 2004; Förster Schreiber *et al.*, 2004; van Dokkum *et al.*, 2004).

Deep submillimetre and radio surveys (Chapman *et al.*, 2003a, 2003b, and references therein) are detecting a population of objects, the so-called sub-millimetre galaxies (SMGs), that appear to be forming stars at rates of up to $1000\,\mathrm{M_\odot\ yr^{-1}}$ (i.e. up to 10 times more than ULIRGs), and that, according to their morphology (Chapman *et al.*, 2003b), luminosity and spectral energy distribution (Egami *et al.*, 2004; Frayer *et al.* 2003), seem to be the high-z counterparts of the low-z luminous (LIRG) and ultraluminous (ULIRG) infrared galaxies (for a review, see Blain *et al.*, 2002). Mid-IR satellites (*Spitzer Space Telescope* as well as millimetre interferometers, IRAM, ALMA) continue to detect high-z LIRGs and ULIRGs in large quantities, as already shown by the data taken with *Spitzer* (Charmandaris *et al.*, 2004; Egami *et al.*, 2004; Le Floc'h *et al.*, 2004). The majority of these galaxies at $z < 1$ are in the LIRG class, and they make a significant contribution to the star formation density at $0.5 < z < 2$ (Pérez-González *et al.*, 2005).

Studies of the evolution of the morphological types of galaxies as a function of redshift indicate that, as in low-z ULIRGs, mergers have played a major role in the shaping of present-day galaxies. These studies (Conselice *et al.*, 2003, 2005) conclude that the fraction of galaxies consistent with undergoing a major merger increases with redshift

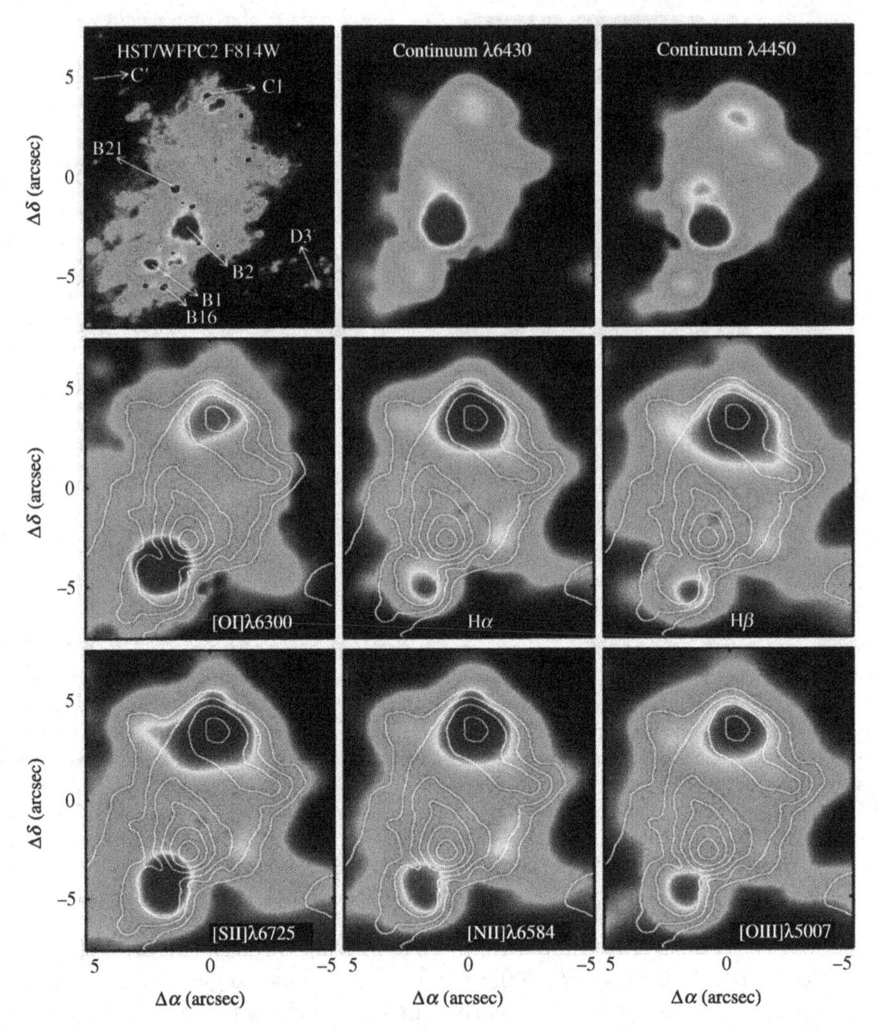

FIGURE 6.5. Continuum and emission line images for NGC 3690 obtained with INTEGRAL. The contours represent the red stellar continuum at 6460 Å. The HST/WFPC2 *I*-band (F814W) image is shown for comparison, with several regions of interest marked on it. Component B2 dominates the stellar continuum emission while components C1 and B1 are the strongest emission line sources (figure reproduced from García-Marín *et al.*, 2006).

for all galaxies, and that at redshifts above 2, more than 80% of the stellar mass is in objects with peculiar morphology. There is also a predominance of irregular and complex morphologies in high-z luminous SMGs, suggesting that major mergers are common in these dusty star-forming galaxies (Chapman *et al.*, 2003b). In addition, galaxies less affected by dust – such as the Lyman break galaxies (LBGs) and UV-selected, high-z star-forming galaxies – do often show irregular and distorted morphologies (Erb *et al.*, 2003, 2004; Giavalisco *et al.*, 1996), indicating that interaction/merging is also involved in these galaxies.

6.2.2 *Structure of the stellar and ionized gas distribution*

The stellar and ionized gas distribution is traced by line-free stellar continuum images and by the light of several optical emission lines, respectively (see Figures 6.5 and 6.6 for

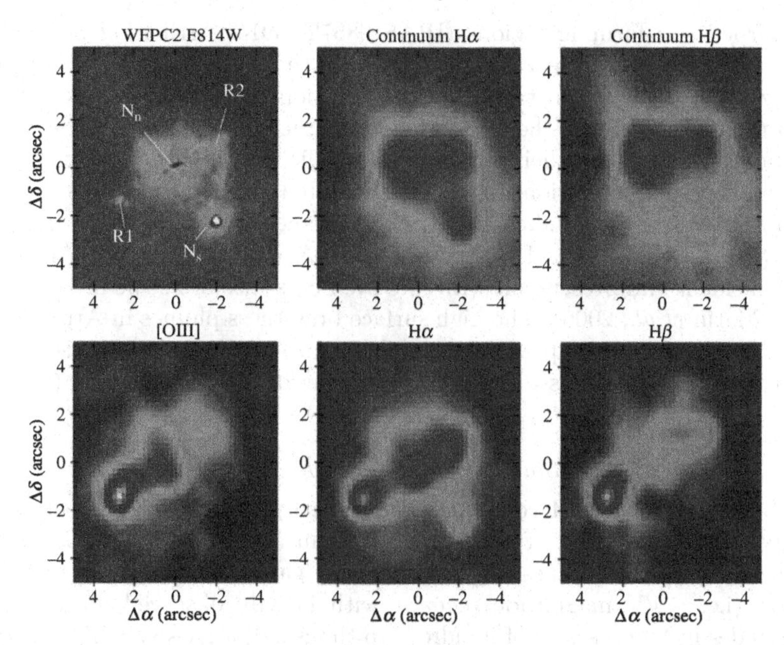

FIGURE 6.6. INTEGRAL images of the ionized gas and stellar light distribution in the central regions of IRAS 12112+0305 as traced by different emission lines and by continuum windows close to Hβ and Hα. The observed [O III] and Hβ emission is dominated by an extinction-free extranuclear star-forming region. The HST *I*-band (F814W) image is shown for comparison (figure reproduced from Colina *et al.*, 2000).

examples). The wavelength dependence of the optical and near-IR stellar light distribution is mostly produced by the absorption due to the large amount of dust towards the nucleus in these galaxies. A clear example of high nuclear extinction is shown in IRAS 12112+0305 where the red bright, compact southern nucleus (N_s) appears as a weak source in the Hβ-continuum light (Figure 6.6) seen through an extinction of about eight magnitudes in the visual (Colina *et al.*, 2000). In other galaxies, such as Arp 220 and IRAS 17208−0014, the presence of a strong central dust lane obscures the true nucleus in the optical. As a result, the optical nucleus, identified as the high surface brightness peak, does not spatially coincide with the near-IR nucleus but is about 1 kpc apart in these galaxies (Arribas *et al.*, 2001; Arribas and Colina, 2003). In several ULIRGs investigated, the internal extinction produces a wavelength dependence of the relative brightness of the heavily obscured nuclear regions with respect to the less obscured external van Dokkum regions of the host galaxy while keeping the overall stellar light distribution, even in interacting pairs.

The observed ionized gas and stellar distributions show decoupled structures with different morphologies in many (U)LIRGs. These differences are due to the presence of extranuclear gas ionized by a central active galactic nucleus (Mrk 273; Colina *et al.*, 1999; Arp 299; García-Marín *et al.*, 2006), the presence of bright, tidally induced extranuclear star-forming regions along the tidal tails (IRAS 14348−1447; Monreal-Ibero, 2004) or far away (9 kpc) from the nucleus of the galaxy (IRAS 12112+0305; Colina *et al.*, 2000) and the presence of circumnuclear gas ionized by a nuclear, dust-enshrouded starburst (Arp 220; Arribas *et al.*, 2001). In addition, the complex, non-axisymmetric structure of the internal extinction in these galaxies produces additional wavelength-dependent effects such as changes in the relative brightness of the nuclear regions with respect to less absorbed star-forming regions (IRAS 08572+3915; Arribas *et al.*, 2000), or to extranuclear regions located several kpc away from the nucleus (IRAS 12112+0305).

The ionized gas distribution traced by the emission lines shows a different structure in some of the galaxies. Since these emission lines trace different ionization mechanisms (shocks, young stars, AGN), their two-dimensional light structure represents the spatial distribution of these different ionizing sources within a given galaxy. This is clearly seen in the two nearest galaxies studied so far, Arp 299 and Arp 220. The peaks of the [O I] λ6300 Å, [O III] λ5007 Å, Hα and [S II] $\lambda\lambda$6717,6731 Å emitting regions in Arp 299 are not spatially coincident and present substructure even on scales of 200 pc or less (Figure 6.5 and García-Marín *et al.*, 2006). The high surface brightness plumes in Arp 220, detected up to 4 kpc from the nucleus, present a complex ionization structure as identified by the changes in the [N II]/Hα emission line ratio measured along the plumes (Colina *et al.*, 2004).

6.2.3 *Star formation and dust extinction*

The distribution of dust, and consequently its absorption effects, presents a complex morphology and a wide range of extinction values measured using optical and/or near-IR hydrogen recombination lines. Within the same galaxy, star-forming regions almost free of dust (i.e. $\leq$1 V magnitude) coexist with heavily obscured nuclear starbursts ($\geq$3 magnitudes in V) on scales of hundreds to thousand of parsecs (Figure 6.7; García-Marín *et al.*, 2006). If corrected by the two-dimensional extinction map, the Hα light distribution agrees with the less-absorbed ionized gas distribution traced by the near-IR Paα emission line (Figure 6.7).

Dust absorption effects and their impact in the observed extranuclear versus nuclear star formation are clearly seen in many ULIRGs, such as IRAS 12112+0305 (Figure 6.6; Colina *et al.*, 2000). The stellar main body of this galaxy is concentrated in three regions: the two nuclei (N_n and N_s) and a third region (R2 hereafter) about 3″ north of N_s. However, the observed dominant line-emitting region (R1) does not coincide with any of the nuclei identified above but is located 5″ (i.e. 7.5 kpc) east of N_s. Moreover, the optically dominant nucleus of the galaxy (N_s) appears as the faintest Hα emission source. These effects are due to the relative extinction of the different regions. While region R1 has an extinction of only 0.9 magnitudes in the visual, in the southern nucleus (N_s) a visual extinction of 7.8 magnitudes is measured from the Hα/Hβ line ratio (Colina *et al.*, 2000). Thus, although N_s is a minor contributor to the observed Hα emission, the reddening-corrected flux emanating from this nucleus dominates the overall Hα luminosity. The corresponding star formation rate would amount to about 74 $M_\odot$ yr^{-1} for a Salpeter initial mass function (IMF), with mass limits of 0.1 and 100 $M_\odot$ (Colina *et al.*, 2000). In many of the (U)LIRGs the extinction is so high ($>$10 visual mag) in the nuclear (IRAS 17208–0014; Arribas and Colina, 2003) or circumnuclear regions (NGC 7469, Díaz-Santos *et al.*, 2007) that an important fraction of the star formation is not detected in the optical, hence the importance of future near-IR and mid-IR IFS studies.

6.2.4 *Detection of AGN*

Traditionally, the level of activity in galaxies has been obtained through diagnostic diagrams (BPT diagrams, after Baldwin, Phillips and Terlevich, 1981) based on the ratios of different optical emission lines. These studies have been restricted to the nuclei of ULIRGs or to extended regions along a given direction (Veilleux *et al.*, 1992). Some ULIRGs, such as Mrk 231, I Zw 1 or Mrk 273, are known to host a powerful AGN. These AGN already show the spectral characteristics of Seyfert 1 (broad emission lines) or

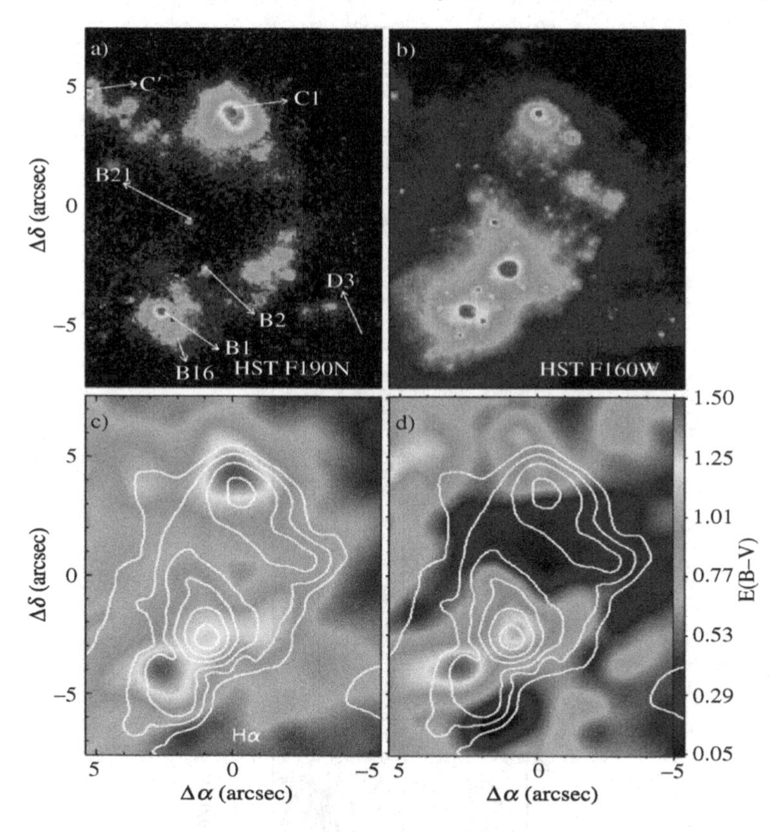

FIGURE 6.7. a) Panel showing the galaxy NGC 3690 the HST/NICMOS continuum-subtracted Paα emission; b) the HST/NICMOS H-band continuum; c) the INTEGRAL Hα extinction corrected emission and d) the extinction spatial distribution obtained with the Hα/Hβ line ratio. The contours represent the red stellar continuum at 6460 Å as obtained with INTEGRAL (figure taken from García-Marín *et al.*, 2006).

Seyfert 2 (strong high excitation [O III] λ5007 Å emission line) nuclei. In most ULIRGs, however, the presence of a dust-enshrouded AGN is not evident in the optical, based only on the nuclear spectrum. Integral field spectroscopy allows us to investigate the two-dimensional ionization structure of the nuclear and circumnuclear gas searching for traces of optically dust-enshrouded AGN. These types of AGN can be traced by the presence of extended extranuclear high-excitation ionized regions similar in size and structure to those detected in nearby AGN. Some ULIRGs, such as Mrk 273 (Colina *et al.*, 1999) and IRAS 11087+5351 (García-Marín *et al.*, in preparation), show the presence of highly ionized [O III] emitting regions at distances of several kiloparsecs from the nucleus of the galaxies. Some other galaxies show a well-defined ionizing cone morphology, as expected from the standard AGN model (Urry and Padovani, 2000 and references therein). One of these cones, detected in the galaxy NGC 3690, is in the process of merging with IC 694 and forms the luminous infrared system Arp 299 (Figure 6.8; García-Marín *et al.*, 2006). According to the standard diagnostic diagrams, three different types of ionization, HII-LINER and AGN-like, are identified. The line ratio maps clearly show the presence of an extranuclear Seyfert-like nebula with a well-defined conical morphology, with its apex in region B1, and located at a projected distance of 1.5 kpc from its apex. This highly ionized conical structure in NGC 3690 is also detected at larger distances from B1 in the

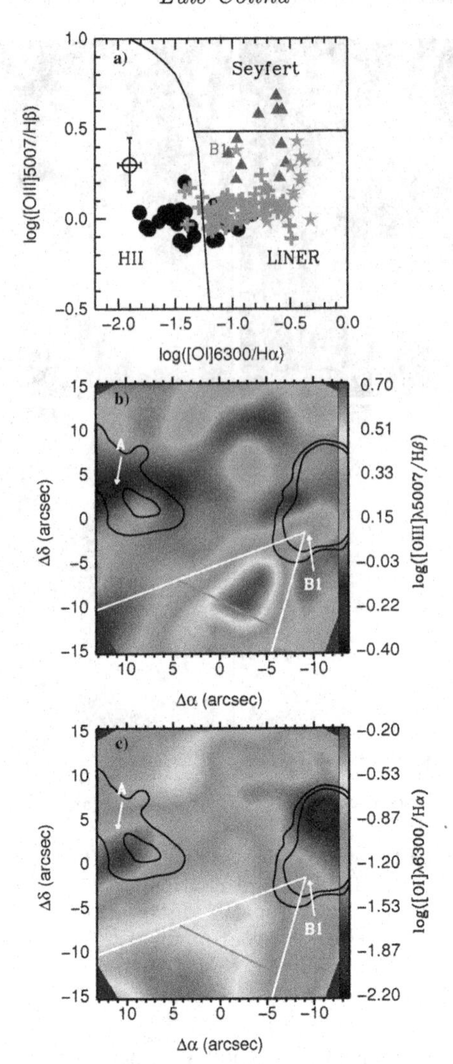

FIGURE 6.8. Diagnostic diagram and line ratio maps derived for the LIRG Arp 299 (IC 694 + NGC 3690) using INTEGRAL with the SB3 bundle. a) [O III] λ5007 Å/Hβ versus [O I] λ6300 Å/Hα diagnostic diagram with all the data displayed. The symbol code is as follows: *triangles* represent values within the ionization cone, and at distances of less than 2 kpc from region B1; *stars* represent the region within the cone at distances of 2–4 kpc from B1; *crosses* mark the interface region between the two galaxies that form the Arp 299 system (IC 694 and NGC 3690); and *filled circles* represent the regions in the two galaxies (defined as the regions within the isocontours). b)–c) [OIII] λ5007 Å/Hβ and [O I] λ6300 Å/Hα maps, respectively. The nuclei of galaxies IC 694 and NGC 3690 are identified with regions A and B1, respectively. The white line delineates the ionization cone with its apex in B1 (figure reproduced from García-Marín *et al.*, 2006).

[O I] λ6300 Å/Hα map (Figure 6.8). However, the ionization associated with the outer regions of the cone is LINER-like rather than Seyfert, indicating that within the cone there are two ionization regimes well separated spatially. Therefore, the two-dimensional ionization maps strongly support photoionization by a radiation cone escaping from a central dust-enshrouded AGN source located in B1. This AGN has also been recently identified at other wavelengths (Della Ceca *et al.*, 2002; Ballo *et al.*, 2004; Gallais *et al.*, 2004).

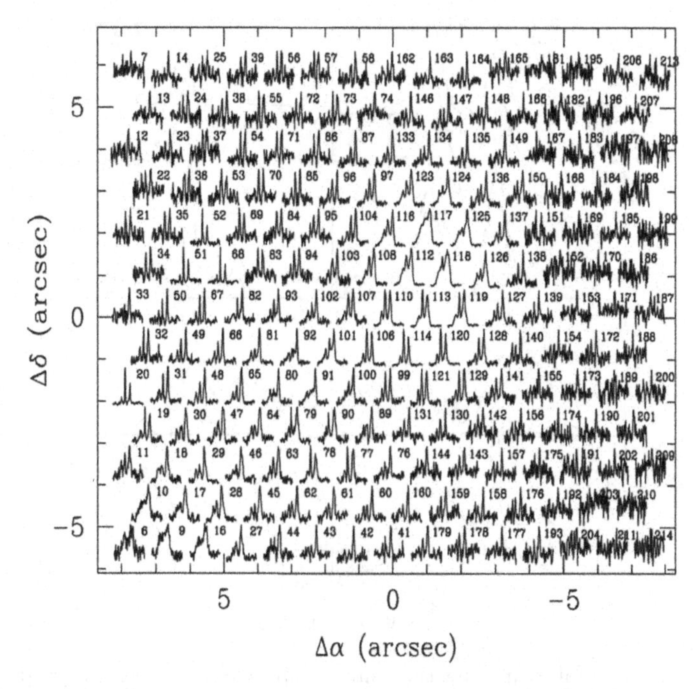

FIGURE 6.9. Spatial distribution of the spectra (centred on the Hα-[N II]) in the circumnuclear region of Arp 220 obtained using the bundle SB2 of INTEGRAL (figure reproduced from Arribas *et al.*, 2001).

6.2.5 *Arp 220: the nearest ULIRG*

The kinematic structure of the ionized gas in the central one to two kiloparsec region, as well as in the extended, several tens of kiloparsec regions, can only be investigated efficiently using IFS with different angular resolutions from sub-arcsec to several arcsec. IFS systems like INTEGRAL and PMAS offering bundles of fibres with different diameters from 0.5″ to 3.0″ are ideal for these types of studies. So far, because of the low surface brightness of the regions, a detailed study of only the nearest ULIRG, Arp 220, covering a field of view of 75.0″ by 40.0″ (i.e. 25 by 18 kpc) has been performed (Arribas *et al.*, 2001; Colina *et al.*, 2004).

(Circum)nuclear regions: rotation and starburst-driven outflows

The complex gas kinematics in the (circum)nuclear one to two kiloparsec region of Arp 220 is already suggested by the dramatic changes detected in the profiles of the Hα-[N II] lines on scales of less than a few hundred parsec using IFS (Figure 6.9; Arribas *et al.*, 2001). To disentangle the different components of the velocity field, a two-dimensional, multi-Gaussian line decomposition analysis has to be performed in such complex emission line profiles (for representative examples, see Figure 6.10; for details, see Arribas *et al.*, 2001). Three kinematically distinct and spatially extended velocity components have been identified in the ionized gas. One narrow (270 km s^{-1} FWHM; full-width-at-half-maximum) component (A in Figure 6.11) is detected nearly everywhere over the central 14 × 12 arcsec^2 region (i.e. 5.2 kpc × 4.5 kpc). The velocity field of this component with a peak-to-peak velocity of 450 km s^{-1} is interpreted as gas located in a rotating disc inclined by about 45 degrees with respect to the line of sight and with the spin axis along the SE–NW direction (position angle 135°). The inclination-corrected rotational

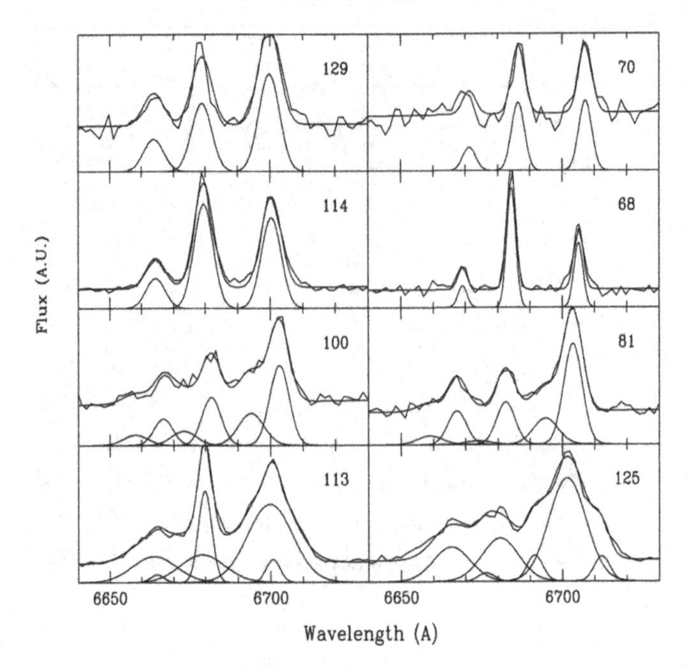

FIGURE 6.10. Some examples showing the complex spectra in the nuclear regions of Arp 220 and the fitted components indicating the presence of spatially separated velocity components. Numbers represent the fibre as displayed in the previous figure (reproduced from Arribas *et al.*, 2001).

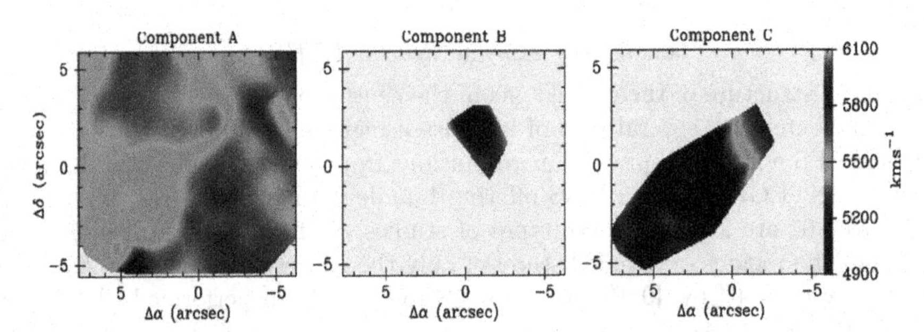

FIGURE 6.11. Velocity field and spatial distribution of the three kinematically distinct gaseous components in the circumnuclear regions of Arp 220. These velocity components are associated with rotational motions (A) and radial flows (figure reproduced from Arribas *et al.*, 2001).

velocities imply a dynamical mass ($M_{\rm dyn}$) of 3.5×10^{10} $M_\odot$ within a radius of 1.5 kpc. This relatively high value indicates a large mass concentration in the nuclear region of Arp 220, as already inferred by the presence of 5×10^9 $M_\odot$ of molecular gas in a nuclear, 0.5 kpc size disc (Scoville *et al.*, 1997).

A broad velocity component (B in Figure 6.11) with an average width (FWHM) of 815 km s^{-1} is identified over an extended region of 1 kiloparsec in diameter, at about 600 pc northwest of the dust-enshrouded nucleus, and blueshifted by 300 km s^{-1} with respect to the system velocity. A third, high-velocity component (C in Figure 6.11) with peak-to-peak velocities of 1000 km s^{-1} is detected mostly southeast of the nucleus and orientated along the kinematic minor axis of the rotating component (for details, see Arribas *et al.*, 2001). The two-dimensional distribution and kinematics of components B and C are consistent with a bipolar gas outflow characterized by an opening angle of about 90 degrees, and orientated perpendicularly to the nuclear rotating disc of gas traced by component A. These overall gas kinematics agree with the basic predictions of the starburst-driven galactic wind scenario proposed by Heckman *et al.* (1990) for galaxies such as (U)LIRGs with powerful nuclear starbursts.

Extended nebulae: tidally induced motions

The extended nebula around Arp 220 covering a region of 28 kpc $\times$ 15 kpc has been mapped entirely using a low angular (2.7″ arcsec diameter fibres) resolution mode, and integrating for a total of close to 11 hours in three different pointings (for the final Hα+[N II] spectra, see Figure 6.12a). These data have produced the first two-dimensional velocity field of the entire ionized nebula in Arp 220 as well as an ionization map of the nebulae and of the stellar light distribution on the same linear scales (see Figures 6.12b and 6.13; Colina *et al.*, 2004).

The two-dimensional velocity field shows its largest gradients (50 km s^{-1} kpc^{-1}) and deviations (+280 km s^{-1} to $-$320 km s^{-1}) from the systemic velocity in the low surface brightness ionized gas associated with the dust lane, the south-eastern stellar envelope and the north-western broad stellar tidal tail at distances of up to 7.5 kpc from the nucleus (Figure 6.13). Therefore, the velocity field of the extended nebula at distances of a few to several kpc from the nucleus and with projected peak-to-peak velocity deviations of 600 km s^{-1} is associated with well-defined stellar structures induced by the merger, i.e. it is dominated by tidally induced flows and has no clear connection with a central starburst-driven galactic wind. Moreover, the kinematics of the nebula, with average velocities of $+8 \pm 79$ km s^{-1} and -79 ± 58 km s^{-1} for the E and W lobe, respectively, are to first order in agreement with the predictions of the tidally induced scenario (McDowell *et al.* 2003), according to which the gas condensations in the lobes could be the products of the merger that are animated by a combination of continuing expansion and residual rotation from the original rotation of the two discs involved in the merger. This scenario has to be explored further by measuring the two-dimensional velocity structure of the gas along the rims of the lobes with better signal-to-noise spectra. Low angular resolution integral field spectrographs mounted on 10 m-class telescopes would be ideal for establishing the role of starburst-driven winds and of merger-induced motions on different linear scales and at different phases in the evolution of the interaction/merger process of galaxies.

6.2.6 *Merger evolution and ionized gas kinematics*

The overall kinematic characteristics of the extended nebulae have been investigated with IFS on a sample of eight ULIRGs (Colina *et al.*, 2005). The sample, although small in size, is representative of ULIRGs, and covers a wide range of morphologies representing different phases of the merging process from well-separated interacting pairs (i.e. IRAS 08572+3915, IRAS 12112+0305), to double-nucleus galaxies (angular separation less than one arcsec, i.e. Arp 220, Mrk 273) and single nucleus (i.e. IRAS 17208$-$0014). Also, the full range of nuclear activity from H II and LINERs to Seyfert 1 and 2 galaxies is covered. Finally, as defined by the ratio of the 25 μm and 60 μm IRAS bands, both warm ($f_{25}/f_{60} \geq 0.2$) and cool ($f_{25}/f_{60} < 0.2$) ULIRGs are included in the sample.

Velocity fields

The velocity fields of the ionized gas in the ULIRGs investigated with IFS show a complex structure (see Figure 6.14), inconsistent in general with large, kiloparsec-scale, ordered motions indicating that rotational motions do not dominate the velocity field. Only one galaxy (IRAS 17208$-$0014) shows a velocity field consistent with that of a rotating

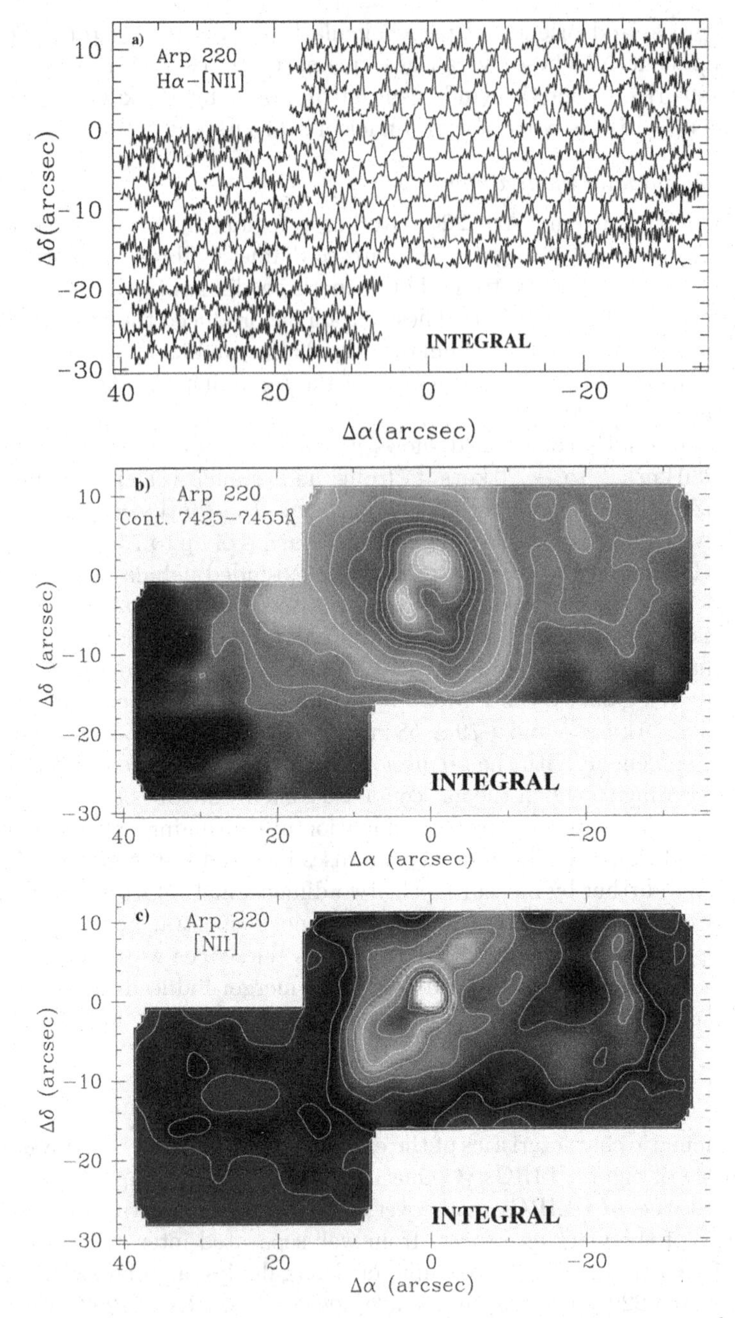

FIGURE 6.12. a) Spatial distribution of the spectra around the redshifted Hα–[N II] spectral region over the entire 75″ × 40″ extended nebulae in Arp 220. Spectra taken with INTEGRAL using the low angular resolution bundle (SB3 with 2.″7 diameter fibres) and three different pointings. b) Image of the stellar light distribution in the central regions and faint extended envelope of Arp 220 as traced by a line-free narrow-band red continuum. c) Image of the ionized gas distribution in Arp 220 as traced by the [N II] λ6484 Å emission line. The bright circumnuclear region, the elongated plumes and the extended diffuse lobes are easily identified. Note the clear differences in the ionized gas distribution with respect to the stellar light presented above (figures reproduced from Colina *et al.*, 2004).

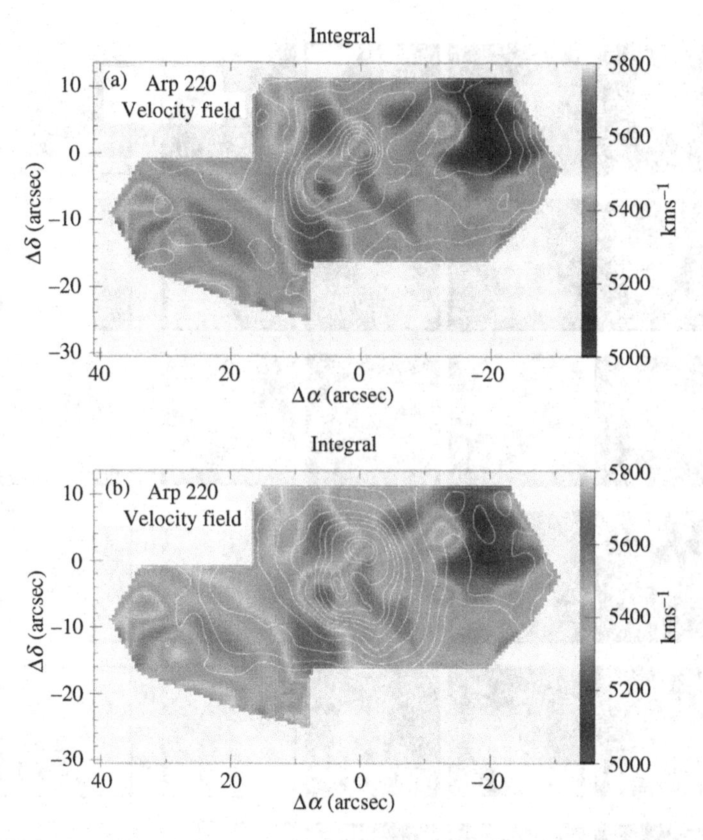

FIGURE 6.13. Velocity field of the extended ionized gas in Arp 220 obtained from the [N II] λ6584 Åemission line. The contours superimposed represent the [N II] emission (a), and the stellar light distribution, respectively (b). Orientation as in Figure 6.12 (figure reproduced from Colina *et al.*, 2004).

disc of about 3 kpc in size (Arribas and Colina, 2003), while one of the three velocity components identified in the central regions of Arp 220 (see Section 6.2.5, *Circum-nuclear regions* and Arribas *et al.*, 2001) could also trace a nuclear rotating disc, already detected in molecular gas (Downes and Solomon, 1998). In addition to IRAS 17208−0014 and Arp 220, there are other galaxies (e.g. IRAS 08572+3915) in which the major photometric and kinematic axes lie close in projection and could be considered as candidates for rotating systems (Arribas *et al.*, 2000). For the rest of the sample (seven galaxies in six ULIRGs), the largest velocity variations, characterized by peak-to-peak differences of 200–400 km s^{-1}, do not show a convincing case for rotation on scales of a few to several kpc. Higher angular resolution AO-based IFS is needed to elucidate the presence of rotation in the inner regions ($\leq$ 1–2 kpc) of ULIRGs in general. Evidence for inward flows (IRAS 15206+3342; Arribas and Colina, 2002), outflows associated with nuclear starbursts (Arp 220; Arribas *et al.*, 2001), flat velocity fields (Mrk 273, IRAS 15250+3609) and motions probably associated with tidal tails (IRAS 12112+0305, IRAS 14348−1447; Monreal-Ibero, 2004) and with the merger process (Arp 220; see Section 6.2.5.2) are also present in these systems. The sample is still small but the results indicate that tidally induced motions dominate the velocity field of the extended ionized nebula on kiloparsec scales. However, there is so far no clear trend between the observed velocity field and the phase of the interaction/merging process traced by the stellar structure, i.e. intermediate phases identified with pairs (i.e. IRAS 12112+0305, IRAS 14348−1447), more advanced

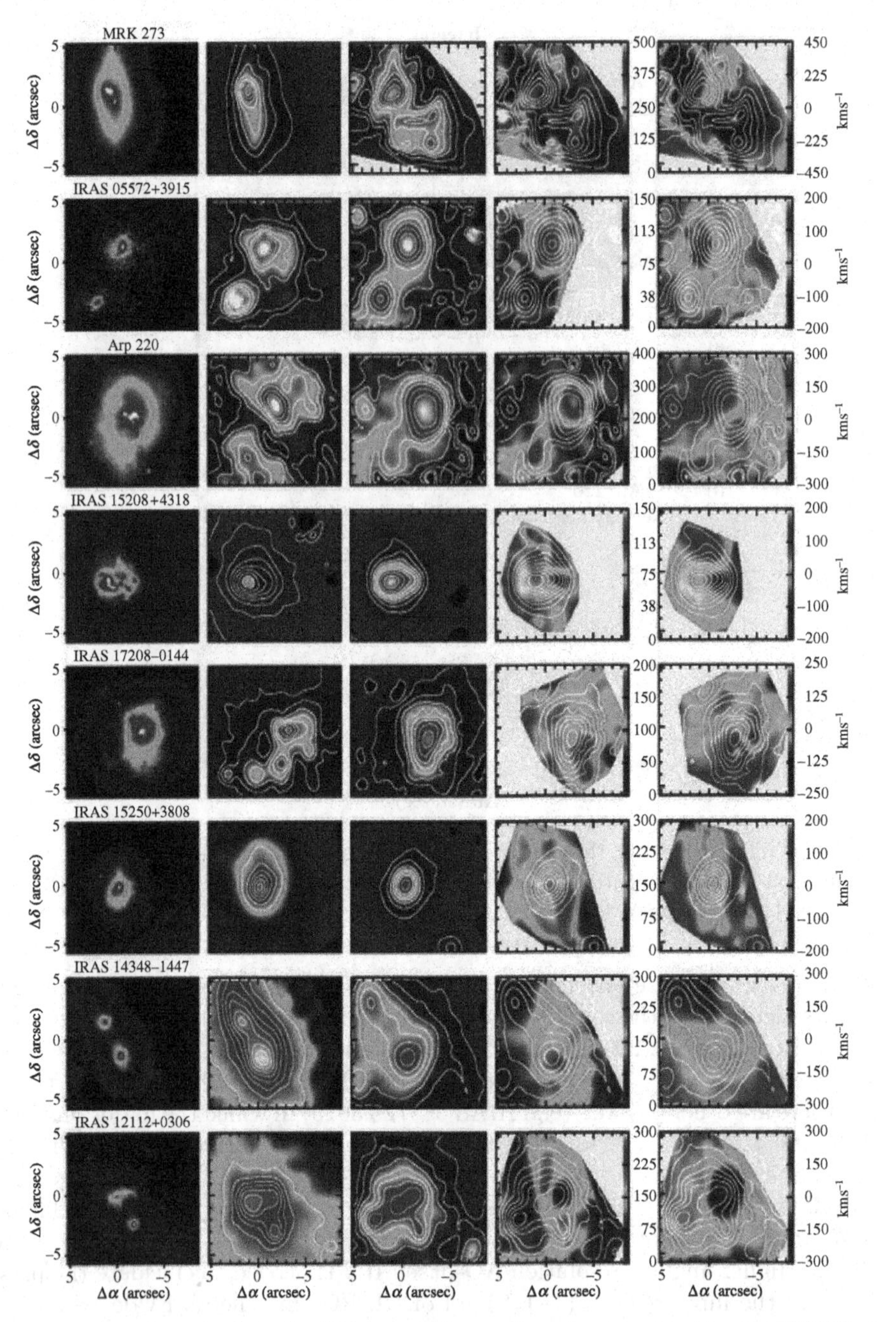

FIGURE 6.14. Panel presenting the IFS-based kinematics of a sample of low-z ULIRGs in different evolutionary states. The stellar and ionized gas distribution are traced by the optical continuum (*second column from the left*), and Hα line (*centre column*). The kinematic properties of the ionized gas are represented by the velocity dispersion map (*second column from the right*) and the velocity field (*right column*). The Hα contours are superposed on the velocity dispersion and velocity field maps. The near-IR (NICMOS F160W filter) *HST* imaging is also shown on the left side for comparison (figure reproduced from Colina *et al.*, 2005).

phases represented by double nuclei (i.e. Arp 220, Mrk 273) and evolved single-nucleus galaxies (i.e. IRAS 17208−0014, IRAS 15250+3609).

Analysis of a larger sample of ULIRGs, also including the lower luminosity LIRGs, is needed in order to characterize the gas kinematics and to establish the role and importance of starburst-driven winds, tidal-induced flows and ordered rotational motions as the interaction/merger process evolves.

Velocity dispersion maps

The two-dimensional velocity dispersion of the ionized gas is in general obtained as 0.425 times the full width at half maximum (FWHM) of the Gaussian profile that best fits the Hα emission line, after subtracting in quadrature the instrumental resolution and correcting for redshift (see Figure 6.14 for final results).

There are several important results worth mentioning. In the three ULIRGs identified as pairs of interacting galaxies (i.e. IRAS 08572+3915, IRAS 12112+0305 and IRAS 14348−1447) the velocity dispersion peak coincides with the dominant near-IR nucleus (IRAS 08572+3915 NW, IRAS 12112+0305 SW and IRAS 14348−1447 SW). However, the velocity dispersion of the gas associated with the secondary nuclei is either indistinguishable (IRAS 08572+3915 SE), or even lower (IRAS 12112+0305 NE and IRAS 14348−1447 NE), than that of the surrounding diffuse extranuclear gas. Advanced interacting/merging systems, identified as double-nucleus within 1.0″ and single-nucleus galaxies, do show different behaviours. There are ULIRGs such as IRAS 17208−0014 in which the peak of the velocity dispersion is offset with respect to the optical nucleus but coincides with the near-IR nucleus. However, other advanced mergers (Arp 220, IRAS 15206+3342, IRAS 15250+3609) do have the peak of the velocity dispersion offset from the near-IR nucleus by up to 2.0″ to 3.0″ (i.e. up to 3–5 kpc). Finally, there is one ULIRG (Mrk 273) in which the velocity dispersion in the nuclear region is indistinguishable from that of the gas ionized by the central AGN and located at about 3 kpc (i.e. 4.0″) from the nucleus.

In summary, the nucleus of the brightest galaxy in binary systems usually has the highest velocity dispersion and therefore seems to trace the dynamical mass of the galaxy. However, the high-velocity dispersions of the extranuclear gas, typically between 70 and 200 km s^{-1}, seem to be associated with tidally induced flows and do not in general trace mass concentrations in these systems but extend shocks, as will be discussed in the next section.

6.2.7 *Large-scale shocks as ionization sources in extended nebulae*

The ionized gas distribution in ULIRGs is characterized by the presence of compact, high-surface-brightness regions and of well-resolved, extended nebulae with typical sizes of between 5 to 15 kpc. In general, the ionization conditions in these extended nebulae are characteristic of LINERs, as obtained from the diagnostic diagrams involving some of the strongest optical emission lines (Figure 6.15). The most likely mechanism to explain the range of observed line ratios is the presence of large-scale, high-velocity shocks with velocities of 150–500 km s^{-1}. Ionization by an AGN or nuclear starburst is in general less likely (see Figure 6.15; Monreal-Ibero *et al.*, 2006). High-speed flows with typical peak-to-peak velocities of 200–400 km s^{-1} have been detected in the tidal tails and extranuclear regions of ULIRGs on scales of few to several kpc away from the nucleus and almost independent of the dynamical phase of the merger (Section 6.2.6, *Velocity fields* and Colina *et al.*, 2005). Moreover, the presence of highly turbulent gas, as identified by large-velocity dispersions of 70–200 km s^{-1} in extranuclear regions (see Section 6.2.6, *Velocity dispersion maps*) also supports the scenario of fast shocks as the main ionization mechanism

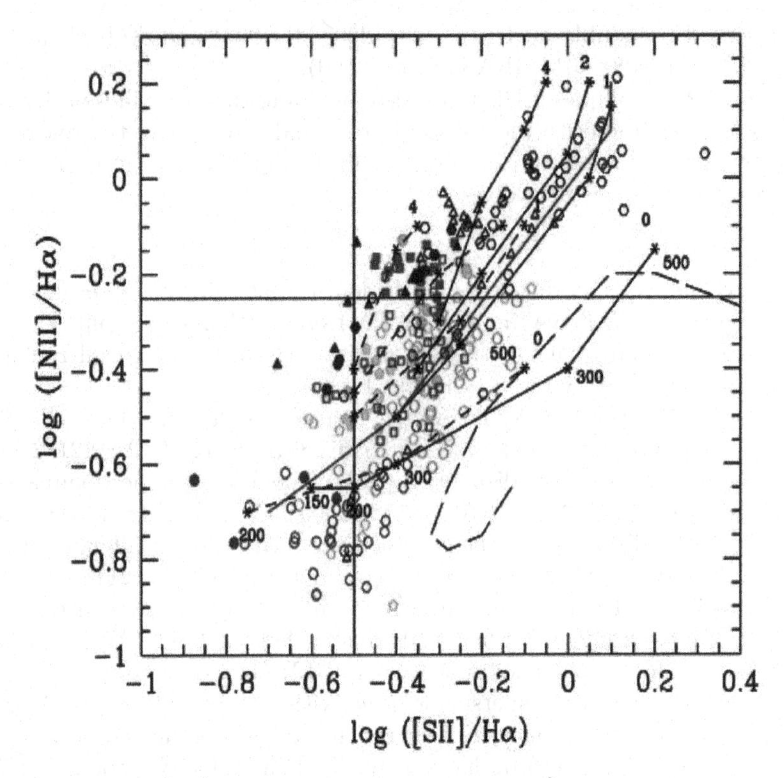

FIGURE 6.15. [N II] λ6584 Å/Hα versus [S II] $\lambda\lambda$6717,6731Å/Hα ratios. This figure can be divided into several regions. In the left bottom corner is located the region occupied by typical H II regions while the top right corner of the plot is the locus for a typical LINER-like spectrum. Different symbols represent values for different galaxies. Values for the fibres associated with the circumnuclear region are indicated with filled symbols, while those for the other fibres appear with open symbols. Models of Dopita and Sutherland (1995) for shocks have been superposed. Those without precursor are indicated with solid lines, while those with precursor are plotted using dashed lines. At the beginning of each line the magnetic parameter $Bn^{-1/2}$ (μG cm$^{3/2}$) is shown. Shocks velocity range from 150 to 500 km s^{-1} for models without precursor and from 200 to 500 for models with precursor. The long-dashed line indicates the [N II] λ6584 Å/Hα versus [S II] $\lambda\lambda$6717,6731 Å/Hα values predicted by photoionization by a power-law model for a dusty cloud at $n_e = 10^3$ cm^{-3} and Z = Z$_\odot$ (Groves *et al.*, 2004). The dotted line indicates the locus for an instantaneous burst model of 4 Myr, Z = Z$_\odot$, IMF power-law slope of -2.35 and $M_{\rm up} = 100\,{\rm M}_\odot$ (Barth and Shields, 2000); dust effects have been included and $n_e = 10^3$ cm^{-3} (figure reproduced from Monreal-Ibero *et al.*, 2006).

in these regions. Further evidence for the dominant role of shocks in the ionization of these regions comes from the positive correlation measured between the velocity dispersion and ionization status of the ionized gas in the extranuclear regions for several galaxies. Figure 6.16 represents such a correlation using the [O I] λ6300 Å/Hα line ratio, i.e. the most reliable diagnostic ratio to detect ionization by shocks (Dopita and Sutherland, 1995), and where each data point represents the value for a specific spectrum (fibre), i.e. different position in the extranuclear nebula but excluding the circumnuclear region, obtained with IFS. The fact that the extranuclear gas in a sample of galaxies covering different dynamical phases of the merging process follows a well-defined relation between the line ratio, and the velocity dispersion reinforces the scenario that tidal-induced shocks are the dominant ionization source on large scales (> 2–3 kpc). However, the fact that the ionization and velocity dispersion of the ionized gas varies from galaxy to galaxy indicates that

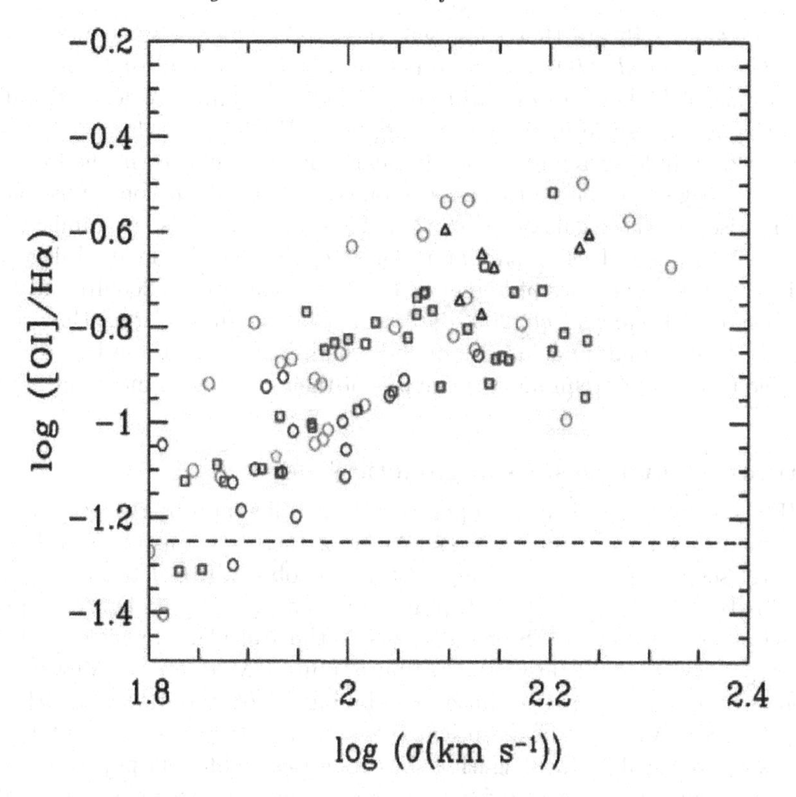

FIGURE 6.16. Relation between the shock-tracer [O I] λ6300 Å/Hα line ratio and the velocity dispersion for the extended, extranuclear gas regions in ULIRGs. Different symbols represent different galaxies. The dashed horizontal line marks the frontiers between the H II region and LINER-type ionization (figure reproduced from Monreal-Ibero *et al.*, 2006).

other mechanisms – such as ionization by a central AGN or by a distributed star formation, and turbulence due to AGN or starburst-related flows – could also be relevant in some ULIRGs. Therefore, the universality of the observed relation is not yet established and needs further investigation with a larger sample of (U)LIRGs.

6.2.8 *(U)LIRGs and high-z star-forming galaxies*

As for the low-z sample of ULIRGs investigated so far, a large fraction of high-z (>2) star-forming galaxies present the morphological features characteristic of interacting/merging systems in different phases of their evolution. There is a predominance of irregular and complex morphologies in luminous SMGs, suggesting that major mergers are common for these galaxies (Chapman *et al.*, 2003b). On the other hand, LBGs and UV-selected, high-z star-forming galaxies do often show irregular and distorted morphologies (Giavalisco *et al.*, 1996; Erb *et al.*, 2003, 2004), suggesting that interaction/merging is also involved in these galaxies. A recent quantitative analysis of the morphology of galaxies detected in the WFPC2 and NICMOS Hubble Deep Field North concludes that the fraction of galaxies consistent with undergoing a major merger increases with redshift for all galaxies, with the highest fraction (40–50%) of objects identified as LBGs (Conselice *et al.*, 2003). Moreover, at redshifts above 2 more than 80% of the stellar mass is in objects with peculiar morphology, and the quantitative structural analysis of these objects indicates that they are involved in major mergers (Concelice *et al.*, 2005). In addition, the optical colours, luminosities, sizes and star formation rates derived for SMGs (Blain *et al.*, 2002;

Frayer *et al.*, 2004) indicate that these galaxies are the high-z analogues of the low-z (U)LIRGs (Arribas *et al.*, 2004). Moreover, typical sub-arcsecond (0.2″) near- and mid-IR observations of high-z galaxies will correspond to a linear scale of about 1.5 kpc, similar to the average scale in the low-z sample of ULIRGs for which IFS is available. Therefore, a good understanding of the physical processes involved in the formation and evolution of a large fraction of the high-z galaxies will only become possible when a direct comparison of these galaxies with their low-z counterparts is established both on the same linear scale and using similar rest-frame wavelength ranges. Initial attempts at investigating the kinematics of high-z SMGs (Swinbank *et al.*, 2006b) and UV-bright galaxies (Förster Schreiber *et al.*, 2006) are now being conducted from the ground. Due to the faintness and small angular size of these high-z systems, these IFS-based studies are only able to provide preliminary estimates of their dynamical mass and evolution.

6.3 Future IFS facilities: extragalactic studies

Several IFS systems have become operational in different observatories as common user instruments over the past few years. Most of these systems are installed in large 10 m-class telescopes and are becoming a standard observational technique for almost all areas of extragalactic research. Future instruments currently under development also include a second generation of IFS systems, such as the Multi Unit Spectroscopic Explorer (MUSE; http://www.eso.org/instruments/muse/) and KMOS for the VLTs. These new instruments will provide new capabilities, combining adaptive optics with either near-IR multi-object IFS (KMOS) or deep optical IFS over a wide field of view (MUSE). Both instruments will be ideal for investigating the properties of different populations of galaxies in the high-z universe, but also for studying the nuclear regions of low-z galaxies on scales of few parsecs. IFS systems will also be installed on board the *James Webb Space Telescope* (JWST) expanding IFS-based science into the mid-IR wavelength range. The JWST Near-InfraRed Spectrograph (NIRSpec) and Mid-InfraRed Instrument (MIRI) will have integral field capabilities covering the near-IR (1–5 μm, NIRSpec) and mid-IR (5–28 μm, MIRI) range in a low background environment. In the following sections, some of the relevant characteristics of these future instruments are summarized and some of the extragalactic and cosmology science that will become available are mentioned.

6.3.1 *Optical wide field IFS on the VLT: MUSE*

MUSE is a second-generation instrument in development for the VLT of the European Southern Observatory (Bacon *et al.*, 2004). MUSE, a panoramic IFS, will be operating in the visible wavelength range from 0.47 to 0.93 μm at a resolution (R) of 2000–4000. Its main operating mode will be the wide field mode combining IFS over a relatively wide field of view (1 arcmin^2) with sub-arcsec spatial resolution (0.3″ to 0.4″ FWHM). MUSE will also have a narrow field mode providing adaptive optics, extremely high angular resolutions (0.03″ to 0.05″ FWHM) over a small field of view (7.5 × 7.5 arcsec^2).

The main scientific driver of MUSE instrument design is the study of the progenitors of normal galaxies out to a redshift of 6. These systems are extremely faint and will only be detected in large numbers through intermediate and deep Lyα line surveys down to a limiting line flux of 3.9×10^{-19} erg s^{-1} cm^{-2}, a factor 100 deeper than currently achieved with narrow band imaging (for more details, see Bacon *et al.*, 2004). In addition, the adaptive optics narrow field mode, with a limiting flux of 2.3×10^{-18} erg s^{-1} cm^{-2} in one-hour integrations, will allow us to investigate the parsec-scale kinematics of the regions surrounding the supermassive black holes in nearby galaxies and the kinematics of the gas and stellar populations in the central kiloparsec regions of low-z galaxies.

6.3.2 Near-IR multi-object IFS on the VLT: KMOS

KMOS will be a cooled multi-object, near-infrared spectrograph able to position 24 integral field units (IFUs) over a circular field of 7.2 arcmin in diameter. Each of the IFU will have a $2.8 \times 2.8\,\mathrm{arcsec}^2$ field of view with a spatial sampling of $0.2''$ and a spectral resolution (R) of 3500–4000 (http://www.eso.org/instruments/kmos/). Since bright optical lines such as the [O II] $\lambda 3727$Å, and the Hα+[N II] complex are shifted into the near-IR windows for redshifts of up to 5, this instrument will be ideal for investigating the physical processes that drive the formation and evolution of galaxies, in particular the mass assembly of galaxies and the connection between active galactic nuclei and galaxy formation. The predicted sensitivity limits will allow us to map the kinematics of the ionized gas in high redshift galaxies ($z \sim 1$ to 3) brighter than AB ~ 20 (K) and AB ~ 22 (J,H) mag arcsec^{-2} with high signal-to-noise ($> 10\sigma$ per pixel) in a 10-hour integration (Sharples *et al.*, 2004).

6.3.3 Near-IR IFS on the JWST: NIRSpec

NIRSpec will be the near-IR spectrograph for the JWST sensitive over the 0.6 μm to 5 μm wavelength range. NIRSpec contains an intermediate spectral resolution ($R \sim 1000$ to 3000) IFS, also capable of obtaining low-resolution ($R \sim 100$) spectra over the entire spectral range in one single exposure. This IFS will provide a field-of-view of $3 \times 3\,\mathrm{arcsec}^2$ with a sampling of $0.1''$, extending therefore the capabilities of ground-based near-IR IFS into the 2 μm to 5 μm window. Expected sensitivities are such that emission lines with fluxes above 2×10^{-21} W m^{-2} will be detected with a signal-to-noise of 10 in a ten thousand seconds exposure with a spectral resolution (R) of 2700 (Arribas *et al.*, 2007). Since the optical Hα emission line shifts into the 2 μm to 5 μm spectral range for redshifts of up to 6.5, this IFS will be able to investigate the two-dimensional ionization and kinematic structure of Lyman break galaxies and Luminous Infrared galaxies at very high redshifts (Arribas *et al.*, 2007; Gardner *et al.*, 2006).

6.3.4 Mid-IR IFS on the JWST: MIRI

MIRI will be the only instrument with IFS capabilities operating in the 5 to 28 μm range (Wright *et al.*, 2004; European Consortium web page at http://www.roe.ac.uk/ukatc/consortium/miri/miri.html). In contrast to ground-based mid-IR instruments, MIRI will be covering this spectral range continuously and will be affected by a thermal background level orders of magnitude below that of the atmosphere. The entire spectral range covered by MIRI will be divided in four different channels designed such that the spectral resolution will be about the same (R of 2400 to 4000). The channels will cover a field of view of $3.6 \times 3.6\,\mathrm{arcsec}^2$ to $7.5 \times 7.5\,\mathrm{arcsec}^2$ with a Nyquist-optimized angular sampling corresponding to pixels of about $0.2 \times 0.2\,\mathrm{arcsec}^2$ to about $0.3 \times 0.6\,\mathrm{arcsec}^2$. The required sensitivities guiding the design of the IFS are $1\text{–}6 \times 10^{-20}\,\mathrm{W\,m^{-2}}$ for a 10σ emission line detection in a ten-thousand second exposure (Swinyard *et al.*, 2004). MIRI will therefore be more than three orders of magnitude more sensitive than the best ground-based mid-IR instrument on a 10 m-class telescope and about 50 times more sensitive than Spitzer, while increasing the spatial and spectral resolution by an order of magnitude with respect to that of Spitzer.

As shown by the IFS-based studies of low-z (U)LIRGs, galaxy interactions can trigger powerful episodes of star formation and accretion on to nuclear supermassive black holes that are deeply shrouded in the interstellar medium of the colliding galaxies. These processes can be studied up to redshifts of 2 to 2.5 through the strong near- and mid-IR emission lines shifted into the 20 to 28 μm range and to very high redshifts ($z > 6.5$)

using the Hα+[N II] optical emission lines shifted into the 5 to 7 μm window (Gardner *et al.*, 2006).

6.4 Concluding remarks

Although still a young technique, IFS is becoming more and more popular and is now used as a standard tool in a whole variety of research topics covering the low- and high-z Universe. Compared to pure broad- and narrow-band imaging and to long-slit spectroscopy, the capabilities of IFS make this technique particularly suitable for the investigation of the complex two-dimensional stellar, gas and kinematic structure of galaxies. The future of IFS for extragalactic studies looks very promising. Almost every large 10 m-class telescope is now equipped with at least one IFS, and all large facilities are now designing future IFS systems including multi-object IFUs and wide-field IFS. Future extremely large telescopes (diameter of 25 m, or more) will also include IFS systems as part of the standard instrumentation. There is, however, an important challenge ahead of us before the power of these systems is fully exploited. These future instruments will produce a huge amount 10–100,000 of spatially separated spectra per exposure. Therefore, emphasis has to be made on the development of new sophisticated software tools to automatically handle the calibration of the raw data and to facilitate their subsequent astrophysical analysis.

6.5 Acknowledgements

Based on observations with the William Herschel Telescope operated on the island of La Palma by the ING in the Spanish Observatorio del Roque de los Muchachos of the Instituto de Astrofísica de Canarias. Also based on observations with the NASA/ESA *Hubble Space Telescope*, obtained at the Space Telescope Science Institute, which is operated by the Association of Universities for Research in Astronomy, Inc. under NASA contract number NAS5-26555.

Much of the work presented here has been supported by the Ministry for Education and Science of Spain under its National Programme for Astronomy and Space Science. I would like to thank my collaborators over the years: Santiago Arribas, Ana Monreal-Ibero, Macarena García-Marín, and Almudena Alonso-Herrero. Finally, a lot of the material presented during the lectures was kindly made available by many colleagues who allowed me to use material given in their PowerPoint presentations. My thanks to Drs Cardiel, Davies, Falcón-Barroso, Hammer, Jarvis, Kuntschner, Lemoine-Busserole, McDermid, Mollá, Swinbank, and Villar-Martín.

REFERENCES

Arribas, S., Colina, L. (2002), *ApJ*, **573**, 576
Arribas, S., Colina, L. (2003), *ApJ*, **591**, 791
Arribas, S., Mediavilla, E. (1994), *ApJ*, **437**, 149–161
Arribas, S., Mediavilla, E., García-Lorenzo, B. (1996), *ApJ*, **463**, 509
Arribas, S., Mediavilla, E., García-Lorenzo, B., del Burgo, C. (1997), *ApJ*, **490**, 227
Arribas, S., Mediavilla, E., del Burgo, C., García-Lorenzo, B. (1999), *ApJ*, **511**, 680
Arribas, S., Colina, L., Borne, K.D. (2000), *ApJ*, **545**, 228
Arribas, S., Colina, L., Clements, D. (2001), *ApJ*, **560**, 160
Arribas, S., Bushouse, H., Lucas, R.A., Colina, L., Borne, K.D. (2004), *AJ*, **127**, 2522
Arribas, S., Ferrint, P., Jakobsen, P. (2007), in *Science Perspectives for 3D Spectroscopy*. ESO Astrophysics Symposia, eds. M. Kissler-Patig, M. Roth & J. Walsh. Springer

Bacon, R., Copin, Y., Monnet, G. et al. (2001), *MNRAS*, **326**, 23
Bacon, R., Bauer, S.-M., Bower, R. et al. (2004), in *Ground-based Instrumentation for Astronomy*, eds. A.F.M. Moorwood & I. Masanori, Proceedings of the SPIE, **Vol. 5492**, 1145
Baldwin, J.A., Phillips, M.M., Terlevich, R. (1981), *PASP*, **93**, 5
Ballo, L., Braito, V., Della Ceca, R. et al. (2004), *ApJ*, **600**, 634
Barth, A.J., Shields, J.C. (2000), *PASP*, **112**, 753
Blain, A. W., Smail, I., Ivison, R.J., Kneib, J.-P., Frayer, D.T. (2002), *Physics Reports*, **369**, 111
Bushouse, H.A., Borne, K.D., Colina, L. et al. (2002), *ApJS*, **138**, 1
Chapman, S.C., Blain, A.W., Ivison, R.J., Smail, I.R. (2003a), *Natur*, **422**, 695
Chapman, S.C., Windhorst, R., Odewahn, S., Yan, H., Conselice, C. (2003b), *ApJ*, **599**, 92
Charmandaris, V., Uchida, K.I., Weedman, D. et al. (2004), *ApJS*, **154**, 142
Cimatti, A., Daddi, E., Renzini, A. et al. (2004), *Natur*, **430**, 184
Colina, L., Arribas, S., Borne, K.D. (1999), *ApJ*, **527**, L13
Colina, L., Arribas, S., Borne, K.D., Monreal, A. (2000), *ApJ*, **533**, L9
Colina, L., Borne, K., Bushouse, H. et al. (2001), *ApJ*, **563**, 546
Colina, L., Arribas, S., Clements, D. (2004), *ApJ*, **602**, 181
Colina, L., Arribas, S., Monreal-Ibero, A. (2005), *ApJ*, **621**, 725
Conselice, C.J., Bershady, M.A., Dickinson, M., Papovich, C. (2003), *AJ*, **126**, 1183
Conselice, C.J., Blackburne, J.A., Papovich, C. (2005), *ApJ*, **620**, 564
Dasyra, K. M., Tacconi, L.J., Davies, R.I. et al. (2006), *ApJ*, **651**, 835
Davies, R. I., Thomas, J., Genzel, R. et al. (2006), *ApJ*, **646**, 754
Della Ceca, R., Ballo, L., Tavecchio, F. et al. (2002), *ApJ*, **581**, L9
de Zeeuw, P.T., Bureau, M., Emsellem, E. et al. (2002), *MNRAS*, **329**, 513
Díaz-Santos, T., Alonso-Herrero, A., Colina, L., Ryder, S.D., Knapen, J.H. (2007), *ApJ*, **661**, 149
Dopita, M.A., Sutherland, R.S. (1995), *ApJ*, **455**, 468
Downes, D., Solomon, P.M. (1998), *ApJ*, **507**, 615
Egami, E., Dole, H., Huang, J.-S. et al. (2004), *ApJS*, **154**, 130
Emsellem, E., Cappellari, M., Peletier, R.F. et al. (2004), *MNRAS*, **352**, 721
Erb, D.K., Shapley, A.E., Steidel, C.C. et al. (2003), *ApJ*, **591**, 101
Erb, D. K., Steidel, C.C., Shapley, A.E., Pettini, M., Adelberger, K.L. (2004), *ApJ*, **612**, 122
Farrah, D., Rowan-Robinson, M., Oliver, S. et al. (2001), *MNRAS*, **326**, 1333
Ferruit, P., Mundell, C.G., Nagar, N.M. et al. (2004), *MNRAS*, **352**, 1180
Flores, H., Hammer, F., Puech, M., Amram, P., Balkowski, C. (2006), *A&A*, **455**, 107
Förster Schreiber, N.M., van Dokkum, P.G., Franx, M. et al. (2004), *ApJ*, **616**, 40
Förster Schreiber, N.M., Genzel, R., Lehnert, M.D. et al. (2006), *ApJ*, **645**, 1062
Franx, M., Labbé, I., Rudnick, G. et al. (2003), *ApJ*, **587**, L79
Frayer, D.T., Armus, L., Scoville, N.Z. et al. (2003), *AJ*, **126**, 73
Frayer, D.T., Reddy, N.A., Armus, L. et al. (2004), *AJ*, **127**, 728
Gallais, P., Charmandaris, V., Le Floc'h, E. et al. (2004), *A&A*, **414**, 845
Ganda, K., Falcón-Barroso, J., Peletier, R.F. et al. (2006), *MNRAS*, **367**, 46
García-Lorenzo, B., Mediavilla, E., Arribas, S., del Burgo, C. (1997), *ApJ*, **483**, L99
García-Lorenzo, B., Mediavilla, E., Arribas, S. (1999), *ApJ*, **518**, 190
García-Lorenzo, B., Arribas, S., Mediavilla, E. (2001), *A&A*, **378**, 787
García-Lorenzo, B., Sánchez, S.F., Mediavilla, E., González-Serrano, J.I., Christensen, L. (2005), *ApJ*, **621**, 146
García-Marín, M., Colina, L., Arribas, S., Alonso-Herrero, A., Mediavilla, E. (2006), *ApJ*, **650**, 850
García-Marín, M. et al., in preparation
Gardner, J.P., Mather, J.C., Clampin, M. et al. (2006), *SSRv*, **123**, 485
Genzel, R., Tacconi, L.J., Rigopoulou, D., Lutz, D., Tecza, M. (2001), *ApJ*, **563**, 527
Giavalisco, M., Steidel, C.C., Macchetto, D. (1996), *ApJ*, **470**, 189
Groves, B.A., Dopita, M.A., Sutherland, R.S. (2004), *ApJS*, **153**, 9
Heckman, T.M., Armus, L., Miley, G.K. (1990), *ApJS*, **74**, 833
Le Floc'h, E., Pérez-González, P.G., Rieke, G.H. et al. (2004), *ApJS*, **154**, 170

Lutz, D., Spoon, H.W.W., Rigopoulou, D., Moorwood, A.F.M., Genzel, R. (1998), *ApJ*, **505**, L103
McDermid, R.M., Bacon, R., Kuntschner, H. *et al.* (2006), *New Astronomy Review*, **49**, 521
McDowell, J.C., Clements, D.L., Lamb, S.A. *et al.* (2003), *ApJ*, **591**, 154
Mediavilla, E., Arribas, S., García-Lorenzo, B., del Burgo, C. (1997), *ApJ*, **488**, 682
Mediavilla, E., Arribas, S., del Burgo, C. *et al.* (1998), *ApJ*, **503**, L27
Mediavilla, E., Guijarro, A., Castillo-Morales, A. *et al.* (2005a), *A&A*, **433**, 79
Mediavilla, E., Muñoz, J.A., Kochanek, C.S. *et al.* (2005b), *ApJ*, **619**, 749
Monreal-Ibero, A. (2004), *Estudio cinemático y de fuentes de ionización de galaxias infrarrojas ultraluminosas con espectroscopa bidimensional*, PhD, Universidad de La Laguna
Monreal-Ibero, A., Arribas, S., Colina, L. (2006), *ApJ*, **637**, 138
Motta, V., Mediavilla, E., Muñoz, J.A. *et al.* (2002), *ApJ*, **574**, 719
Motta, V., Mediavilla, E., Muñoz, J.A., Falco, E. (2004), *ApJ*, **613**, 86
Mueller Sánchez, F., Davies, R.I., Eisenhauer, F. *et al.* (2006), *A&A*, **454**, 481
Pérez-González, P.G., Rieke, G.H., Egami, E. *et al.* (2005), *ApJ*, **630**, 82
Rieke, G.H., Low, F.J. (1972), *ApJ*, **176**, L95
Sánchez, S.F., García-Lorenzo, B., Mediavilla, E., González-Serrano, J.I., Christensen, L. (2004), *ApJ*, **615**, 156
Sarzi, M., Falcón-Barroso, J., Davies, R.L. *et al.* (2006), *MNRAS*, **366**, 1151
Scoville, N.Z., Yun, M.S., Bryant, P. (1997), *ApJ*, **484**, 702
Sharples, R.M., Bender, R., Lehnert, M.D. *et al.* (2004), in *Ground-based Instrumentation for Astronomy*, eds. A.F.M. Moorwood & I. Masanori, Proceedings of the SPIE, **5492**, 1179
Soifer, B. T., Sanders, D.B., Madore, B.F. *et al.* (1987), *ApJ*, **320**, 238
Solomon, P.M., Downes, D., Radford, S.J.E., Barrett, J.W. (1997), *ApJ*, **478**, 144
Swinbank, A.M., Smith, J., Bower, R.G. *et al.* (2003), *ApJ*, **598**, 162
Swinbank, A.M., Smail, I., Bower, R.G. *et al.* (2005), *MNRAS*, **359**, 401
Swinbank, A.M., Bower, R.G., Smith, G.P. *et al.* (2006a), *MNRAS*, **368**, 1631
Swinbank, A.M., Chapman, S.C., Smail, I. *et al.* (2006b), *MNRAS*, **371**, 465
Swinyard, B.M., Rieke, G.H., Ressler, M. *et al.* (2004), in *Optical, Infrared and Millimeter Space Telescopes*, eds. J.C. Mather, Proceedings of the SPIE, **Vol. 5487**, 785
Tacconi, L.J., Genzel, R., Lutz, D. *et al.* (2002), *ApJ*, **580**, 73
Tecza, M., Baker, A.J., Davies, R.I. *et al.* (2004), *ApJ*, **605**, L109
Tully, R.B., Fisher, J.R. (1977), *A&A*, **54**, 661
Urry, M., Padovani, P. (2000), *PASP*, **112**, 1516
van Dokkum, P.G., Franx, M., Förster Schreiber, N.M. *et al.* (2004), *ApJ*, **611**, 703
Veilleux, S., Kim, D.-C., Sanders, D.B. (1992), *ApJS*, **522**, 113
Villar-Martín, M., Sánchez, S.F., De Breuck, C. *et al.* (2006), *MNRAS*, **366**, L1
Wright, G.S., Rieke, G.H., Colina, L. *et al.* (2004), in *Optical, Infrared and Millimeter Space Telescopes*, eds. J.C. Mather, Proceedings of the SPIE, **5487**, 653

7. Tutorials: How to handle 3D spectroscopy data

SEBASTIÁN F. SANCHEZ, BEGOÑA GARCÍA-LORENZO AND ARLETTE PÉCONTAL-ROUSSET

7.1 Introduction

Integral field spectroscopy (IFS) is a technique to obtain both spatial (x,y) and spectral (λ) information of a more or less continuous area of the sky simultaneously on the detector. Only a few instrumental concepts allow 3D information on 2D detectors to be obtained, and all of these are based on field splitters such as fibre bundles, lens array, or image slicers (see Figure 7.1) to sample the field of view. Each sampled element is then dispersed using a classic spectrograph and produces a spectrum on the detector. Depending on the field splitter used, the geometry of the spectra on the detectors may be very different. This diversity leads to the creation of very specific reduction techniques and/or packages, i.e. one per instrument built (e.g. P3d, Becker, 2001). Combined with the inherent complexity of 3D techniques, such software diversity has reduced the use of IFS for decades to a handful of specialists, mainly those involved in the teams building such instruments.

Conscious that this would be a handicap IFS specialists Walsh and Roth (2002) have started to standardize techniques and tools for integral field units (IFU). Recently, the Euro3D Research Training Network (RTN), whose aim was to promote 3D spectroscopy all over Europe (Walsh and Roth, 2002), made a great effort to create a standard data format (Kissler-Patig *et al.*, 2004) for storing and exchanging 3D data, developing an application programming interface, API (Pécontal-Rousset *et al.*, 2004), to ease the use of such a data format and creating a visualization tool (Sánchez, 2004) usable by any existing IFU. All these tools are freely distributed to the community (http://www.aip.de/Euro3D). However, there is still a lack of software packages of general use for most of the data reduction and data analysis processes (reduction steps, line fitting, gas and stellar kinematics, etc.) of 3D data. The astronomical community faces a plethora of different pipelines and tools specific to each instrument. We have developed a set of tools and procedures useful for most of the IFS data; these will be explained in these tutorials, as they were presented in the Instituto de Astrofísica de Canarias (IAC) Winter School. Many of these tools are still in the development process, and we are distributing the current version without any guarantee that they will work in the optimal way. The aim is to provide simple and standardized tools, and also to teach the user the main logical steps he or she will face in trying to reduce and then analyse his or her 3D data.
The layout of the chapter will be as follows:

- Section 7.2 describes the most widely used data formats to store 3D spectroscopy data.
- Section 7.3 describes some of the existing tools to analyse 3D data before the Euro3D network.
- Section 7.4 describes the installation requirements of the different packages used in these tutorials.
- Section 7.5 introduces `E3D`, the Euro3D visualization tool (Sánchez, 2004).
- The sky background subtraction is described in Section 7.6.
- Section 7.7 introduces some notions of emission-line fitting and gas kinematics.
- Crowded field spectroscopy and 3D modelling is described in Section 7.8.

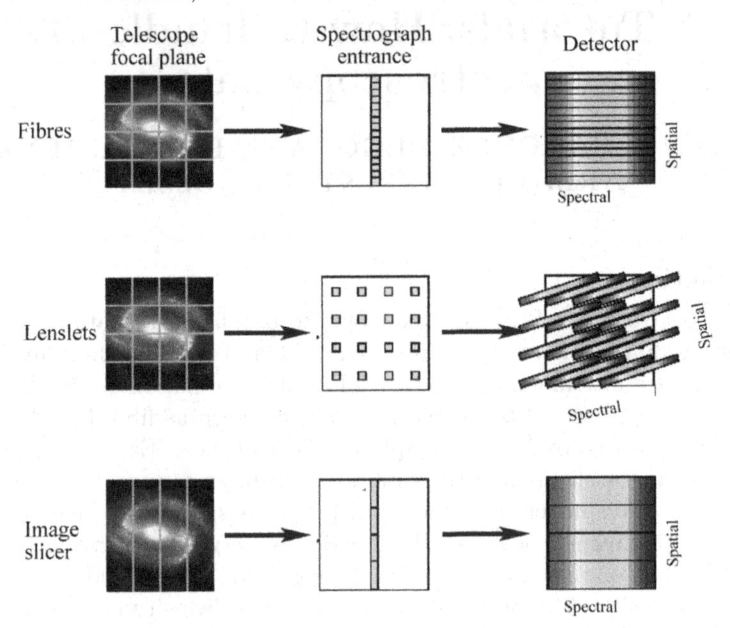

FIGURE 7.1. Layout of the main techniques used to perform integral field spectroscopy (Allington-Smith and Content, 1998).

- Section 7.9 describes the empirical correction of the differential atmospheric refraction.
- Section 7.10 describes the data reduction process with fibre-feed spectrographs.

7.2 Data formats for storing IFS reduced data

IFS data can be stored using different data formats, as far as it is possible to store the spectral information associated with the position on the sky, to which it belongs, and the shape of the spaxel (which stands for sampled spatial element; see Kissler-Patig *et al.*, 2004). The various existing data formats were designed by software developers to meet instrument specifications, and are, in general, only valid for a specific instrument. The exchange and comparison of data from different integral field instruments was therefore difficult to achieve.

After several discussions and brainstorming sessions, the Euro3D network created a standard format, the Euro3D format (Kissler-Patig *et al.*, 2004), whose aim is to share data and analysis tools. This data format is based on FITS (Flexible Image Transport System), a widely used data format in the astronomical community all over the world. It consists mainly in a binary table, where the position of the spatial sampled elements (spaxels) on the sky are stored on the same row as the associated spectra. Although this format is now intended to be a standard for any reduced 3D data, other formats (pre-Euro3D network) are still in use. Two of them are widely used for IFS: datacubes (3D image) and row-stacked spectra (RSS hereafter).

Datacubes are only valid to store reduced data from instruments that sample the sky plane in a regular grid or for interpolated data. The data are stored in a 3D FITS image, were the x and y axes stand for the two spatial dimensions, and the Z axis corresponds to the dispersion axis.

RRS format is a 2D FITS image where the x and y axes contain the spectral and spatial information regardless of their position in the sky. It requires an additional file (a FITS or ASCII table), where the position of the different spatial elements on the sky is stored. RRS is widely used by IFUs where spectra are well-separated on the detector (e.g. INTEGRAL, GMOS or Gemini Multi-Object Spectrograph).

7.3 Tools before the Euro3D network

The analysis of 3D data is not an easy task due to the huge amount of information to process. Before the Euro3D network, there was no dedicated software to analyse 3D data, in part due to the lack of a standard format for storing IFU data as we have already mentioned. Scientists working with 3D used to combine well-known and widely used spread astronomical software like IRAF (Image Reduction and Analysis Facility), MIDAS (Mid-Infrared Asteroid Spectroscopy), STARLINK, etc., with home-made software developed using their favourite programming language (fortran, C, IDL, etc.). A common procedure was, for example, to have a look to 3D data using IRAF (with the DS9 visualization tool), then fit spectral lines using DIPSO, and create images with a home-made FORTRAN program.

In this section, we will briefly describe some ways to visualize and analyse 3D data using traditional astronomical packages, but many other software packages may also be used. The user should always keep in mind that these tools were not specifically designed to handle IFS. However, some of them have recently included a few tasks to process 3D data due to the increasing popularity of 3D techniques in the last years.

7.3.1 *DS9 visualization tool*

DS9 (http://hea-www.harvard.edu/RD/ds9/) is an astronomical imaging and data visualization application. It handles FITS images and binary tables, multiple frame buffers, region manipulation, many scaling algorithms and colour maps, mosaic images, tiling, blinking, geometric markers, arbitrary zoom, rotation, pan, and a variety of coordinate systems.

- To display an RSS image with DS9, first select the File menu and then select Open.
- To display datacubes (i.e. 3D images) with DS9, do the same as above (i.e. select the File menu and then the Open option). Once you have choosen the file to be loaded, a pop-up window comes up entitled 'Data Cube', with a scroll bar that allows you to scroll through spectral pixels (i.e. wavelength). Moving that scroll bar, DS9 displays the slice corresponding to the selected wavelength (monochromatic image).
- To display datacube stores in multi extension FITS file, use the File menu and then the 'Open Other' option and the 'Open Multi Ex as Data Cube' option.

DS9 can be also run from the shell, entering ds9 followed by the name of the FITS image (e.g. ds9 test.fits). For multiple extension FITS files, use the metacube option of DS9 (e.g. ds9 -medatacube test.fits).

7.3.2 *Checking 3D data using IRAF*

IRAF (http://iraf.noao.edu/) is a popular tool to reduce, visualize and analyse astronomical data. IRAF needs a visualization tool such as DS9 (Section 7.3.1) to display data. It is possible to visualize datacubes slice by slice through the IRAF command window using the *display* command. The user should know the structure of its data FITS file. Let's say the datacube contains more than two axes (i.e. NAXES>2 in the FITS header). Use

```
display test.cube.fits[*,*,n],
```

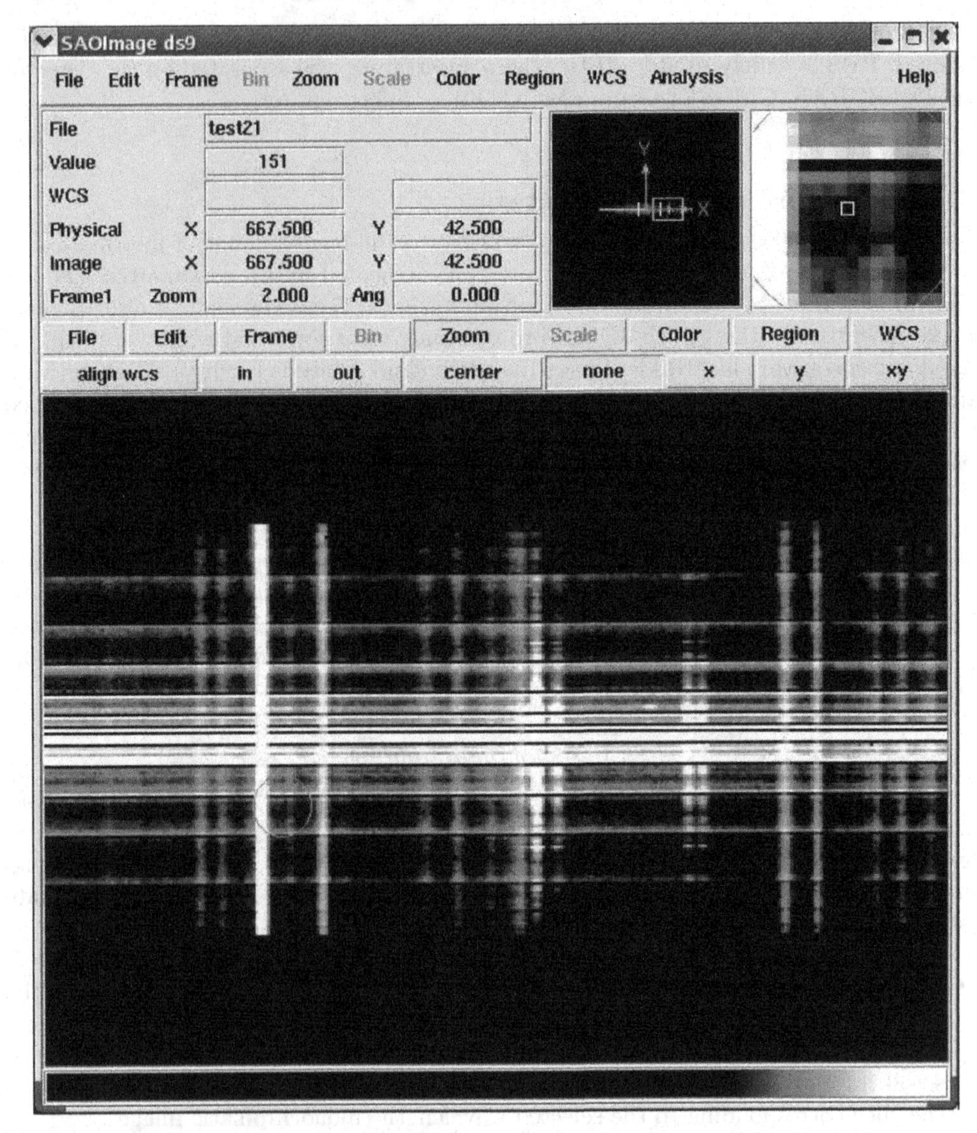

FIGURE 7.2. DS9 displaying an RSS image. The different rows correspond to the various spectra of a particular area of an object (an active galaxy in this case). The horizontal axis is wavelength. Vertical lines covering all the spectra are sky lines. A few emission lines from the observed object can also be identified in this frame.

n being the value of the third dimension to display. If it is a multiple extension format, use

```
display test.cube1.fits[n][*,*],
```

where n is the extension number to display.
You can also inspect the different spectra using the *splot* command:

```
splot test.cube.fits[10,10,*].
```

The RSS images (2D images) may also be visualized using the *display* command:

```
display test.ms.fits.
```

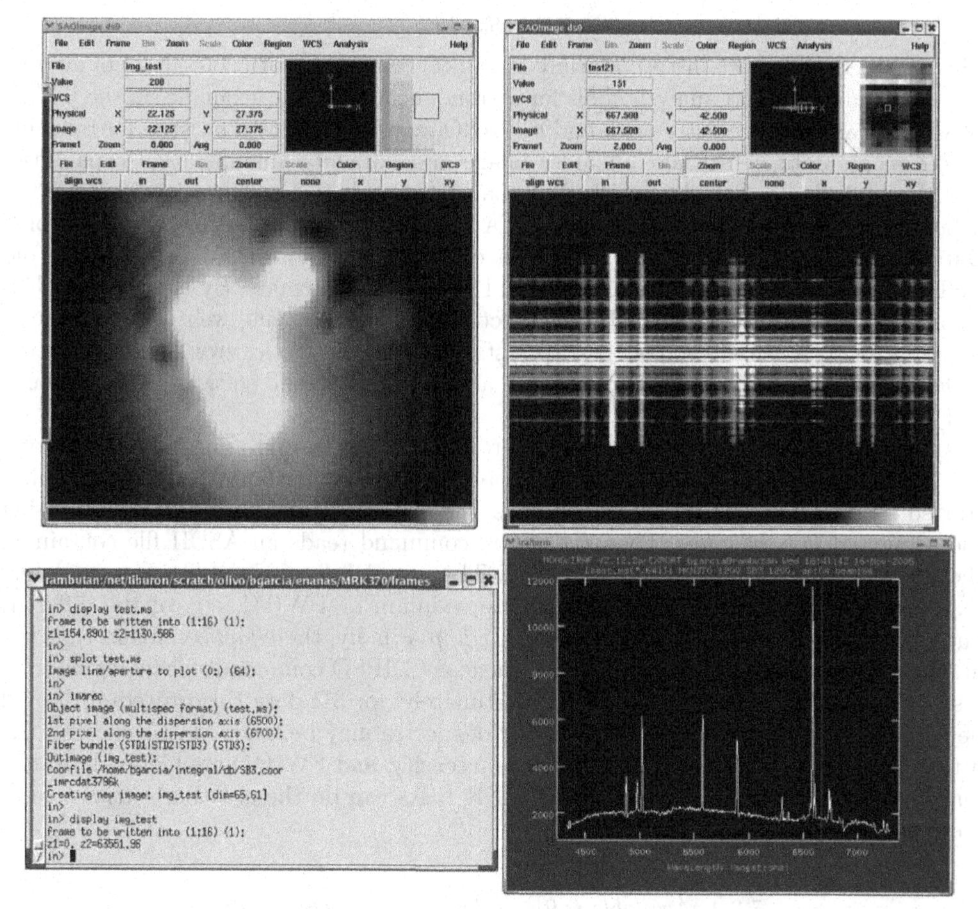

FIGURE 7.3. Visualizing 3D data using IRAF and imarec. The snapshot of the console shows the different IRAF tasks used to visualize RSS data.

To have a look at the spectra, use the command *splot*:

```
splot test.ms.fits
```

and select a spectrum. It is possible to overplot other spectra using the options of the *splot* command.

However, IRAF has no routine to visualize monochromatic (data from a given wavelength) or filtered images from RSS data. The INTEGRAL team[1] has developed a task called *imarec* (which stands for image reconstruction) to extract and view monochromatic images from RSS images and ASCII files containing the positions of the spectra. *Imarec* is based on standard IRAF routines and a FORTRAN program (more information on IRAF routines developed for INTEGRAL data can be found at http://www.iac.es/proyect/integral/). *Imarec* prompts the user for some parameters and produces a filtered image that can be displayed in DS9. Figure 7.3 shows the different steps to visualize RSS data in IRAF.

[1] INTEGRAL (http://www.iac.es/proyect/integral/) is a 3D instrument at the William Herschel Telescope (WHT) (Arribas *et al.*, 1998).

7.3.3 *Starlink*

The STARLINK project (http://starlink.rl.ac.uk/) compiled (until 2005) a set of interactive data reduction and analysis tools for astronomical data. For example, in the STARLINK software package list we can find FIGARO, a great reduction package for reducing spectroscopic data, which also includes image and datacube manipulation capabilities (for more information see http://starlink.rl.ac.uk/star/docs/sun86.htx/sun86.html). By default, FIGARO accesses data files in STARLINK's NDF (Extensible N-Dimensional Data Format) data format, but it can also access other data formats, like FIGARO's old DST (Data Storage Technology) format, FITS and IRAF formats. By the way, IRAF is a software facility of the STARLINK project. Check the following web page for a complete list of STARLINK software: http://starlink.rl.ac.uk/static_www/soft2_LSF.html. Unfortunately, the STARLINK project has already finished and software packages such as FIGARO are unsupported.

As an example of the STARLINK software capabilities for analysing 3D data, we will use DIPSO, a friendly spectrum analysis program, for fitting lines. Once the user has started the STARLINK and DIPSO sessions, the plotting device should be defined using the command *dev xwindow.* Then the *alasrd* command reads an ASCII file containing the spectral data and relevant information. The initial values of the fitting parameters (e.g. central wavelengths, full-width-at-half-maximum or FWHM, etc. of the different line components) are defined using the task *elfinp.* Finally, the *elfopt* command runs the fit. Figure 7.4 shows both the complete sequence of DIPSO commands for fitting lines on a spectrum and a plot of the results. Unfortunately, for 3D data this procedure should be done once per spectrum and the number of spectra may be huge. Moreover, after the fitting, maps of the line parameters (such as intensity, and FWHM) are highly desirable, but neither DIPSO nor any other STARLINK tasks can do these. Other software tools must be used to create such maps.

7.3.4 *Available software for specific instruments*

Currently, a large number of integral-field spectrographs (IFSs) are operated around the world and, most of the time, the team who built the instrument provides the community with dedicated data reduction software; however, few of them provide tools to analyse the data.

One example is the set of XOasis/XSauron and XSnifs software (see, for example, http://www.ing.iac.es/Astronomy/instruments/oasis/xoasis.html), the data reduction and processing packages designed to handle data from the OASIS (Optically Adaptive System for Imaging Spectroscopy), SAURON (Spectrographic Areal Unit for Research on Optical Nebulae) and SNIFS (Supernovae Integral Field Spectrograph) instruments built at the Centre de Recherche Astrophysique de Lyon (CRAL). OASIS (http://www-obs.univ-lyon1.fr/labo/oasis) is a multi-mode spectro-imager, previously operated on the CFHT, and now installed at the WHT. SAURON (http://www.strw.leidenuniv.nl/sauron) is another IFU installed at the WHT and optimized for studies of the stellar kinematics, gas kinematics and line-strength distributions of nearby spheroids. The last one, SNIFS; http://snfactory.lbl.gov, is currently operated at the University of Hawaii (UH) telescope at Mauna Kea and dedicated to the survey of Type Ia supernovae (part of the SNFactory project, whose aim is to measure the expansion history of the Universe and explore the nature of Dark Energy).

All of these software codes share the same development basis (using C, Fortran and tcl/tk), and were designed to be run through a pipeline, or via shell scripts, or in interactive mode using a more user-friendly interface (see Figure 7.5). Through a menu-driven

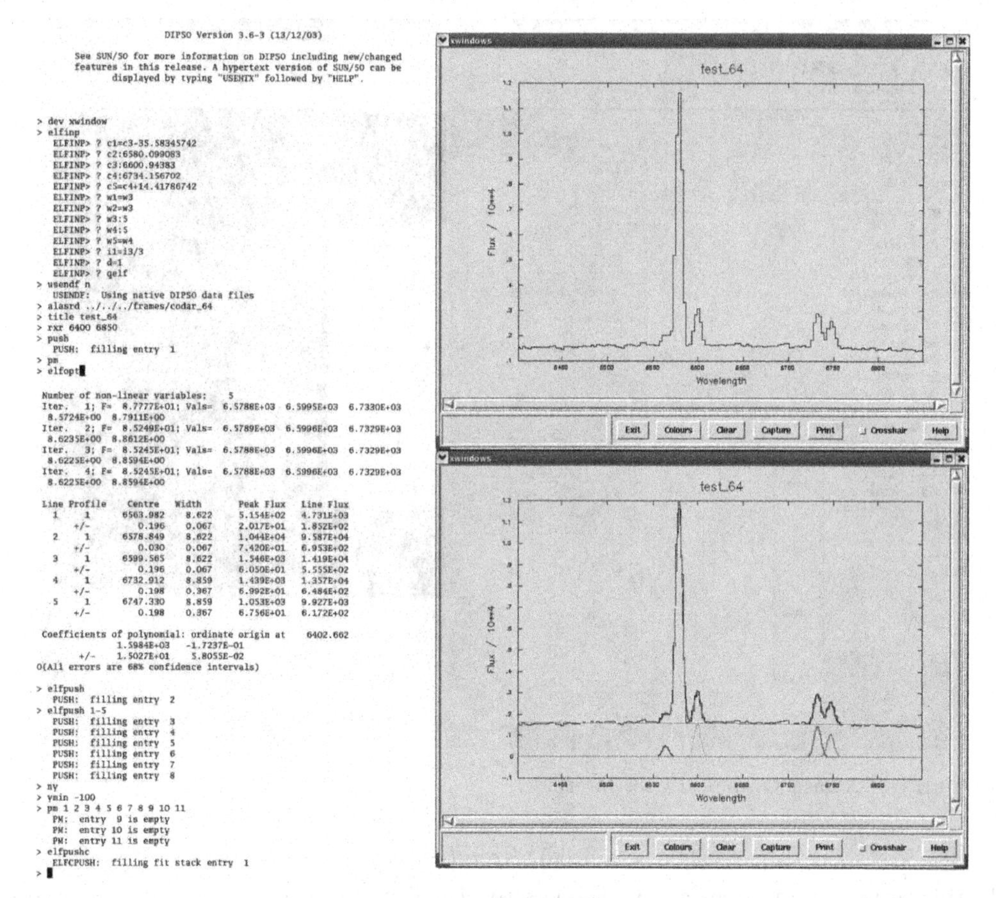

FIGURE 7.4. Line fitting using DIPSO (from the STARLINK software).

graphical user interface (GUI), they provide the user with all the necessary data reduction steps, plus a set of generic graphical tools to view or plot data, assign keywords, etc. They also provide analysis tools like image reconstruction (maps), spectra summation over an aperture, line fitting (developed externally and currently being incorporated), merging of datacubes, etc. They all handle different data formats, and are currently debugged to fully support Euro3D. Binaries of the XOasis software are made available to the user community and may be downloaded from the Oasis Webserver, while others may be obtained through the consortia who develop the instruments. The last ones usually contain more sophisticated analysis tools like line-of-sight velocity distribution (LOSVD) fitting, etc.

Another example is IDA. INTEGRAL team developed IDA, the Integral Data Analysis tool (García-Lorenzo *et al.*, 2002). IDA was written in interactive data language (IDL) and offers a widget-based interface. IDA uses an interactive display tool, ATV (for more information about ATV, see http://www.physics.uci.edu/~barth/atv/), which allows control of images similar to other display tools like DS9. IDA plots individual spectra, creates and displays maps, overplots spectra on recovered images, zooming in and zooming out, and calculates velocity fields through the cross-correlation technique. Although IDA was written to work with INTEGRAL data, it can be also used for similar data from other IFSs through minor software changes. Figure 7.6 shows some of the IDA capabilities for analysing INTEGRAL data.

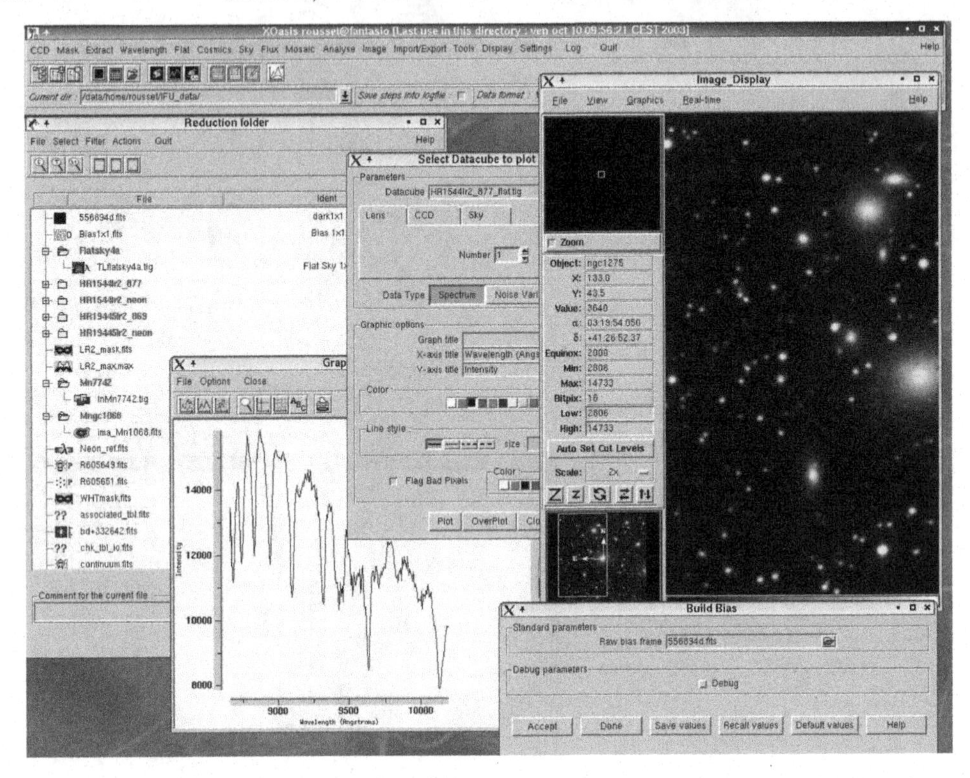

FIGURE 7.5. XOasis Graphical User Interface.

7.3.5 *Exercises*

In this section, we present a few exercises to visualize 3D data using traditional astronomical software. Hereafter, we use the following filename convention:

- RSS → NAME_RSS.fits (with associated position table NAME_RSS.pt.txt);
- datacube → NAME.cube.fits;
- Euro3D format data → NAME.e3d;
- Position Table → NAME.pt.

Using IRAF

To run IRAF, type `cl` or `ncl` at any *xterm* or *xgterm*:

```
cl> imred NAME_RSS.fits
cl> specred NAME_RSS.fits
```

It is possible to zoom in, zoom out, fit lines, plot different lines, with all the options of the gtools. For more help, see help/splot.txt and help/gtools.txt

Using DS9 through IRAF

To display datacubes slices to slices.

```
cl> disp NAME.cube.fits[*,*,500]
```

To display RSS spectra:

```
cl> disp NAME_rss.fits
```

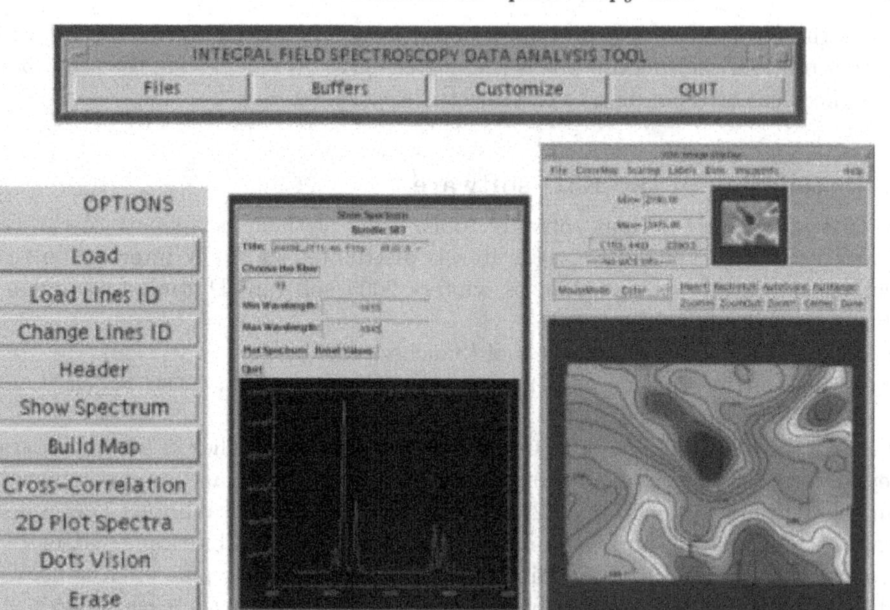

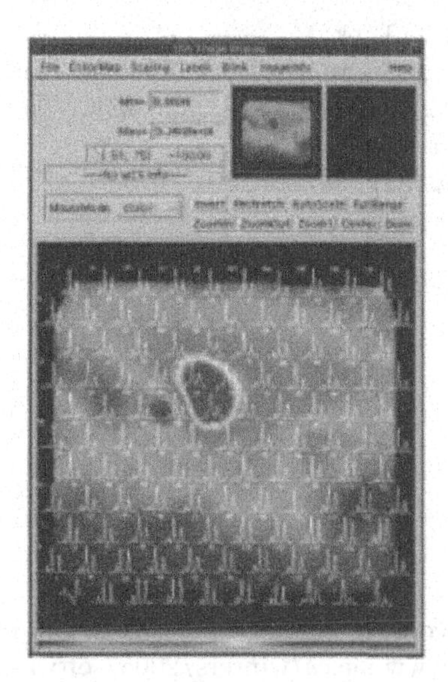

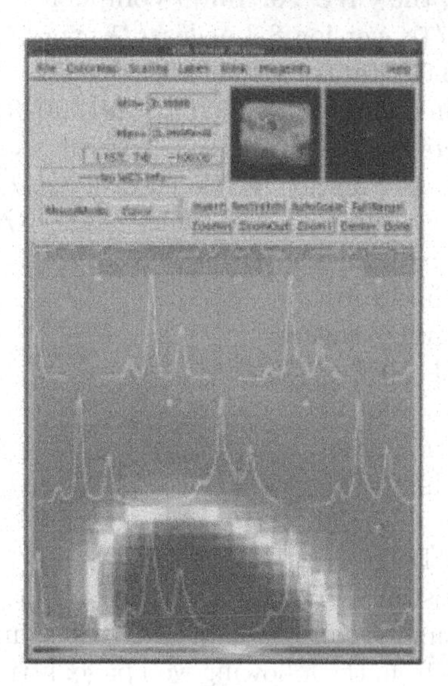

FIGURE 7.6. The INTEGRAL Data Analysis (IDA) tool capabilities for visualizing and analysing 3D data.

To display spectra from a datacube.

cl> disp NAME.cube.fits[10,*,*]
cl> splot NAME.cube.fits[10,10,*]

Using DS9 outside of IRAF (datacubes)

To display datacubes, using DS9 GUI:

Menu: File → Open

Once the file is loaded, a pop-up window comes up, with a scroll bar allowing the user to scroll over the spectral pixels (i.e. wavelength). While moving the scroll bar, DS9 shows the slice corresponding to the selected wavelength.

7.4 Installation of the E3D software

The tutorial will use various software tools, some of them being part of the Euro3D network activities. The software (as it was distributed at the Winter School) can be downloaded from: http://www.iac.es/winters/2005/software3D.html. In summary, the software tools distributed are:

- `E3D`, the Euro3D visualization tool (Sánchez, 2004);
- `FIT3D`, a fitting tool for 3D spectroscopy (Sánchez *et al.*, 2004; Sánchez *et al.*, in preparation);
- `R3D`, a software package for reducing fibre-based IFS (Sánchez *et al.*, in preparation).

They all have a C-core written using the I/O Lyon-C Library and its development environment (Pécontal-Rousset *et al.*, 2004). The PGPLOT graphical library (Pearson, 1995) is used also for plotting purposes and, in some cases, `Tcl/Tk` libraries for GUIs. In many cases, `perl` is also used as a scripting language, or as a prototyping environment.

Here is the list of libraries and binaries required to build or run such tools:

- `gcc`, the GNU C compiler
- `g77`, the GNU Fortran77 compiler[2]
- `Tcl/Tk`, version 8.3 at least. `http://www.tcl.tk`
- `PGPLOT`, (`http://www.astro.caltech.edu/~tjp/pgplot`), installed with the following drivers: `XSERVER, XWINDOWS, POSTSCRIPT, GIF` and `XTK`. The required `drivers.list` is distributed with the software.
- LCL I/O library, version 1.2, `http://www.aip.de/Euro3D`

and the following `perl` modules (`http://www.cpan.org`).

- Astro::FITS::CFITSIO.
- Math::Approx.
- Math::Derivative.
- Math::FFT.
- Math::Matrix.
- Math::Spline.
- Math::Stat.
- PDL.
- PGPLOT.
- Statistics::OLS.

We recommend the user installs the source code of all these packages. You can download this from the following web page: http://www.iac.es/winters/2005/software3D.html. When available, pre-compiled binaries are also included for both Linux or Mac OSX, to ease the installation for users who are not very familiar with software building, and who are using standard Linux/Mac OSX distributions.

7.4.1 *Installing the LCL library*

The Lyon-C library (LCL; Pécontal-Rousset *et al.*, 2004) is a `C`-coded I/O library that handles multi-extensions fits files in many ways (1D, 2D and 3D images, ASCII or binary tables and the Euro3D format). It also includes a simple and useful command-line argument-handling system. It is distributed through the Euro3D web page (http://www.

[2] The new `gcc` compiler does not include anymore `g77`, so you must install it separately.

aip.de/Euro3D/). A copy of version 1.2 is included there (E3D_io_LCL-1.2.tar.gz). To built it, just follow these instructions:

(i) First create a directory in your computer, that we name hereafter as `yourDIR`. Copy the compressed tar file into this directory `yourDIR`.
(ii) Uncompress and untar the file.
(iii) Go to the created directory, e.g. `cd E3D_io_LCL-1.2`
(iv) Run the configure program: `./configure`
(v) Run the make program: `make`
(vi) Define the environment variables: `IFU_PATH` and `IFU_DEFAULT_FMT`; e.g. for a tcsh shell:

```
setenv IFU_PATH yourDIR³;
setenv IFU_DEFAULT_FMT Euro3D;
```

(vii) To check the library, go to the directory:

```
cd checklib/exec
```

and run the test program:

```
./check_E3D_io
```

If you do not have any error messages, then the install is OK.

7.4.2 *Installing PGPLOT*

PGPLOT is a fortran graphic library (Pearson, 1995) that has a `C` and a `perl` wrapper, which allow the user to use it also when coding with these languages. From http://www.astro.caltech.edu/~tjp/pgplot/ you may download the latest version of PGPLOT. The instructions on how to install it are described in the `INSTALL.txt` file included in the tarball.

7.4.3 *Installing E3D*

E3D (Sánchez, 2004) is a visualization tool for IFS created through the Euro3D RTN. We distribute in these proceedings the lastest version, v1.3b: http://www.iac.es/winters/2005/software3D.html in two formats: the source code (`e3d-1.3b.tar.gz`), and a binary distribution, for both Linux (`E3D_v1.3b_linux.tgz`) and Mac OSX (`E3D_MacOs.tgz`). Binary distributions include most of the required libraries, although some of them have to be installed.

Installing E3D from the source code

To install, follow the instructions below:

(i) Copy the tar file into the Euro3D directory (which is referred to using the previous environment variable `$IFU_PATH`).
(ii) Using the command `tar xvfz e3d-1.3b.tar.gz`, extract the archive file `e3d-1.3b.tar.gz` under `$IFU_PATH`. It will create the directory `e3d-1.3b` (or a similar one depending on the distribution number you are installing).
(iii) Change directory to the newly created one (`cd e3d-1.3b`), and execute the configure and make scripts:

```
./configure
make
```

[3] The `IFU_PATH` should refer to `yourDIR` and not to `yourDIR/E3D_io_LCL-1.2`.

(iv) Go to the scripts directory (i.e. `cd scripts`) and execute the install script (`./install.pl`). This script will customize the `tk_e3d.tcl` according to your directory tree.
(v) Add the path to the directory `user/bin` (under the E3D directory) to your environment variable PATH.
(vi) Test your E3D installation by executing the script `tk_e3d.tcl`, and then loading the file `test.e3d` that you will find under the `scripts` directory.

Installing the binary distributions of E3D

Follow the instructions below:
(i) Copy the tar file into the Euro3D directory (which is referred to using the previous environment variable `$IFU_PATH`).
(ii) Uncompress the tarfile.
(iii) Go to the newly created directory.
(iv) Execute the `./install_static.pl`, and follow the instructions.
(v) Test your E3D installation by executing the script `tk_e3d.tcl`, and then loading the file `test.e3d` that you will find under the `scripts` directory.

7.4.4 *Installing FIT3D*

FIT3D is a fitting tool (Sánchez *et al.*, 2004; García-Lorenzo *et al.*, 2005) to fit emission lines and continuum spectrum, in order to analyse gaseous emission and stellar kinematics. It consists of a core coded in `C` for the low-level fitting routines, and a wrapper written in `perl` to facilitate the creation of the configuration files, describing the model to fit. (This is an idea it shares with DIPSO, to create the fitting model using a configuration file.) Several emission lines and continuum backgrounds can be fitted simultaneously, with different kinematics, setting bounds to the parameters of the emission lines, linking parameters between lines for creating kinematics systems (i.e. several lines sharing the same kinematics properties), and deblending different emission line components. It uses datacubes and row-stacked spectra as input data, and produces FITS files as output. In some cases, it requires E3D for the creation of the interpolated maps.

Based on the LCL development environment, it shares the same requirements with E3D. To install it, just use the same installation instructions as above. At the website: http://www.iac.es/winters/2005/software3D.html we distribute two versions of FIT3D: the source code (`fitting_tool-1.0.tar.gz`), and the binaries distributions for Linux (`FIT3D_v1.0_linux.tgz`) and Mac OSX (`FIT3D_v1.0_MacOsX.tgz`). Except for the `C`-core, `FIT3D` uses `perl` routines that require the modules listed above to be installed (see Section 7.4).

7.4.5 *Installing R3D*

R3D is a package for reducing IFS data obtained with any fibre-based IFU (Sánchez, 2006). Although valuable efforts were made by the Euro3D network to produce standardized tools to analyse data, there is still a lack of generic packages for data reduction (which was beyond the scope of the network). The astronomical community encounters a wide variety of pipelines specific to given instruments, whose procedures and parameters differ strongly, and whose outputs are difficult to compare. Following preceding efforts (e.g. P3d, Becker, 2001), we have developed R3D, a reduction package able to reduce any fibre-feed IFU data.

We first coded the algorithms in Perl, using the Perl data language (PDL, Glazebrook *et al.*, 2001), in order to speed up the algorithm testing phase. Once the algorithms were tested with both simulated and real data, we recoded the most time-consuming routines in C to improve their performance. The Perl version of R3D is fast enough to produce valuable science frames in a reasonable elapsed time, although for instruments with many fibres (like VIMOS or Visible Imaging Multi-Object Spectrograph), or for some slow processes like cross-talk correction algorithms, we strongly recommend the C-version. A major advantage of using Perl and C is that they are freely distributed and platform independent, and therefore R3D can be installed under almost any architecture without major efforts and at no cost. The instructions on how to download and install R3D can be found at http://www.caha.es/sanchez/r3d/, where we keep the different versions updated.

Installing the `perl` *version of R3D*

R3D uses `perl` routines that require the modules listed above to be installed (see Section 7.4). We distribute a tar file including the routines and some documentation (`R3D.tgz`). To install them, uncompress the tar file and follow the instructions included in the `INSTALL.txt` file.

Installing the `C` *version of R3D*

Based also on the LCL development environment, it has the same requirements as E3D. To install it, just follow the same installation instructions. We distribute two versions of R3D: the source code (`r3d-0.1.tar.gz`) and a Linux binary distribution (`R3D_c_linux-0.1.tgz`). Except for the C-core, `FIT3D` uses `perl` routines that require the modules listed at the beginning of Section 7.4.

7.5 E3D: the Euro3D visualization tool

E3D is a generic visualization tool for IFS data, fully independent of the instrument used to obtain such data. It was created on behalf of the Euro3D network (Sánchez, 2004), and thus handles the Euro3D data format as default format but is able to work with others formats too (like datacubes and RSS data).

E3D has a C-coded core, and may be divided in three logical blocks:

- The `Euro3D.o` library is a type of application programming interface (API) of low-level functions, including routines calling the Euro3D I/O library, routines to access the shared memory server (SHM), and basic functions needed to plot and analyse the data. These functions may be invoked by any external program, which needs to read/write Euro3D data, or uses the SHM.
- The `tk_e3d` `Tcl/Tk` interpreter. This is a stand-alone program embedding the `Tcl/Tk` interpreter with Euro3D routines. These routines wrap the previous C-functions in `Tcl`, and can be used, within the tcl interpreter, to load/save Euro3D files, plot single or co-added spectra, plot monochromatic/polychromatic maps, interpolate these, save maps in FITS format, etc.
- A number of stand-alone tools, which may be used from the command line, may be helpful in accessing Euro3D files. Perhaps the most interesting routine is `any2Euro3D`, converting single-group IFS data from any instrument into the Euro3D Format.

Together with this C-core, we provide a `Tcl/Tk` GUI, which calls the Euro3D-Tcl routines (`tk_e3d.tcl`). The GUI is organized around three different windows corresponding to three different ways of investigating the data:

- The Stacked Spectra Inspector, or main window, includes the main menu with various options to handle the input and output data and to pop up the other windows. Once loaded, the data are automatically plotted in its canvas, using the stacked-spectra representation. Moreover, a command-line prompt has been included in this window

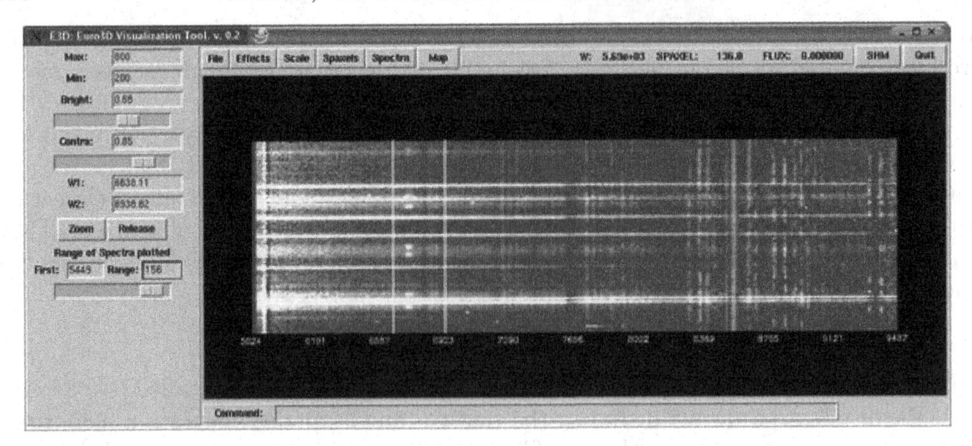

FIGURE 7.7. E3D Stacked Spectra Inspector: this is the E3D main GUI window, from which all the others can be created. It displays all the spectra contained in the 3D file stacked row by row.

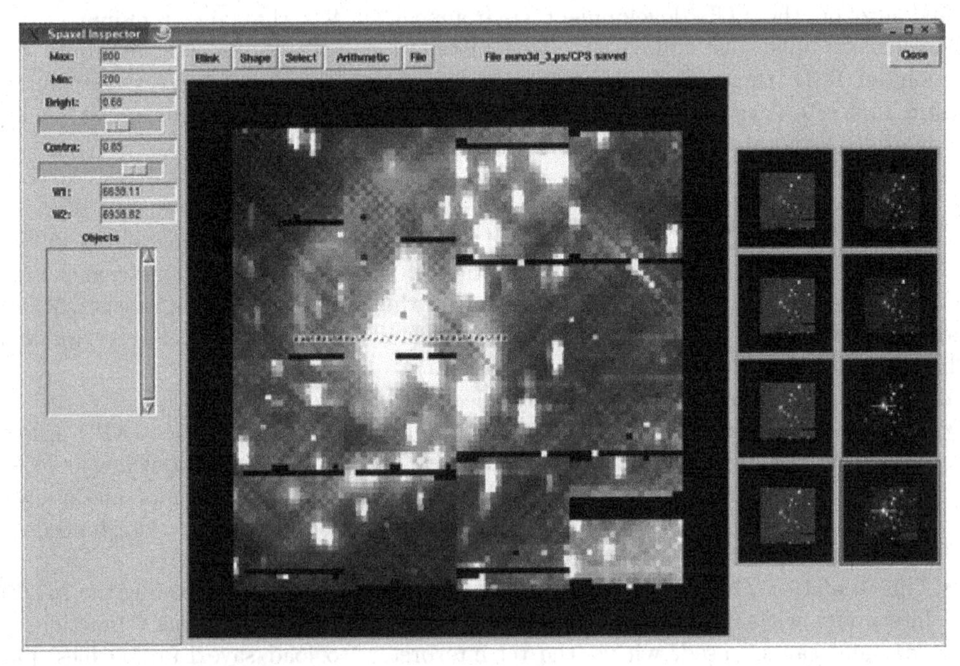

FIGURE 7.8. E3D Spaxels Inspector: this is the GUI provided for plotting monochromatic or polychromatic maps. It is possible to select different spaxels, to be displayed subsequently on the Spectral Inspector.

to directly call any Euro3D-'tcl flavor' routines. Figure 7.7 gives a snapshot of this window.

- The Spaxels Inspector provides the user with a main canvas for viewing monochromatic/polychromatic datacube slices or maps, which were selected from the main window or from the Spectral Inspector (see below). It includes eight minor canvasers for buffering the most recently created maps. It also includes a menu to handle spaxels/maps options, including spaxels selection, object creation and different methods to view the data. Figure 7.8 shows a screen shot of such a window.

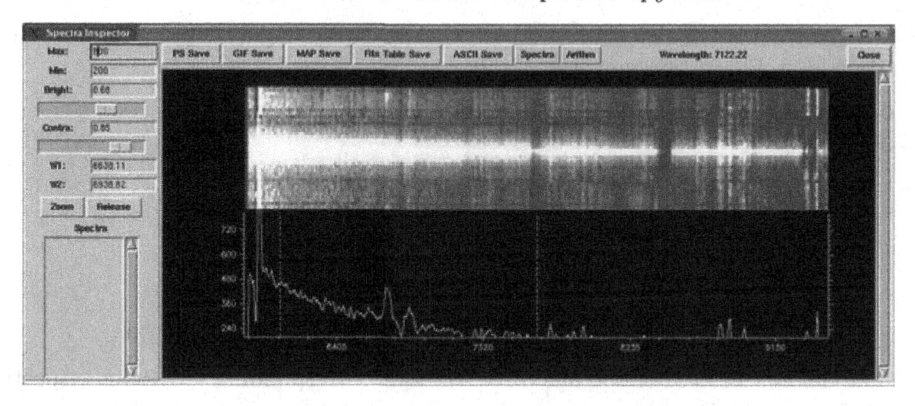

FIGURE 7.9. E3D Spectral Inspector: this is the GUI provided for plotting spectra.

- The Spectral Inspector has a canvas for viewing/plotting the spectra corresponding to the spaxels selected from any of the two previous windows. It also includes a menu with different options to handle the spectra. Figure 7.9 gives a snapshot of this window.

7.5.1 *Working with* `E3D`*: the GUI*

We describe here, with a few examples, the capabilities of `E3D`. As we quoted above, `E3D` can handle data in the Euro3D format, its default format, or use the RSS and datacube formats. All of these are converted internally to the same structures, so, once loaded, they are all handled in the same way.

- Loading data in the Euro3D format:
  - In the top menu of the main window, select `Load` in the `File` menu. A pop-up window appears.
  - Change from default `fits` extension to `e3d`.
  - Select the file `NAME.e3d`.
- Loading data in the RSS format:
  - In the top menu of the main window select `Import RSS` in the `File` menu. A pop-up window appears.
  - Select the `NAME_RSS.fits` RSS file.
  - Select the `NAME_RSS.pt` position table.
- Loading datacubes:
  - In the top menu of the main window select `Import cube` in the `File` menu. A pop-up window appears.
  - Select the file `NAME.cube.fits`.

You can now follow the exercises. In order to be able to run the exercises, first change to the directory `tut1` and open the `E3D` GUI running the `tk_e3d.tcl` command.

Exercise 1: Try to load your own data. *First identify what format your data are in (read the reduction tool notes if necessary).*

If you need to build a position table, you have to know that the position tables in `E3D` are `ASCII` files with four columns, where the first row gives the geometry of the dataset and the remaining rows contain the positions, as below:

```
TYPE    SIZE1   SIZE2   TYPE_ID
INDEXi  Xi      Yi      TYPE_IDi
...
```

TYPE is the shape of the spaxels and can be: `R` for a rectangle shape, `S` for a square, `C` for a circle, and `H` for a hexagon. Here is an example of such a table:

```
H   0.2  0.2  1
1   0.0  0.0  1
2   0.5  0.0  1
3   0.5  0.5  1
...
```

7.5.2 *Quick look over the various windows of the* E3D *GUI*

As quoted before, the GUI of `E3D` has three windows corresponding to various inspections of the data:

The main window or stacked-spectra inspector

This is the window which popped up first when running `E3D`, and where the data are displayed as stacked spectra (RSS format), using the same order in which they are stored on disk (i.e. no preferential spatial direction, *a priori*). Using this window, it is possible to browse the spectra.

Perform the following instructions:

- Change the directory to `tut3` and run `E3D`.
- In the top-level menu of the main window, select `Import cube` in the `File` menu. A pop-up window appears.
- Select the file `NAME.cube.fits`.

By default, the main window displays only 256 spectra (i.e. not the total number of spectra) as a 2D image, where each row represents a single spectrum. E3D automatically sets the cuts of intensities for the look-up table, determining the median and the standard deviation of the intensities plotted. Sometimes, the estimate is not good enough and it is better to set the values manually. To change these maximum and minimum values, just fill the `Max` and `Min` entries in the left-hand menu. Note: A range of [−1200] would be acceptable to visualize the data. For a better contrast it is sometimes useful to tune the `Brightness` and `Contrast` settings too, edit the corresponding entries, or just move the corresponding scroll bars.

Exercise 2: *Change the maximum, minimum, brightness and contrast settings until you get a proper visualization of your data.*

To browse the spectra (in cases where you have more than 256 spectra) you can move the scroll bar in the bottom part of the left-hand menu, below the label `Range of Spectra plotted`. It is also possible to change the number of plotted spectra, editing the values of the two entries labelled `First` (the first spectrum plotted) and `Range` (the number of spectra plotted).

Exercise 3: *Let's play with this browsing capability.*

It is possible to `zoom` over a given spectral region, by selecting this region and clicking twice over the main display with a left click of the mouse. The selected wavelength range appears in the `W1` and `W2` entries in the left-hand menu. You can also edit these entries directly, if you want a more accurate selection of the wavelength bounds. In order to return to the default range click on the `Release` button.

E3D always uses the same method to select elements. The left button of the mouse is used to select a window and the right button for sending this selection to another window (inspector). We will experiment with this later on.

Exercise 4: *Play with the zooming capabilities until you get used of them.*

The combination of zooming capability and range selection is useful to look at specific wavelength and spatial ranges; e.g. for emission line objects.

Exercise 5: *Try to locate some of the emission lines in this object.*

The spaxels inspector

The spaxels inspector is the window used to visualize the spatial distribution of the intensities for each spaxel, at a selected wavelength range. It allows you to select individual (or groups of) spaxels/channels to view the spectrum/spectra corresponding to them.

On the `main window` menu, click on the `Open` option of the `Spaxels` submenu. A new window pops up: this is the spaxels inspector. It consists of a top menu bar, a left-hand menu, with similar entries to the one of `main window`, and then other entries specific to this inspector. There is also a main display, to display slices (or maps), and eight small buffer windows, where the last eight displayed slices are stored. `Row.No` indicates the row number in the RSS representation of the spaxel over which the cursor is located. `Flux`, `X` and `Y` show the corresponding flux, and the coordinates of the spaxel. The flux is the average value of the intensities of the spectra for the selected wavelength range.

As a general remark, E3D displays average values when a multiple selection (several spectral channels or spaxels) is done. This feature has the advantage that, for subtracting wavelength ranges, it is not necessary to know the width of the subtracted range. It is however good to note this width somewhere, when saving the data, because it may be useful for future computations. `N.SPAXELS` is the number of currently selected spaxels and `MEAN` and `SIGMA` show the mean flux value and its corresponding standard deviation over the selected channels. We will explain below how to select spaxels.

To display a slice, go back to the main window (by clicking on it) and select a wavelength range. Once selected, right-click on the mouse. The reconstructed slice, computed by averaging the intensities over all the spectra at the selected wavelength range, is displayed in the spaxels inspector.

Exercise 6: *Select various wavelength ranges, and look at the differences between the slices.*

One of the strange (and useful) features of the spaxel inspector is the compatibility of selecting a narrowband image and creating a movie effect by clicking on the display on the `main window`. Keeping the central mouse button pressed, move the mouse along the spectral direction (browsing the wavelengths). Once the button is released, the most recently selected range is stored in the spaxels inspector.

Exercise 7: *Play with the movie effect to see the the object shifted from the blue to the red due to the differential atmospheric refraction (DAR) effect, and the areas dominated by the emission lines.*

It is possible to go back to one of the previous displays (from the last eight displayed at least) by right-clicking on the mouse on the corresponding buffer (small display), at the bottom of the spaxels inspector window.

To select a spaxel (or a group of them), just left-click on the mouse and keep it pressed while moving over the spaxel you want to select. A hashed polygon will appear over the selected spaxels. Note that the `N.SPAXELS`, `MEAN` and `SIGMA` entries change when selecting the spaxels. More selection options will be described below. To clear a given selection, just click on the `Clear` option of the `Select` submenu in the Spaxels Inspector top menu bar.

Exercise 8: *Play with the spaxels selection capability.*

The spectra inspector

The spectra inspector is used to display and plot selected spectra (possibly in a different order to the ones that are stored on disk). On the `main window` menu, click on the `Open` option of the `Spectra` submenu. The new window that pops up is the spectra inspector. It consists of a top menu bar, a main display and a left-hand menu. Most of the entries in the left-hand menu are similar to those of the main window, except for the `spectra` box list, which will be explained below.

The easy way to select a group of spectra is to come back to the `spaxels inspector` and make a selection of spaxels, as described above, then to right-click on the display of the `spaxels inspector`. The spectra corresponding to the selected spaxels will be displayed. By default, the `spectra inspector` shows, on the upper part of the display, a stacked-spectra image of the selected spectra in the sequential order used for the selection and, on the bottom, a plot of the average of these spectra.

Exercise 9: *Practise using the spectra inspector. Change the maximum and minimum, the brightness and the contrast in order to visualize the selected spectra properly.*

As in the main window, it is possible to zoom over spectral regions and/or select them, using the same settings as above. You can also display the average slice of a selected number of spectra by right-clicking on the mouse.

7.5.3 *Main-menu options*

The top menu-bar of the `main window` presents the various options that affect the dataset (except for the sky subtraction, which is part of the `spectra inspector`). It also shows the setup of the visualization parameters.

The `File` menu

This contains all the features to input or output the data on disk. The `Euro3D data format` (Kissler-Patig *et al.*, 2004) is the native and default format of this visualization tool, so the `load/save` options are intended for use with that format. Other data formats, like the RSS and datacube formats, can be imported/exported using the `import/export` menu option.

When exporting data to a datacube, `E3D` always performs an interpolation using the method currently set in the main window (configuration). If you import a datacube and export it again, it does not keep the previous format of the datacube, but creates a new datacube with the interpolated data. A feature to keep the original format for datacubes may be implemented in the future. We will explain how to bypass this potential problem, if necessary.

It is also possible to import/export slit-spectra, without taking into account the spatial information, or store a cut of the current selected spectral region. These options are not for 3D data.

`Spaxels` *and* `Spectra` *menus*

These are used to open the `Spaxels inspector` and the `Spectra inspector`. They allow also a switch between two selection modes (option `Select`): one based on wavelength ranges or another based on the distribution of the spectra in the RSS. The `plot` options just send the selected regions to the respective inspector.

Exercise 10: *Open both the spaxels inspector and the spectra inspector. Select a wavelength range and right-click on the mouse. Have a look to the spaxels inspector. Now, change the selection mode from spaxels to spectra (by clicking on the tag 'Select' in the* `Spectra` *menu). Select a set of spectra (two left-clicks and then one right-click on the mouse), and look at the spectra inspector. Go back to the default setting.*

The `configuration` *menu*

This allows you to change the current settings for the visualization tool:

- `ColorMap`, used to change the current colour look-up table.

Exercise 11: *Experiment with the different look-up tables.*

- `Scale`, to change the colour-map scaling (also known as intensity transform table).
- `Spectral Representation`, used to change the way the spectra are displayed in the `spectra inspector` (combine display and plot, plot only, etc.).

Exercise 12: *Experiment with the different spectral representations: select a group of spaxels and display their corresponding spectra in the spectra inspector. Try the different modes until you get used to them.*

- `Interpolation`. This brings up a pop-up menu where it is possible to modify the interpolation method and its parameters. E3D has five different interpolation routines implemented, which were borrowed from PLPLOT (with permission of the authors under the condition of not copyrighting the code). It is possible to switch between these different interpolation methods by pulling down the `FUNCTION` menu in the interpolation window. The values of the `Grid Option` parameter and the output `pixel size` may be overwritten. Some of the routines require an additional parameter, which require changes from function to function (for their exact meaning, see the E3D documentation).

Exercise 13: *Import the* `NAME_RSS.fits` *file in E3D, select a given wavelength range and display the results in the* `spaxels inspector`. *Select the* `map` *option in the* `Shape` *menu to display an interpolated map of the slice. Change the interpolation parameters and methods to see the effect on the interpolated map. As far as we know, the optimal interpolation is the Natural Neighbour method. However, it is a slow one. Choose a very small value for the* `Grid Option`, *e.g. 1e-9. The smaller it is, the better it is; however, it is also the slower.*

- `Spaxel Selection`: This brings up a pop-up window with a menu that allows you to change the method of selecting spaxels, switching from `single` to `aperture`. The default method is `single`, and allows for the selection of spaxels one by one in the `spaxels inspector`. The `aperture` method allows you to select all the spaxels within a given radius (selected in the pop-up window).

  There is another selection method, the `pseudo-slit` method. By hitting the `s` key over the `spaxels inspector`, it is possible to switch between the `single` (or aperture) selection and the `pseudo-slit` selection. When this mode is selected, you just need to select two points on the `spaxels inspector` display (with a left-click on the mouse) and all the spaxels over the straight line joining these two points will be selected. As in the previous method, you can view the corresponding spectra on the `spectra inspector` display by right-clicking on the mouse. The `pseudo-slit` selection is useful for comparing your data with slit spectrum.

Exercise 14: *Experiment with the various spaxel selection methods. Select spectra one by one, switch to aperture selection, and to the pseudo-slit selection. See the differences in the* `spectra inspector` *and experiment until you get used to the different methods.*

The DAR tools menu

E3D includes some tools for correcting for the differential atmospheric refraction. These tools can be divided into two groups:

- tools for determining the shift and angle of the image along the wavelength;
- tools to correct for this shift.

The first tool computes the theoretical DAR correction (Filippenko, 1982). It relies on HEADER keywords not included in the standard Euro3D format (see the E3D user guide), and in a proper orientation of the position table in the sky (east left, north up). This tool is designed as the `Th.Det.DAR` in the DAR menu. Two other empirical determinations of the DAR have been included. These tools rely on the presence of a clear peak in the spatial distribution at any wavelength (e.g. a point-like source). They are designed as `PM.Det.DAR`, which determines the centroid of the image based on the principal momentum, and `peak.Det.DAR`, which determines the peak of the emission by a simple maximum search.

Once the DAR correction is determined, it is possible to perform the correction, choosing the `Correct DAR` option. It is important to understand what the program is doing when performing a DAR correction: first of all, at each wavelength, the data are interpolated (using the interpolation method set in the configuration), then the intensity corresponding to the original spaxels grid, at the new position, is determined using the off-set of the centroid computed previously (for more detail, see for example Arribas *et al.*, 1999).

Any DAR correction implies an interpolation in the spatial domain. To resample the data in the original shape is just a mathematical trick, since the recovered data come through an interpolation. We therefore do not recommend using the DAR tools included in `E3D` for science applications; however, it can be useful for a quick check of the data and the DAR effects. We will explain below how to better correct for DAR.

Exercise 15: Experiment with DAR correction (`dir: tut3`). *Run E3D and import the cube* `star_no_dar.cube.fits`. *Open the spaxels inspector and, using the movie effect, check the DAR effect. Click on the peak correction on the main window menu bar (DAR tools) and perform the correction. Store the result in another file (e.g.* `star_dar.cube.e3d`*). Note that the correction is not optimal.*

The command-line entry

`E3D` is based on a `Tcl/Tk` interpreter embedded with the E3D commands (see the E3D user guide). The GUI is no more than a front end to that set of commands. It is therefore possible to invoke any of these commands by entering them directly in the `command` line entry, at the bottom of the main window.

It is also possible to run a script stored on disk by entering `source scriptfile` at the command line prompt, where `scriptfile` is the name of the file.

Moreover, each time you execute `tk_e3d.tcl`, the program creates a log file to record the various actions you have performed. This log is stored in the file `e3d.log`. You can therefore cut and paste from this file to run a previous set of commands.

Exercise 16: Experiment with the command line entry. *Run E3D, and import the cube* `VIMOS.cube.fits`. *Tune the maximum and minimum values to see the data. Enter the following command lines:*

```
recolor_image heat 0
recolor_image iraf 0
recolor_image rainbow 0
```

See the effect on the display.

Exercise 17: Experiment with the scripts (`dir: tut3`). *Run E3D and enter the following words on the command line:* `source script.tcl`. *See what happens. Open the file* `script.tcl` *with your favourite text editor (e.g. vi, emacs), and have a look to the sequence of commands you have executed.*

7.5.4 *Menu options of the Spaxels Inspector*

The top-level menu of the `spaxels inspector` gathers the options that affect the spaxels-oriented representation and selection of data.

The `File` *menu*

This handles the various input and output of data (or plots) driven by the spatial distribution. The data files can be slices (i.e. collections of coordinates with their corresponding fluxes) or maps (2D images). The slices may be stored in FITS or ASCII tables (four-column tables, with id, x, y and the flux), while the maps can be stored in FITS images. The maps are derived from the currently displayed dataset, interpolated according to the current configuration settings. A particular case is the `save object` option, that we will explain below.

The plots are Postscripts or GIF hardcopies of the current display.

The `Shape` *menu*

This allows the user to change the way the data are represented in the spaxels inspector, including methods combining the original spaxels shape and the image of interpolated data (i.e. the map).

Exercise 18: *Import the file* `INTEGRAL_SB2.rss.fits`, *and the position table* `INTEGRAL_SB2_pt_219.txt` *(*`dir:tut4`*). Open the spaxels inspector and browse different wavelength ranges. Switch between the different options in the* `Shape` *menu to experiment with the various representations. Store the plots/slices obtained using the different output options in the* `File` *menu. In the case of 2D FITS images, visualize them using an external tool, e.g. DS9. View the ASCII tables using an editor, or use the Unix command* `more` *to look at them. For the plots, display them with ghostview or xv.*

The `Select` *menu*

First, this allows you to clear the current selection and start a new one using the `Clear` option, but you may also:

- `plot spectra`: This has the same behaviour as right-clicking on the mouse.
- `plot profile`: This plots the intensity profile of the slice currently displayed along the selected spaxels. It is particularly useful when combined with the `pseudo-slit` selection.

Exercise 19: *Hit the* `s` *key on top of the spaxels inspector display. Look at the information labelled displayed on top of it. It should say* `Pseudo-Slit Selection`. *Select a cut that goes through the centre of the image, and select the* `plot profile` *option. The intensity profile along the pseudo slit will appear in the spectra inspector, fitted with a Gaussian.*

- `create object`: An `object` stands for a collection of selected spaxels. Once selected, a new entry will appear on the `Objects` list box in the left-hand menu. A maximum of nine different objects can be created; if this number is exceeded, the objects are overwritten. `Objects` are the best way to store your previous selections. To get back to any `object`, just click on it in the `Objects` list box. The spectra corresponding to the selection of spaxels stored in this object will be displayed in the `spectra inspector`.

Exercise 20: Play with the object creation (`dir: tut4`). *On the spaxels inspector select the central spaxels and store the selection as an object. Clear the selection and repeat the process with the outer part of the central square, and with the external circle of spaxels. Open the spectral inspector and select alternatively each of the objects. Look at the different selections of spectra displayed.*

It is possible to store a selected group of spaxels (or object) in a separate Euro3D file, using the `save object` option. Just click on one of your objects (or select a group of spaxels) and select this menu option. If you want to see what the result looks like, you can load it in the main window.

- `reverse object`: this option is used to reverse the spaxels selection stored in an object (i.e. select the ones that were not selected, and vice versa). Click this option and then on one of your objects.

Exercise 21: Separate sky and objects fibres (`dir: tut4`). *Import the* `INTEGRAL_SB2.rss.fits` *file, and display a slice in a given wavelength range using the spaxels inspector. One by one select all the spaxels in the outer ring, i.e. those corresponding to the sky fibres. Once selected, store the selection as an object twice. Reverse the selection of one of these objects, so that you have the central square selected. Apply the two available selections and store the results with the* `save object` *option in two different files:* `obj.e3d` *and* `sky.e3d`. *Load these files in E3D to see the result.*

Exercise 22: Mask regions creating objects (`dir: tut3`). *Import the* `VIMOS.cube.fits` *file in E3D. Open the spaxels inspector. Select a given wavelength range and display the corresponding slice on the spaxels inspector. You may see four* `bad` *regions corresponding to dead fibres. Select them and store the selection as an object. Reverse the selection and store the object as* `good_data.e3d`. *Load the newly created Euro3D file and display the result in the spaxels inspector. You have removed the bad data.*

7.5.5 *The* `Tools` *menu*

This gathers the basic tools that operate over the slices:

- `Blink`: Allows blinking between slices stored in the eight-length small display buffer.

Exercise 23: Play with the blinking facility (`dir: tut4`). *Import the* `INTEGRAL_SB2.rss.fits` *file and display different slices at various wavelength ranges (hint: select continuum dominated regions and emission line regions). Then, click on the* `blink` *button and select, by left-clicking the mouse on the small displays, the slices you want to blink* *in between. Click again on the* `blink` *button to start the blinking, and once again when you want to stop it.*

- `Arithmetic`: This allows the user to perform simple arithmetic computations between two buffered slices. It brings up a small pop-up window whose parameters are: the two slices to be processed (using their buffer number) and the number of the buffer in which to store the result. You select the type of the computation by pulling down the `Arithm` menu at the bottom of the window.

Exercise 24: Map of emission lines, continuum subtracted (`dir: tut4`). *Load the* `obj.fits` *(created before), and display two slices each corresponding to the wavelength region of [OIII] and of a continuum region nearby. Subtract the continuum to the [OIII] slice. This will be a raw estimation of the continuum-free emission-line map.*

Keyboard shortcuts

There are a few keys that may be used for different operations:

- The 's' key switches from the pseudo-slit selection mode to the spaxel-to-spaxel (or aperture), selection.
- The 'z' and 'u' keys allow the user to zoom over the image. Hit the `z` key, with the mouse cursor located on the upper corner of the area to zoom and once more with the mouse cursor over the lower corner. A zoom over the selected region will be displayed in the spaxels inspector. To unzoom, hit the `u` key over the spaxels inspector to go back to the display of the whole image.

Exercise 25: Play with the zooming capability (`dir: tut3`). *Import the* `VIMOS.cube.fits` *file in E3D. Open the spaxels inspector. Select a given wavelength range and display the corresponding slice on the spaxels inspector. Zoom in and out over different regions until you get used of it.*

7.5.6 *Menu options of the Spectra Inspector*

The top menu of the `spectra inspector` gathers the options related to the spectra-oriented representation and selection of data.

The `File` *menu*

This handles the capabilities of saving spectra or hardcopy of the plots. Hardcopies may be saved in Postscript or GIF, spectra as 2D RSS images. A single spectrum (like the average spectrum) may be saved using FITS or ASCII tables. In `E3D`, a spectrum saved as a table is formatted as follows: the file contains three columns, which are respectively the Id (i.e. the spectral index), the wavelength and the flux. Please remember that the stored spectrum is the mean of the currently selected (and plotted) spectra, not the sum of them.

Exercise 26: Testing the saving capabilities of the spectra inspector (`dir: tut4`). *Import the* `INTEGRAL_SB2.rss.fits` *file and display a slice using the spaxels inspector. Select different groups of spaxels and display the corresponding spectra in the* `spectra inspector`. *Save the selected spectra as a Postscript and a GIF file (hardcopies), then as a 2D RSS image, a FITS table and an ASCII table. Visualize the outputs using external tools, such as ghostview, xview, iraf or DS9. Repeat the operation until you get used to it.*

The `Spectra` *menu*

This menu handles the storage in memory (or buffering) of average spectra. The capability is similar to the `object` creation in the `spaxels inspector` (see above). These spectra may be plotted, one-by-one or simultaneously, or may be used to perform basic computations, like those exhibited in the `spaxels inspector` using slices.

Clicking on `Save Spectrum` stores a new spectrum in the memory (up to a maximum of nine), and a new entry appears in the list box labelled `Spectra`. In order to display any of the buffered spectra, just click on the corresponding entry in the list box.

It is possible to view several spectra on the same display by selecting the tag labelled `single/multiple plot` on the pull-down menu, and then selecting the spectra to be plotted using the `spectra` list box. Each spectrum will be plotted using a different colour in order to distinguish them. Click on the `Clear Multiple selection` option to clear the screen and create a new selection.

Exercise 27: Test the spectra buffering capability (`dir: tut4`). *Import the* `INTEGRAL_SB2.rss.fits` *file and display a slice using the spaxels inspector. Select different groups of spaxels and plot the corresponding spectra in the* `spectra inspector`. *Store each of these spectra in a separate slot. Select each of the spectra to view them in the* `spectra inspector` *display. Change from single to multiple selection and repeat the selection to see the difference. Store them as Postscripts and/or GIF files, and visualize them using external tools.*

Other menu options are:

- `Arithmetic` allows you to perform basic arithmetic computations using buffered spectra. A window pops up (similar to the slice arithmetic one), requesting the selection of two spectra (from the list box) and the type of arithmetic operation to perform (pull-down menu `Arithm` on the bottom of the window). The result will be stored in the first available slot.[4]

Exercise 28: Get used to the spectra arithmetic capability (`dir: tut4`). *Import the* `INTEGRAL_SB2.rss.fits` *file and display a slice using the spaxels inspector. Select various groups of spaxels and plot the corresponding spectra in the* `spectra inspector`. *Store each of these spectra in a separate slot. Select spectra to view them in the display of the* `spectra inspector`. *Perform different arithmetic operations with the selected spectra and visualize them using the single or multiple display options to compare the results.*

- `Sky Subtraction` allows to subtract one of the buffered spectra from the dataset. This option is mostly used to perform a sky subtraction. To use it, just select this option in the menu, and then select the spectrum to be subtracted (i.e. the sky spectrum).

Exercise 29: Sky subtraction (`dir: tut4`). *Import the* `INTEGRAL_SB2.rss.fits` *file and display a slice using the spaxels inspector. Select the spaxels in the outer ring (the sky-fibres ones), and visualize the average* `sky` *spectrum in the* `spectra inspector`. *Store this spectrum in one of the memory slots, and perform a sky subtraction using this spectrum as the* `sky` *spectrum. Once the subtraction is performed it is possible to save the whole dataset on disk. Go back to the main window and save it as an RSS or a Euro3D dataset. It is also possible to remove the* `sky` *spaxels by storing the sky spaxels selection as*

[4] There is a known bug when trying to use spectra from the first position of the slot.

an object, performing a `reverse object` *operation, and then storing the object (e.g. as* `sobj.fits` *for RSS format or* `sobj.e3d` *for Euro3D datacube), as described in the Exercise above.*

The `Fitting` *menu*

`E3D` includes some basic fitting routines. More sophisticated routines will be discussed later on in a dedicated section. We will describe here the fitting routines used to derive the FWHM of a profile (cut on a slice using the `spaxels inspector`). Two fitting capabilities are provided:

- `Fit Single Emission line`: This fits the plotted spectrum with a single emission line (a Gaussian function), plus a continuum (as a threshold). The four free parameters of the fitting are the intensity of the Gaussian, the FWHM, the central wavelength and a constant for the continuum. The result of the fit will be overplot in red (in contrast to the data which are plotted in white). The output of the fitting routine is printed in the xterm.

Exercise 30: Fit an emission line (`dir: tut4`**).** *Import the* `sobj.fits` *or* `sobj.e3d` *file and select the slice whose wavelength range is centred on the [OIII] emission line, without any other emission line pollution, using the spaxels inspector. Plot the corresponding spectrum using the spectral inspector. Fit this spectrum using the* `Fit Single Emission line`*.*

- `Full Automatic Kinematics Analysis`: This performs a kinematics analysis using a single emission-line fitting (as described above) per spectrum. Being an automatic procedure, the result is not fully reliable but must be considered as a quick estimation of the kinematics on the object. It fits each spectra in the dataset with a single Gaussian function (the emission line) plus a constant (the continuum). Once the fitting is done, it creates four maps using the output of the fit, i.e. a map of the line intensity, a map of the FWHM, a map of the central wavelength and a map of the continuum emission. The fit is performed over the currently selected wavelength range.

Exercise 31: Full kinematics estimate (`dir: tut4`**).** *Import the* `sobj.fits` *or* `sobj.e3d` *file and select a slice centred on the [OIII] emission line, without pollution of any other emission line, using the spaxels inspector. Plot the corresponding spectrum using the spectral inspector. Fit this spectrum using the* `Fit Single Emission line`*. Once this first fit is performed, click on the* `Full Automatic Kinematics Analysis`*. The routine loops over the spectra of the dataset. Once finished, four new maps will be created in the spaxels inspector, one for each free parameter of the model (see above). Check visually the different maps.*

7.6 Sky subtraction

The sky background correction is usually applied by subtracting a sky spectrum from the object spectra. The sky spectrum is usually obtained by computing the mean of several spectra belonging to sky spaxels on 3D observations.

For a proper sky background subtraction, the sky spectrum must be wavelength calibrated. Figure 7.10 shows a typical sky background spectrum in optical wavelengths. Multitude of bright emission lines can be identified in the optical sky spectrum, being worse in the infrared spectrum. [OH] emission lines may be observed between 6500 Å and 2.62 μm, with the brightest lines above 1.5 μm. For example, in the H band, about 70

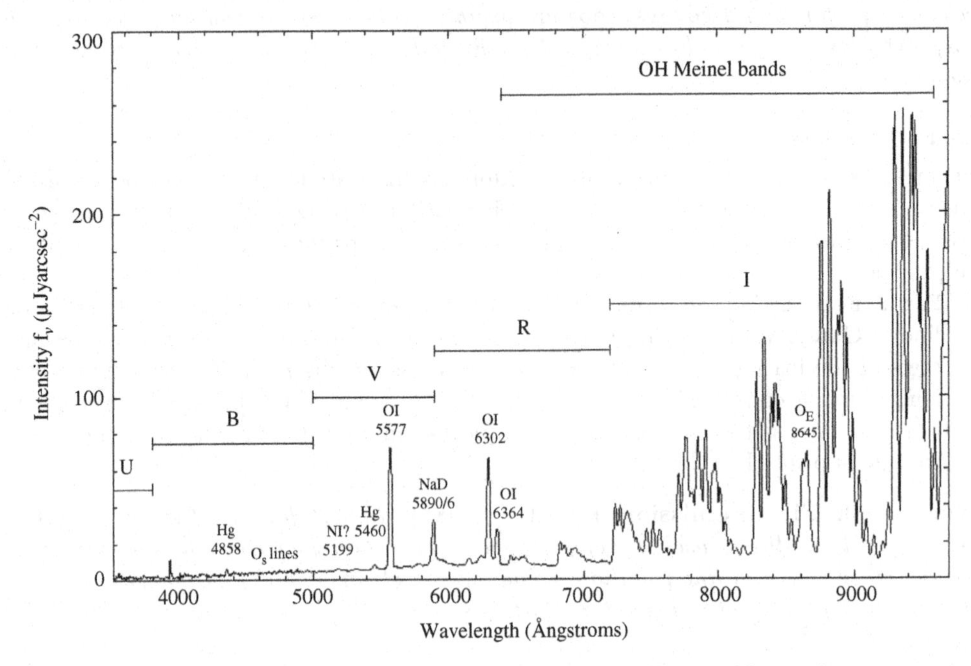

FIGURE 7.10. Typical sky spectrum in the optical wavelength range obtained on a moonless night at the Observatorio del Roque de los Muchachos (ORM) on La Palma island (Spain). Most of the emission features identified in the sky spectrum are due to airglow. The Meinel rotation-vibration bands of [OH] dominate the sky background spectrum redward of 6500 Å. The typical broad band filters for astronomical observations are marked (figure from Benn and Ellison, 1998).

distinct [OH] lines can be identified which are spectrally unresolved at a resolution of $\sim$17,000. The problem of subtracting such number of sky lines is a difficult one because the absolute and relative intensities of [OH] lines are strongly temporal and spatially dependent. For this reason, the accuracy of the sky subtraction is much better in the continuum than in atmospheric emission lines, and the large number of [OH] lines makes an accurate sky background subtraction in the red difficult.

But the accuracy of the sky background also depends on instrumental features. Typical instrumental problems such as flexures in the spectrograph, aberrations in the optics or variations in the spectral efficiency of the detector also cause bad features when subtracting the sky spectrum.

Therefore, atmospheric and instrumental factors complicate such an *a priori* easy task of subtracting the sky spectrum from the object spectra. Figure 7.11 illustrates troubles found when subtracting the sky background on 3D data.

Depending on the science to be done, sometimes it is very important to subtract the sky background with a high level of accuracy. In some cases – for example in infrared observations – the object may have a surface brightness much fainter than the sky (for example to study faint objects). Because IFS is usually used for observing extended objects, most of the time it is necessary to sample the sky background far enough from the target object (sky background exposures) using a telescope off-set. The problem is worse when IFS is used with adaptive optics, when the point-spread function (PSF) delivered by the telescope and the adaptive optics system should be the same in both target and background exposures for a proper and accurate sky subtraction (for a description of

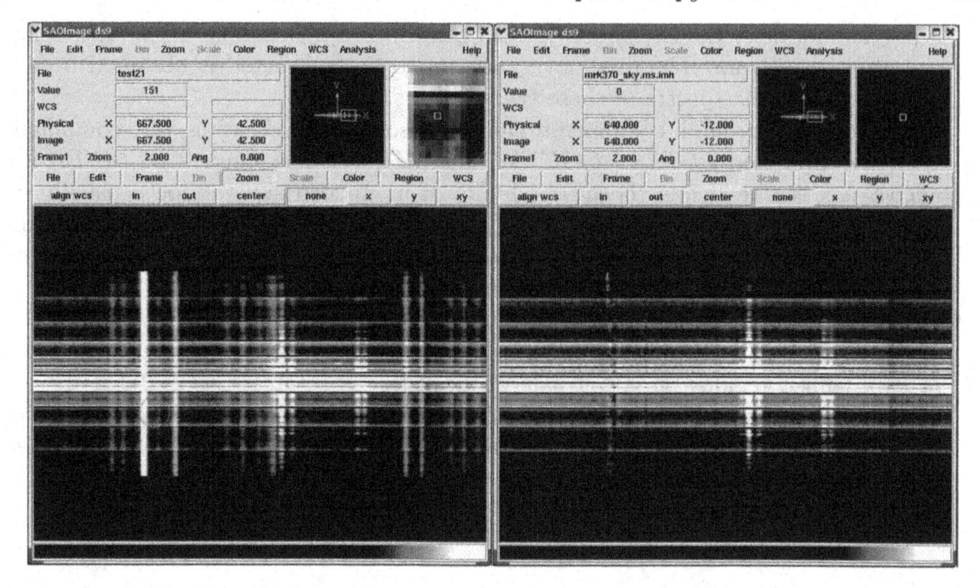

FIGURE 7.11. The DS9 displays viewing the same RSS image before (left) and after (right) the sky background subtraction. On the left display, several emission lines are present. After the sky background subtraction (on the right), some residuals remain where the brightest sky lines were located.

different ways of removing the sky background, see Allington-Smith and Content, 1998). Therefore, the accuracy of the sky background subtraction often depends on observing strategies.

Software to subtract the sky background

Most of the traditional astronomical software includes tasks that allow subtraction of the sky background. However, none of these tools was specifically designed to work with 3D data. Therefore, a few modifications of the original packages (or scripts involving several packages) are needed to subtract the sky background from IFS data.

- IRAF (http://iraf.noao.edu/) includes a few tasks to remove the sky background from object spectra such as *SKYTWEAK* or *SKYSUB*. *SKYTWEAK* is part of the *onedspec* IRAF package and subtracts sky background from 1D spectra after tweaking the sky spectra. The IRAF *specred* package includes the previous *SKYSUB* task. Users select a subset of spaxels from a multispec format image (RSS image) corresponding to sky spectra. They are combined (mean, median or sum) into a master spectrum that is subtracted from another subset of spectra defined as the object spectra.

  Other methods have been implemented for estimating and removing the sky background spectrum through IRAF tasks. For example, the *SVsvdfit* task – an IRAF add-on package developed at the Smithsonian Astrophysical Observatory Telescope Data Center – is used to calculate and remove the background spectrum from spectra without a simultaneous measurement of the sky, using eigenvector fitting techniques. The technique is based on the iterative subtraction of continuum estimates and eigenvector sky models derived from singular value decompositions of sky spectra and sky spectra residuals (Mink and Kurtz, 2001).
- MIDAS (http://www.eso.org/projects/esomidas/): The command *SKYFIT* is the MIDAS task for removing the sky spectrum from long-slit data. This program makes

a polynomial fit of the sky spectrum using two different windows above and below the object spectrum. The final sky spectrum to be subtracted is obtained from a combination of the spectra of these two windows.

The sky background component can be also estimated using the commands *DEFINE/SKY* and *EXTRACT/SKY* within MIDAS. The command *DEFINE/SKY* allows the user to define the off-set limits of up to two sky windows, usually on both sides of the object spectrum. The command *EXTRACT/ECHELLE* performs an average extraction of the sky background using these off-sets. The two extracted images can be subtracted for sky background correction.

- FIGARO (http://starlink.rl.ac.uk/): The *POLYSKY* task is used in FIGARO to subtract the sky from long-slit data by polynomial fitting in the spatial direction of two defined regions on either side of an object of interest. FIGARO users can also use the command *SCNSKY*, which uses the minimum median algorithm to create a sky spectrum from an image without any lines clear of stars. FIGARO also provides the *SKYLINER* task removing a sky spectrum normalized by the height of the 5577 Å [OI] emission line.

7.7 Fitting and modelling

Fitting and modelling is often useful to model the spectra and/or the 2D distribution of the flux (or even the entire dataset) to extract more information from the data. Here we will discuss the peculiarities of fitting and modelling 3D data.

7.7.1 *Emission-line fitting*

We give here a brief description on how to fit and deblend emission lines (like ionized gas emission lines). For this purpose, we will use our own fitting routines, handling both RSS images and datacubes (`FIT3D`, Sánchez *et al.*, 2004); however, please note that these are still under development and distributed as they are. These routines consist of a core coded in `C` and a set of `perl` wrappers.

`fit_spec_back` , *the* `C`*-coded core fitting routine*

This program fits the data with the given model computing a minimization of the reduced χ^2 and using a modified Levenberg–Marquardt algorithm. The user provides a spectrum (ASCII table in the E3D style[5]) and a configuration file where the model to fit is described and formatted as follows:

- first, a header line with four elements:
  - two integers, the first one always left to 0 (to identify the file), and a second giving the number of individual functions included in the model;
  - two floats, the first one being the goal value for the reduced χ^2 and then the goal value for the variation of this value between two consecutive iterations. When one of these two limits is reached, the fitting process stops.
- for each function:
  - a first line indicating its name;
  - nine further lines (one per parameter, nine being the maximum number of parameters allowed in `FIT3D`), with five numbers each.
- each of these five elements gives:
  - a first guess of the parameter value (a float value);

[5] This is a three-column ASCII file containing: ID, WAVELENGTH and FLUX.

- an integer flag indicating if this parameter is to be fitted (=1) or if it will stay fixed (=0);
- the lower bound for the parameter;
- the upper bound for the parameter;
- an integer flag indicating if this parameter is free (= −1) or if it is linked to the same parameter of another function of the model (= to the number of the function in the configuration file to which it is linked, 1 being the first one).

• The linking method allows the fitting of lines that share some properties (e.g. lines that are kinematically coupled) or the inclusion of lines for which the line ratio is known (e.g. the line ratio between [OIII] 5007 Å and [OIII] 4959 Å). There are two ways to link parameters:
  - additive: the parameter of this function is equal to the parameter of the function indicated by the element number (5) plus a constant. In this case element number (3) indicates the constant to be added and element number (4) must be set to 0.
  - multiplicative: the parameter of this function is equal to the parameter of the function indicated by the element number (5) multiplied by a constant. In this case element number (3) indicates the multiplicative factor, and element number (4) must be set to 1.

Although many functions are described with fewer than nine parameters, the nine lines are mandatory (but possibly filled with zeroes). The currently available functions in `FIT3D`, to build a model for Observatoire de Lyon, are:

- `gauss1d` one-dimensional Gaussian, whose parameters are (in the following order): (1) the central wavelength, (2) the peak intensity and (3) the σ of the Gaussian function (describing the width, FWHM = $2.345\,\sigma$).
- `eline` a Gaussian emission line, whose parameters are (in the following order): (1) the central wavelength, (2) the integrated flux, (3) the σ of the Gaussian function (FWHM = $2.345\,\sigma$) and (4) the velocity in km s^{-1}.
- `poly1d` a n-order polynomial function ($n < 9$), whose parameters are the coefficients of the polynomial function.
- `back` a spectrum background, included in the list of files `BACKGROUND_LIST` (one file per line), whose parameters are: (1) order of the background file in the list, (2) the intensity scaling factor, (3) the σ of a Gaussian function to convolve the background spectra to match the resolutions and/or the velocity dispersions, and (4) the velocity in km s^{-1} to shift the spectrum.

The following example illustrates a model based on four functions:

- two emission lines;
- a zero-order polynomial function;
- a background spectrum.

The constraints included are as follows:

- The lines and the background are kinematically coupled, defining a system (see below).
- Two lines have the flux intensity linked.

As shown in this example, one of the powerful features of this tool is the possibility of defining systems. A system is a set of emission lines which share the same velocity and velocity dispersion. The dispersion is included as a dispersion in terms of wavelength and not of velocity, which may be a problem for large wavelength ranges. This co-moving set of emission lines may be fundamental for deblending several components in complex systems. The continuum can also be coupled kinematically (or not) to a system.
For example:

```
    0   4    0.2   0.001
eline
 5007   1      0        0   -1
  1.0   1    0.2       20   -1
  4.6   1      0        0   -1
12000   1      0        0   -1
    0   0      0        0   -1
    0   0      0        0   -1
    0   0      0        0   -1
    0   0      0        0   -1
    0   0      0        0   -1
eline
 4959   1      0        0   -1
 0.33   1   0.33        1    1
  4.6   1      0        0    1
12000   1      0        0    1
    0   0      0        0   -1
    0   0      0        0   -1
    0   0      0        0   -1
    0   0      0        0   -1
    0   0      0        0   -1
...
```

```
...
poly1d
 0.01   1    -10     10   -1
    0   0      0      0   -1
    0   0      0      0   -1
    0   0      0      0   -1
    0   0      0      0   -1
    0   0      0      0   -1
    0   0      0      0   -1
    0   0      0      0   -1
    0   0      0      0   -1
back
    1   0      0      0   -1
 10.0   1    0.2    200   -1
  4.6   1      0      0    1
12000   1      0      0    1
    0   0      0      0   -1
    0   0      0      0   -1
    0   0      0      0   -1
    0   0      0      0   -1
    0   0      0      0   -1
```

The result of the fitting process is displayed on the screen and stored in the following output files:

- `out.fit_spectra`, an ASCII file including a header line with the number of fitted functions (i.e. the number of function included in the model), followed by a line per function. Each line contains 19 values:
  - the label of the model, as described above;
  - the values of its nine parameters;
  - the nine corresponding error estimates.
- `out_config.fit_spectra`, a configuration file where the output values of the fit are used as first-guess values. This file may be useful when fitting iteratively similar spectra with the same model.
- `out_mod_res.fit_spectra`, a five-column ASCII table. The columns are:
  - the wavelength;
  - the original flux at this wavelength;
  - the derived model;
  - the residual; and
  - the residual continuum.

The syntax is as follows:

`fit_spectra SPECTRUM.TXT CONFIG_FILE [BACKGROUND_LIST NBACK]`

`fit_spec_back.pl` *, a* `perl` *wrapper to create configuration files*

This program creates the configuration file needed by `fit_spec_back`. It requires an input spectrum in the ASCII format (see above), an ASCII file listing the emission lines to be fitted (see `tut4/emission_lines.txt`), the wavelength range to be fitted, a file containing the list of backgrounds to use (if not applicable, type `none`), the number of backgrounds files, and a file defining the wavelength bounds of the spectral regions to mask.

The emission lines file is an ASCII file with two columns, one for the wavelength of the emission line, and the other one for the name (or label) of the line; for example:

```
5006.84   [OIII]5007
4958.91   [OIII]4959
4861.32        Hbeta
```

The file defining the spectral regions to mask is a two-column ASCII file with respectively the lower bound (minimum wavelength) and the upper bound (maximum wavelength) of the region to mask. The syntax is:

```
fit_spec_back.pl SPECTRUM.TXT LINE_FILE START_WAVELENGTH
END_WAVELENGTH BACK_LIST NBACK MASK_LIST [CONFIG]
```

Exercise 32: Fitting and deblending emission lines (`dir: tut4`)**.** *Using E3D, import the* `sobj.fits` *or* `sobj.e3d` *file and display a slice with a wavelength range centred in the [OIII] emission line, avoiding other emission lines. Plot the corresponding spectrum. Save this spectrum as an ASCII table (e.g. as* `spec_cen.txt`*). Exit from E3D. Enter on the shell command-line:*

```
fit_spec_back.pl spec_cen.txt emission_lines.txt 4800 5300 none 0 none
```

Then answer the following questions:

```
Number of systems to include:2
Redshift of the system number 0:0.03
Systemic Velocity=8993.77374 km/s
Flexibility of the velocity (km/s):500
GAUSSIAN sigma of the line (GAUSS,min,max): 3 0.1 40
Redshift of the system number 1:0.03
Systemic Velocity=8993.77374 km/s
Flexibility of the velocity (km/s):500
GAUSSIAN sigma of the line (GAUSS,min,max): 40 0.1 1000
HeII 4686.0 at 4826.58 found
Include? (y/n):n
[OIII]5007 5006.84 at 5157.0452 found
Include? (y/n):y
Flux (Flux, min, Max)?:2000 100 5000
Included
[OIII]4959 4958.91 at 5107.6773 found
Include? (y/n):y
Flux (Flux, min, Max)?:2000 100 5000
Included
Hbeta 4861.32 at 5007.1596 found
Include? (y/n):y
Flux (Flux, min, Max)?:2000 100 500
Included
[FeII] 4889.62 at 5036.3086 found
Include? (y/n):n
[FeII] 4905.34 at 5052.5002 found
Include? (y/n):n
[FeII] 5111.6299 at 5264.978797 found
```

```
Include? (y/n):n
System 0 contains 3 lines
HeII 4686.0 at 4826.58 found
Include? (y/n):n
[OIII]5007 5006.84 at 5157.0452 found
Include? (y/n):n
[OIII]4959 4958.91 at 5107.6773 found
Include? (y/n):n
Hbeta 4861.32 at 5007.1596 found
Include? (y/n):y
Flux (Flux, min, Max)?:2000 100 5000
Included
[FeII] 4889.62 at 5036.3086 found
Include? (y/n):n
[FeII] 4905.34 at 5052.5002 found
Include? (y/n):n
[FeII] 5111.6299 at 5264.978797 found
Include? (y/n):n
```

The data are plotted with white lines, the model in blue and the model continuum subtracted in red. The green line shows the residuals. The first fit is very bad since we have not included any background. We will now model the background as a polynomial function. Enter the following:

```
cp tmp.config fit_system.config
emacs fit_system.config &
```

The file `fit_system.config` *is the configuration file used for the model. It included a header line and the different parameters for each function included in the model (see the file* `README_fitting_tool.txt` *or Section 7.1.1). We are going to add a new function to the model, changing the header line from:*

```
0 4 0.2 0.001
```

to

```
0 5 0.2 0.001
```

and adding the description of the new parameters at the end of the file:

```
     0  4     0.2  0.001
poly1d
   100  1  -1e12    1e12  -1
   0.1  1  -1e12    1e12  -1
     0  0      0       0  -1
     0  0      0       0  -1
     0  0      0       0  -1
     0  0      0       0  -1
     0  0      0       0  -1
     0  0      0       0  -1
     0  0      0       0  -1
```

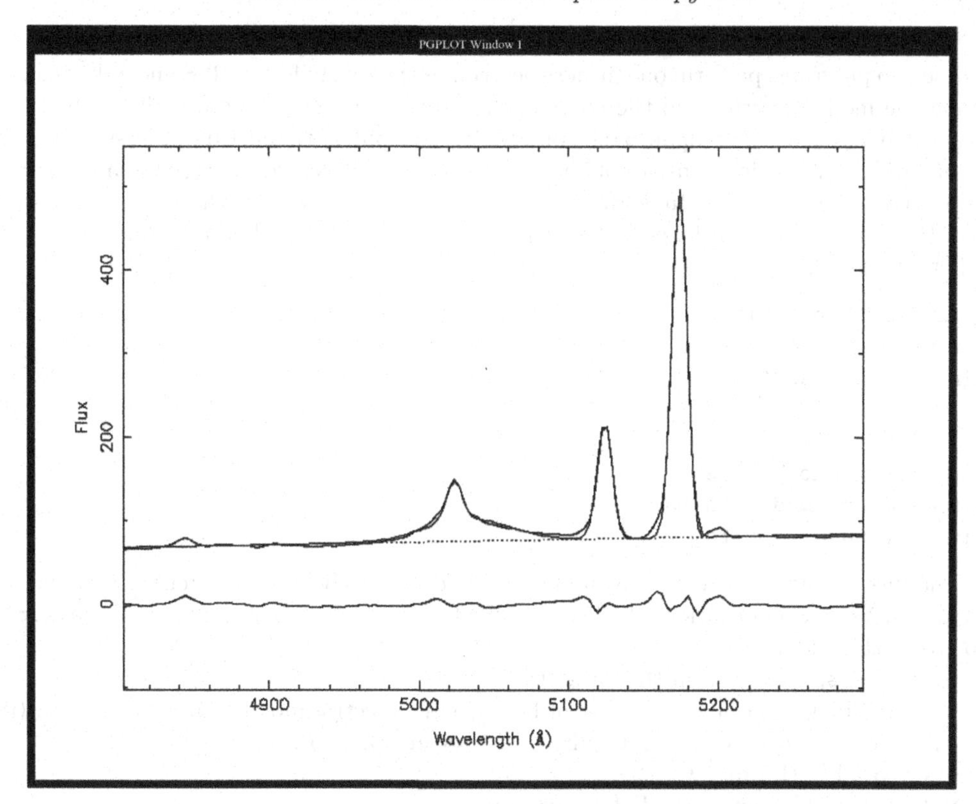

FIGURE 7.12. Example of the use of **FIT3D**: snapshot of an iteration of **fit_spec_back** where a quasi-stellar object (QSO) spectrum is fitted with three emission lines plus a continuum. The lines show the input data, overplotting the fitted model. The dot line is the continuum. The residuals is the line close to zero.

Again, execute the fitting procedure using this new config file as input:

```
fit_spec_back.pl spec_cen.txt emission_lines.txt 4800 5300 none 0 none
fit_system.config
```

The output of the fitting process is illustrated in Figure 7.12. Play with the parameters, or include a new system to describe the broad line until you get an acceptable fit.

7.7.2 *Gas kinematics*

Gas kinematics is derived from the individual fits (one or several systems) of all the spectra of the dataset. For each spaxel (i.e. one row of the position table), the user is provided with the velocity and velocity dispersion of the systems, the intensity of each line of the system, the continuum level and the estimate of the error for each parameter. In the case of grid-distributed positions (i.e. datacubes), the output of the kinematics analysis would be a set of 2D images (or maps), one per parameter. In other cases, it may be required to interpolate the outputs of the process to create 2D images, but it is also possible to handle ASCII of FITS table as slices, as stated above.

Different tools are made available for gas kinematics. FIT3D includes all the subroutines required to handle the problem in a consistent and simple way. We describe some of these tools here.

Fitting the spectra: `kin_back_rss.pl` *and* `kin_back_cube.pl`

These two programs perform one fit per spectrum respectively for an RSS and a datacube, using the model described in the configuration file. The fitting is interactive, using the previously described `fit_spec_back` routine. The output of this fitting process is stored in an ASCII file, which consists of the concatenation of each `out.fit_spectra` log file produced by the routine. In addition, a RSS (or a datacube) of the original dataset (i.e. the trimmed section fitted), the derived model and the residual are stored in three different FITS files, i.e.:

`kin_back_rss_org.fits`,
`kin_back_rss_mod.fits` and
`kin_back_rss_res.fits`

or

`kin_back_cube_org.fits`,
`kin_back_cube_mod.fits` and
`kin_back_cube_res.fits`.

The two programs require as input a 3D dataset (RSS or datacube format), a `fit_spec_back` configuration file (as described before). Please run `fit_spec_back.pl` to create this file, using:

- a typical spectrum from the dataset;
- an ASCII file containing the list of background spectra and the number of them (if no background spectrum is required, just set `nback` to 0);
- the wavelength range to fit;
- the output filename (ASCII format); and
- a PGPLOT device for plotting the output.

The syntax for calling these tools from the command-line is as follows:

```
kin_back_rss.pl rss.fits config_file back_file nback
start_wavelength end_wavelength out_file DEVICE
```

or

```
kin_back_cube.pl cube.fits config_file back_file nback
start_wavelength end_wavelength out_file DEVICE
```

Creating maps: `mapgrid_back.pl` *and* `map_interp_back.pl`

These programs create flux intensity, velocity and velocity dispersion maps, and their corresponding error maps, using the outputs of the previous kinematic analysis programs. The maps are stored in 2D FITS files, reflecting the original disposition of the spaxels in the sky for the first one (a case of a regular-grid datacube) or interpolating the data for the second program (for RSS or non-regular gridded IFU). They require as input the position table of the dataset (or an ASCII slice obtained with `E3D`; see Section 7.5), the files output from the previous step (kinematics analysis programs) and a label to prefix the output FITS files. In the case of `map_interp_back.pl`, three additional parameters are required to define the kind of interpolation to perform:

- the interpolated function to be used (as a number from 1 to 5);
- an additional parameter required by some interpolation functions;
- the final pixel size.

These three parameters are described in Section 7.5.3, and in the `E3D` user guide. Based on our experience, we recommend use of the natural neighbour interpolation,

INTERPO_MODE=3, with a GRID_FUNC=1e-12 and a pixel size of 1/3 of the original spaxel size.

The names of the output maps indicate their contents. Their filename starts with the given prefix, followed by four possible labels:

- flux, for the flux intensity;
- vel, for the velocity;
- disp, for the velocity dispersion (both of them in $\mathrm{km\,s^{-1}}$);
- cont for the background continuum.

The filenames end with a running number (starting from 0) indicating the emission line that they refer to. Each of these maps has its corresponding error map, which includes an e in the name before the label type; for example, out_sobj_kin_flux_00.fits and out_sobj_kin_eflux_00.fits are the flux intensity map and its corresponding error map for the first emission line analysed in a kinematic study, using out_sobj_kin as prefix.

The syntax for calling these tools from the command-line is the following:

```
mapgrid_back.pl slice.txt/POS_TABLE out.model PREFIX_OUT
```

or

```
map_interp_back.pl slice.txt/POS_TABLE out.model PREFIX_OUT
INTERP_MODEGRID_FUNC GRID_PIX
```

Exercise 33: Kinematics analysis. *Using* E3D, *import the* sobj.fits *or* sobj.e3d *file. Select the wavelength range between 4800–5300 Å (to speed up the process) and store it as a RSS file (*sobj_rss.fits*). We will use the same configuration file as for the previous exercise. It is highly recommended to fit a single spectrum before performing the whole semi-automatic analysis. Execute the program* kin_back_rss.pl*:*

```
kin_back_rss.pl sobj.rss.fits fit_system_tunned.config null 0
4800 5300 out_sobj_kin.txt /xs
```

This program will fit all spectra of the dataset, one by one, using fit_spec_back, *and the configuration file* tmp.config. *Each time a fit is performed it is possible to execute the following options:*

- s *to store the results and jump to the next spectrum;*
- a *to perform a non-interactive automatic fit over the whole dataset;*
- r *to repeat the fit.*

Since the program opens an emacs *terminal with the current configuration file,*[6] *it is possible to modify it and repeat the fit until a suitable result is obtained. The program stores the results from each individual fit in a log file (*out_sobj_kin.txt *in the example), and the fitted model and residual spectra in the FITS files* kin_back_rss_mod.fits *and* kin_back_rss_res.fits.

Perform an automatic fit using a *option to speed up the exercise. It is possible to check the results with IRAF and* ds9 *(or a similar tool), using the commands:*

```
disp kin_back_rss_org.fits 1
disp kin_back_rss_mod.fits 1
disp kin_back_rss_res.fits 1
```

[6] If you are not an emacs user or you do not have it installed in your system, please edit kin_back_rss.pl and change the preferences to use your favourite editor.

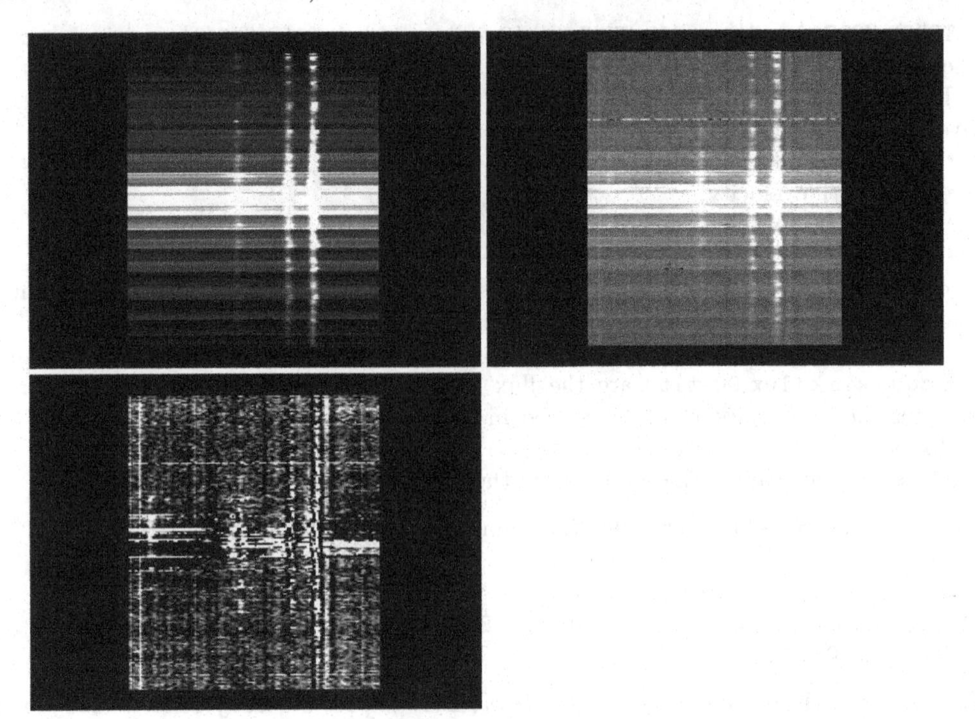

FIGURE 7.13. Example of the kinematics analysis. The top-left panel shows the derived model using the kinematics analysis in Exercise 34. The top-right panel shows the original RSS dataset, and the bottom-left panel the residuals after subtracting the model.

as illustrated in Figure 7.13. As explained above, it is possible to map the fitted parameters, creating 2D images and interpolating data if required (in case of RSS data). To do so, run the following:

```
map_interp_back.pl sobj.pt.txt out_sobj_kin.txt out_sobj_kin 3
1e-12 1
```

Once the maps are created, it is possible to visualize them using IRAF:

```
disp out_sobj_kin_flux_00.fits 1
disp out_sobj_kin_vel_00.fits 1
```

The results are shown in Figure 7.14.

In case of velocity dispersion, it is necessary to subtract, to the observed values, the instrumental component. The instrumental component consists of all the dispersions in the lines that are not physically associated with the studied source but due to the instrument, night conditions or any other effect. The standard procedure is to perform a kinematic analysis over a nearby emission line (over the non-sky-subtracted data), and to subtract the obtained velocity dispersion to the observed object (with a quadratic subtraction).

It is important to note here that the velocity dispersion is fitted assuming a dispersion in Å, that it is not the proper approach for wide wavelength ranges. Kinematics analyses are normally performed over a narrow wavelength range, centred on the emission lines of interest; this is why the current approach is used. It is also easier to decontaminate this method from instrumental effects in the velocity dispersion.

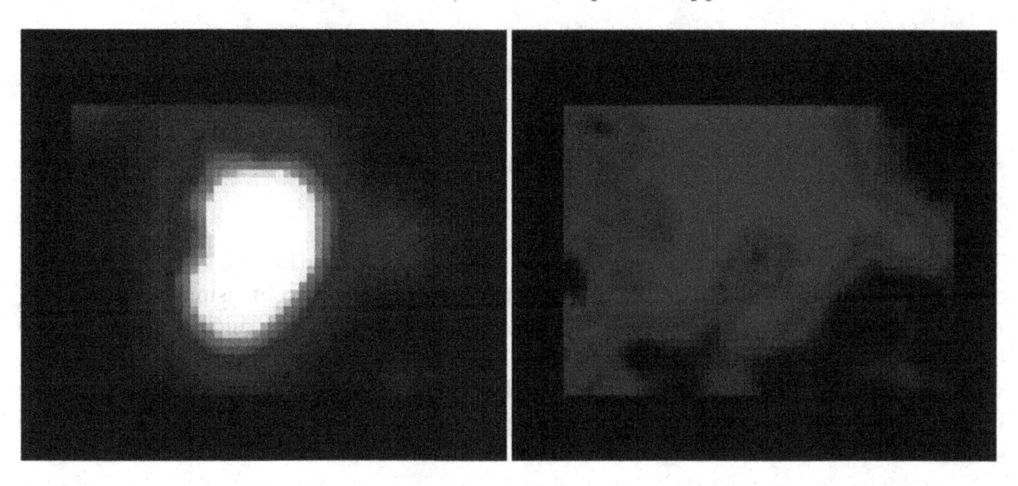

FIGURE 7.14. Example of the use of FIT3D. It shows the flux intensity map (left panel) together with a velocity map (right panel) derived with the automatic line fitting performed in Exercise 34.

7.8 Decoupling spectra of different objects: crowded field spectroscopy

3D data can be seen as a collection of narrow-band or monochromatic images of the same object at different wavelengths. Assuming a smooth variation in the morphology of the objects with wavelength, it is possible, using 3D data, to apply techniques of decoupling or deblending, normally used in the analysis of 2D imaging, and to extend them to 3D (García-Lorenzo *et al.*, 2005; Sánchez *et al.*, 2006a). This method is particularly interesting:

- when we have low signal-to-noise images of objects, all with the same morphology in the image (e.g. QSO lenses; Wisotzki *et al.*, 2003);
- when we have a crowded field where we want to deblend the spectra (e.g. the core of globular clusters or galaxy clusters; Sánchez *et al.*, 2006b); or
- when we want to deblend two well-defined components on the same object (e.g. QSO-hosts or bulge-to-disk decomposition, or point-like sources over bright backgrounds, like in XPNs, Sánchez *et al.*, 2004).

There are different implementations of this technique in the literature (Roth *et al.*, 2004; Sánchez *et al.*, 2004; Wisotski *et al.*, 2004; Fabrika *et al.*, 2005; García-Lorenzo *et al.*, 2005; Sánchez *et al.*, 2006a, 2006b). We will not describe all of them, but take a particular case, namely the extension of `galfit` to 3D, or `galfit3d` (Sánchez *et al.*, 2006a, 2006b). The `galfit` (Peng *et al.*, 2002) program is a program for modelling 2D images of one or several objects using analytical models for galaxies and/or point-like sources, convolving them with a PSF.

In the case of `galfit3d`, we have extended the technique to model regular-gridded 3D data (i.e. datacubes), which may be interpolated versions of non-regular ones (e.g. García-Lorenzo *et al.*, 2005), and decouple the spectra of different objects in the field and/or deblend the spectra from different components of the same object (Sánchez *et al.*, 2006a, 2006b). A major advantage of IFS data is that it is possible to impose links over the parameters used for each n-fits (one per wavelength), on the basis of a smooth variation of the morphological parameters, or we can force them to a particular value if we have a source of external information (e.g. high-resolution Hubble Space Telescope images; García-Lorenzo *et al.*, 2005; Sánchez *et al.*, 2006b). In both cases the fitting process is iterative, with a first loop of non-acceptable results, and a set of iterations where the morphological parameters are constrained to different values, using as input parameters the output of the previous fits (Sánchez *et al.*, 2006a).

7.8.1 *Components of* `galfit3d`

`galfit3d` is a collection of `perl` scripts, including:

galfit3d.pl

This program performs a modelling of an input datacube creating, for each wavelength, a narrow-band image of the width of a spectral pixel, and fitting it using `galfit` (Peng *et al.*, 2002). The output of the fitting is stored in a log file, which consists of the concatenation of the different outputs (e.g. `galfit.01`) produced by `galfit` for each wavelength. A quality log file is also stored, with an integer flag that indicates if the fitting process has finished the minimization (1) or ended in abnormal break (0). The syntax is:

```
galfit3d.pl INPUT_CUBE.fits MOD_CUBE.fits RES_CUBE.fits
GALFIT_CONF LOGFILE QUALITY.txt
  GALFIT_CONF: A) object.fits, B) output.fits
```

galfit3d_force.pl

This program works like `galfit3d.pl`, but using as guessed values for the fit the output values of a previous iteration (`INPUT_LOGFILE`). Each parameter can be smoothed using a polynomial function, or forced to be a given value. This is indicated with the `INPUT_NPOLY` configuration file, which is a three-column ASCII file whose elements per line are:

- the number of the model in the `galfit` configuration file whose parameter needs to be smoothed;
- the number of the parameter of this model to be smoothed;
- the order of the polynomial function to fit (if value is an integer) or fixed value (if value is a float).

The syntax is:

```
galfit3d_force.pl INPUT_CUBE.fits MOD_CUBE.fits RES_CUBE.fits
GALFIT_CONF INPUT_LOGFILE INPUT_NPOLY OUTPUT_LOGFILE QUALITY.txt [PLOT]
```

extract_par_g3d.pl

This program allows the extraction of the fitted values for a parameter of a given model along the spectral pixels and the storage of them in an ASCII output file, plotting this parameter along the spectral pixels using PGPLOT with the defined device. It requires as input the output log files obtained with `galfit3d.pl` or `galfit3d_force.pl`, the name of the output file where the results are to be stored, the number of the model and the number of the parameter to extract. The syntax is:

```
extract_par_g3d.pl INPUT_LOGFILE QUALITY.txt OUTPUT_FILE
model_number par_number DEVICE [min max]
```

extract_spec_g3d.pl

This program allows the extraction of the integrated spectrum corresponding to an object (or component of an object) modelled with a 2D model, and the storage of it as ASCII spectrum (see Section 7.5). It requires as input the output log files obtained with `galfit3d.pl` or `galfit3d_force.pl`, the name of the output file where the results are to be stored, the number of the model in the `galfit` configuration file, the zero point of the magnitude indicated in that configuration file, the starting wavelength and the step between wavelengths for the fitted datacube. The syntax is:

```
extract_spec_g3d.pl INPUT_LOGFILE QUALITY.txt OUTPUT_FILE
model_number mag_zeropoint CRVAL CDELT
```

7.8.2 *Installation of* `galfit3d`

`galfit3d` is still under development. We distribute in these proceedings a beta version as it is (no warranty). A large number of simulations is still needed before producing reasonable outputs for the most usual cases. It requires having `galfit` and `IRAF` installed on your computer. We recommend that the user modifies the source code of `galfit` in order to increase the accuracy of the produced output, not to be truncated to the second decimal of magnitude (see `galfit` documentation). As a set of `perl` scripts, it requires the modules described in Section 7.4 to be installed too. Once these modules are installed, please update the following lines in the `perl` scripts with your own paths:

```
require /data/sanchez/code/R3D/my.pl;
$galfit="/data/sanchez/galfit/galfit";
$cl="/usr/local/bin/cl";
```

Exercise 34: 3D modelling. *It is beyond the scope of this tutorial to make a complete description of the deblending and modelling options for 3D spectroscopy. We will therefore perform a very simple modelling on GMOS data. The data presented here are those of a nearby QSO, where we would like to deblend the bright QSO from the extended Host galaxy. To do so, the model mimics the point-like source using a Gaussian convolved with a PSF, and the Host galaxy using a Sersic profile, with the Sersic index equal to 1, i.e. a Disk galaxy. We will guide the fit by forcing a scale-length, ellipticity and position angle for the galaxy, and we will not perform a second iteration. Run the program:*

```
galfit3d.pl input.cube.fits input_cube.mod.fits
input_cube.res.fits galfit.qso_fix input_cube.fit.log
input_cube.fit.quality
```

The program will loop over all the slices in the dataset and fit them with the given model. The output of the fit is stored in two files: `input_cube.fit.log` *(the log file), with the direct outputs from galfit, and* `input_cube.fit.quality` *(a file containing a flag indicating if the fit ended normally or crashed). It also produces two datacubes: one with the fitted model and another with the residuals from the fitting technique. It is possible to inspect the progress of the fit, and visualize any of the fitted parameters, by executing:*

```
extract_par_g3d.pl input_cube.fit.log input_cube.fit.quality
junk.junk 1 3 /xs 22 18
```

It is also possible to visualize each of the individual fits using IRAF, by typing:

```
cl < display.cl
```

Once the fit is finished, it is possible to extract the deblended spectra of each components by executing:

```
extract_spec_g3d.pl input_cube.fit.log input_cube.fit.quality
spec_nuc.txt 1 25 6600 2.3
extract_spec_g3d.pl input_cube.fit.log input_cube.fit.quality
spec_host.txt 2 25 6600 2.3
```

7.9 DAR correction using external tools

As mentioned before, E3D contains its own tools to perform the DAR correction. However, in most cases, they cannot produce results accurate enough, and are only valid for a quick correction. Let us remember that any DAR correction requires the spatial resampling of the data, which produces a loss of the initial spatial configuration of the spaxels. (This is not critical, but is important to know.)

We discuss here other tools that perform empirical DAR corrections for datacubes. 'Empirical' means that the DAR correction is determined directly from the data, which is a unique feature of 3D spectroscopy. For doing this you need a bright object with a well-defined intensity peak at any wavelength. The empirical correction is determined by looking to the spatial coordinates of the intensity peak for each wavelength and estimating the shift from one to the other. In many cases the empirical correction works better than the theoretical one, based on the Filippenko (1982) formulae. The reason is that the theory behind these formulae assumes that the layer of the atmosphere with equal refraction index is flat and perpendicular to the zenith of the telescope. However, it is well known that this in only a first-order approximation since these layers depend strongly on the orography of the observatory. The next sections describe these tools.

7.9.1 imcntr_cube.pl

This program finds the centroid of an object (using a first-guess position) at each wavelength, and smoothing the centroid position variation from wavelength to wavelength using polynomial fittings. It stores the results in a two-column ASCII file. It requires as input the datacube used for the analysis, the first-guess centroid of the object, the order of the polynomial functions to fit among the x and y coordinates, and the name of the output file. In addition, it is possible for the analysis to exclude a given number of spectral pixels both at the beginning and at the end of the wavelength range. The syntax is:

```
imcntr_cube.pl input_cube.fits X Y NP_X NP_Y centers_list.txt
[N_START] [N_END]
```

7.9.2 cube_shift.pl

This program applies the DAR correction previously computed, recentring an object along each wavelength. It requires as input the names of the input and output datacubes, the list of centroid positions, the final coordinates of the object and the name of an output IRAF script. This script can be used further to apply the same correction on another datacube, by modifying the relevant values. This option is interesting for datacubes where there is no bright object to trace the centroid positions, but for which it is possible to use other observations of a bright nearby star (a type of DAR calibration). The syntax is:

```
cube_shift.pl input_cube.fits output_cube.fits centers_list.txt
X0 Y0 shift.cl
```

7.9.3 cube_shift_delta.pl

This program performs a global and spatial shift of a datacube. It requires as input the names of the input and output datacubes and the relative shifts in the x and y axis. The syntax is:

```
cube_shift_delta.pl input_cube.fits output_cube.fits delta_x delta_y
```

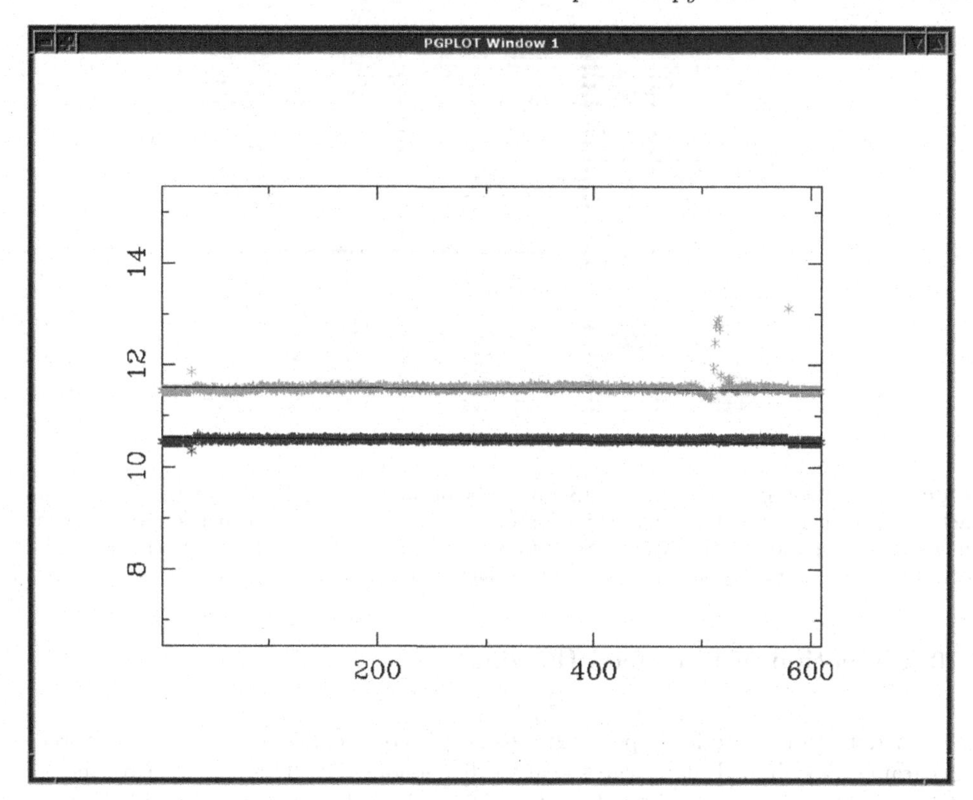

FIGURE 7.15. Example of the output of the script **imcntr_cube.pl**: the x (top) and y (bottom) coordinates of the centroid of a star as a function of the spectral pixel numbers, for a datacube.

To use these scripts, it is first necessary to edit the files and change the line

```
require /data/sanchez/perl/MY/my.pl;
```

to fit your own path. It is also necessary to run a `mkiraf` before running the programs in the directory where you have the dataset, since these programs are simply IRAF wrappers.

Exercise 35: Empirical DAR correction (`dir: tut3`). *Using IRAF with DS9 (or any similar tool), display the channel 300 of a file, e.g.* `disp star_no_dar.cube.fits [*,*,300]` *and determine the centroid of the star in the field, using* `imexam` *or any similar tool. The centroid is near the coordinates (10.5,11.5). To determine the centroids along the wavelength, run:*

```
imcntr_cube.pl star_dar.cube.fits 10.5 11.5 2 2 centers_list.txt
10.5 11.5 7.5 14.5
```

The output of this program is illustrated in Figure 7.15, with the distribution of the x *and* y *coordinates of the centroid of the star as a function of the spectral pixel numbers (representing the wavelength). The list of centroids is stored in the* `centers_list.txt` *file, used as input in the correction process:*

```
cube_shift.pl star_dar.cube.fits star_no_dar.cube.fits
centers.lis 10.5 11.5 shift.cl
```

The file `star_no_dar.cube.fits` *contains the DAR corrected datacube.*

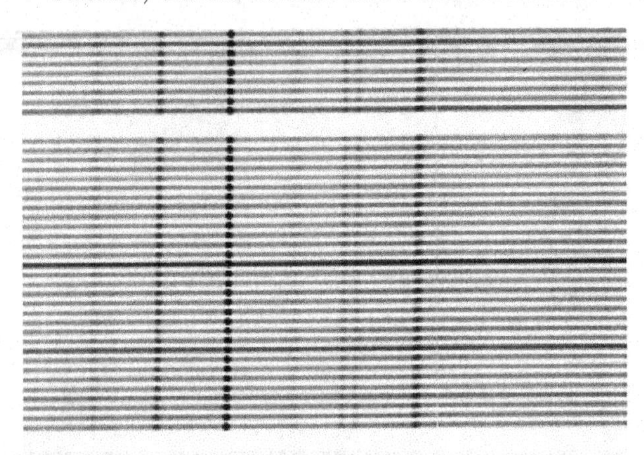

FIGURE 7.16. View of a section of IFS raw data obtained with PMAS in the PPAK mode. Each dark line corresponds to the projection of a spectrum along the dispersion axis, which, in this case, corresponds to the X-axis. Spectra are separated by 7 pixels across the Y-axis, and projected in a few pixels, contaminating the adjacent spectra.

7.10 Reduction of fibre-feed IFS data

7.10.1 *The raw data*

The raw data from fibre-feed spectrographs consist of a collection of spectra, stored as a 2D frame and aligned along the so-called dispersion axis. This axis can be the x- or y-axis, or neither, although it tends to be almost aligned with one of them. Figure 7.16 shows an example of IFS raw data, corresponding to PMAS (Potsdam Multiaperture Spectrophotometer; Roth *et al.*, 2005) in the PPAK mode (Kelz *et al.*, 2006), illustrating this distribution. Each spectrum is spread over a certain number of pixels along the perpendicular direction to the dispersion (cross-dispersion axis). The shape of the profile along the cross-dispersion direction is similar to an asymmetrical Gaussian. There is therefore a contamination from each spectrum to the adjacent ones. This is what is well known as the cross-talk (see Figure 7.16). The profile changes for each spectrum along the cross-dispersion axis and, in general, is more asymmetrical on the edges of the pseudo-slit than in the centre. It also changes along the dispersion axis. Well-designed instruments have well-defined and narrow profiles, and a limited cross-talk. However, there is always a compromise between the number of fibres 'packed' in the spectrograph, the size of the charge coupled device (CCD) and the width of the profiles and cross-talk. As we quoted above, the spectra are not perfectly aligned along the dispersion axis, due to the configuration of the instrument, its setup, the optical distortions, the instrument focus and the mechanical flexures (if any). In addition, the transmission changes from fibre-to-fibre due to physical differences among them, differences in the tension suffered by the fibres at different positions, and slight mis-alignments of the fibres in the pseudo-slit. These effects also affect the width of the projected profiles across the cross-dispersion and the dispersion axis, and introduce small differences in the spectral dispersion from fibre-to-fibre.

7.10.2 *IFU data reduction*

The data reduction of all fibre-feed IFUs shares similar steps, due to the similarities of their raw data. We assume, as a starting point for the reduction, that the data have been bias subtracted and flat-field corrected (when possible) with an external tool. Once this pre-processing is complete, the data reduction consists of:

- identification of the position of the spectra on the detector for all pixels along the dispersion axis;
- extraction of each individual spectrum;
- distortion correction of the extracted spectra and determination of the wavelength solution (i.e. dispersion correction); and
- re-arrangement of the spectra in the sky position.

Each of these reduction steps has been treated with a set of command-line routines in `R3D` (Sánchez, in preparation), a package specially written for reducing fibre-feed IFS data. We briefly describe below the different steps required to perform a reduction.

Identification of the position of the spectra on the detector

To identify the position of the spectra on the detector you need a continuum exposure (e.g. a dome-flat), which, in the case of instruments that suffer from flexures (i.e. displacement of the spectra in the CCD at different positions of the telescope), should be taken with the telescope pointing to the same location as your science exposure. The location of the spectra is found to be at a given column in the CCD by comparing the intensity at each row along the column (or a co-added set of rows around it) with those of n adjacent pixels, checking for those pixels that they verify the maximum criteria, using `peak_find`.

This program requires as input the continuum raw data frame, a flag indicating the principal direction of the dispersion axis (0 for x, 1 for y), the width around the central column where the intensities (for low-level signal-to-noise frames) are to be co-added, a flag indicating if the results are to be plotted (1) or not (0), the number of subframes to plot in the screen, the number of adjacent pixels to check for the maximum criterion, the minimum distance between adjacent spectra (projected in the cross-dispersion axis), the fraction of the intensity peak to use as a threshold for detecting intensity peaks (i.e. spectra), and the name of an ASCII output file containing the location along the cross-dispersion axis of the detected peaks. The syntax is:

```
peak_find RAW.fits Spectral_axis[0/1] Coadd_width plot nplot
nsearch DMIN IMIN(% of the MAX) OUTFILE [colum_search]
```

Then, the tracing of the peak intensities along the dispersion axis is done by looking for maxima around each original location within a given window, using `trace_peaks_recursive`. This program requires as input the continuum raw data frame, a flag indicating the dispersion direction (0 means along x, 1 along y), the file containing the location of the intensity peaks (i.e. the output of the previous program), the width of the window where the intensities along the dispersion axis are to be co-added around each column, the width (in pixels) indicating where to look for a new centroid of the peak along the cross-dispersion axis, a flag indicating if the results are to be plotted (1) or not (0), the number of subframes to plot in the screen, the number of adjacent pixels to check for the maximum criterion, and the output file, a 2D FITS file where the x-axis corresponds to the original dispersion axis and the y-axis indicates the number of traced spectra. The value stored in each pixel is the location of the peak of the centroid in the original frame of the corresponding spectrum at the corresponding spectral pixel. This file is called 'trace'. The syntax is:

```
trace_peaks_recursive RAW.fits Spectral_axis[0/1] PEAKS_FILE
coadd_width search_width plot nplot nsearch TRACE.fits [y_shift]
2nd_search_width]
```

Extraction of individual spectra

After tracing the location of the spectra on the CCD, the next reduction step is to extract, for each spectrum, the flux corresponding to each spectral pixel along the dispersion axis. The simplest method to perform this extraction is to co-add the flux within a certain aperture around the 'trace' of the spectra in the raw data, and store it in a 2D image. This procedure is named aperture extraction, and it has been implemented in `R3D`. The x-axis of the resulting image corresponds to the original dispersion axis, while the y-axis corresponds to the ordering of the spectra along the pseudo-slit. This is the so-called row-stacked spectra representation or RSS (Sánchez, 2004). The extraction is performed using `extract_aper`.

This program requires as input a raw data frame, a flag indicating the direction of the dispersion axis (0 for x, 1 for y), the trace frame, the aperture width for the extraction and the name of the output file. The syntax is:

```
extract_aper RAW.fits Spectral_axis[0/1] TRACE.fits Aperture
OUTPUT.fits [SHIFT]
```

Distortion correction

Most of the spectrographs do not disperse the light homogeneously along the cross-dispersion axis. The dispersion is distorted, being larger on the edges of the slit than in the centre. This distortion is sometimes called the 'C' distortion, due to its shape on the CCD. In the case of fibre-feed spectrographs, additional distortions are introduced due to the way the fibres are placed in the pseudo-slit. Once the spectra are extracted, these distortions are self-evident. The left panel of Figure 7.17 shows an example of extracted spectra, corresponding to an arc exposure obtained with PMAS in the PPAK mode. The 'C' distortion and the discontinuities in the dispersion solution along the cross-dispersion may be clearly seen. These discontinuities and the slight differences in the dispersion from fibre to fibre do not allow a 2D modelling of the distortion map with an analytical function to be created, neither at the level of the extracted spectra, nor at the level of the raw data. The distortions have to be corrected fibre to fibre before finding a common wavelength solution.

`R3D` performs this correction by a two-step procedure, using arc calibration lamp exposures (like the one in Figure 7.17). First, the peak intensity of a single emission line is traced along the cross-dispersion axis and shifted to a common reference by a linear shift. This is done by the program `dist_cor`. This program requires as input the aperture extracted file, the output corrected file, the output first-order distortion file, a flag indicating whether to smooth (1) or not (0) the solution along the cross-dispersion axis, the reference pixel in the dispersion axis to look for the emission line peak, the width of the window in pixels to look for this peak, the number of adjacent pixels to check for the maximum criterion, and a flag indicating whether to plot the results (1) or not (0). The syntax is:

```
dist_cor EXTRACTED.fits CORRECTED.fits DISTORTION_CORRECTION.txt
SMOOTH[0/1] start_index delta_index nsearch plot [n_fib] [nsigma]
[center_index]
```

After this the intensity peak of a set of selected emission lines is traced, and a polynomial distortion correction is determined in order to recentre all the lines to a common reference. This is done using the program `mdist_cor_sp`. This program requires as input the output of the first-order correction, an aperture to look for emission lines, a $n\sigma$ threshold over the average intensity to look for emission line peaks, the order of the

FIGURE 7.17. Example of extracted spectra from an arc exposure obtained with PMAS in the PPAK mode, before correcting for the distortion in the dispersion along the cross-dispersion axis (left panel) and after (right panel). The x-axis corresponds to the dispersion axis.

polynomial function to fit the distortion correction (spectrum by spectrum), the output file with the distortion-corrected frame, the output file containing the 2D map of the corrections, the range of pixels in the dispersion direction to consider for the analysis, and a flag indicating if results will be plotted (1) or not (0). The syntax is:

```
mdist_cor_sp EXTRACTED.fits APERTURE NSIGMA NPOLY OUTPUT.fits
DISTORTION.fits NX_min NX_max plot [NSIGMA2] [BOX] [NY_CEN]
```

Wavelength solution

To derive the wavelength coordinate system, the emission lines of an arc exposure are first identified using an interactive routine. The distortion-corrected spectra of the arc are then transformed to a linear wavelength coordinate system by a 1D spline interpolation, assuming a polynomial transformation between both coordinate systems. The required transformation is stored in an ASCII file to be applied on the science data at a later stage. This procedure is done using the `disp_cor.pl` script.

This program requires as input the output of the distortion correction analysis, the starting wavelength and the wavelength step needed for the final spectral pixels, the aperture to look for emission lines, the row number of the central spectrum over which to perform the analysis, width of pixels in the cross-dispersion axis for co-adding the spectra (to increase the signal-to-noise ratio), the order of the polynomial function to fit the wavelength solution, the output file containing the spectra transformed to a linear wavelength coordinate system, the output file containing the dispersion or wavelength solution applied, and a flag indicating if results will be plotted (1) or not (0). The syntax is:

```
disp_cor.pl EXTRACTED.fits CRVAL CDELT APERTURE NY_SPECTRA
WITDH NPOLY OUTPUT.fits DISPERSION.txt plot
```

Additional steps

Once wavelength is calibrated, the data need to be corrected from fibre-to-fibre differential transmission, combined with other exposures (using with off-sets), corrected for the differential atmospheric refraction, sky-subtracted and flux-calibrated. R3D includes tools to perform some of these operations, but it is beyond the scope of this quick reference guide to explain all of them in detail here. Moreover, they may strongly depend on the instrument used for the acquisition.

Exercise 36: Reducing data with R3D[7] (dir: tut6). *We are going to perform a simple data reduction on PMAS Roth et al. (2005) data in the PPAK mode (Kelz et al., 2006) using* R3D. *First, we determine the number of spectra and their location along the cross-dispersion axis for the central column of the CCD:*

```
peak_find.pl DomeFF_Cont_Galaxy.fits 0 2 1 4 2 3.5 0.10
DomeFF_Cont_Galaxy.peaks
```

Then, we trace the location of the peak intensity of the spectra along the cross-dispersion axis of each pixel in the dispersion axis:

```
trace_peaks_recursive DomeFF_Cont_Galaxy.fits 0
DomeFF_Cont_Galaxy.peaks 2 2 1 4 1 DomeFF_Cont_Galaxy.trc.fits
```

Once the location of the spectra in the CCD (trace) has been determined it is possible to extract them:

```
extract_aper.pl IRAS18131_A.fits 0 DomeFF_Cont_Galaxy.trc.fits 5
IRAS18131_A.ms.fits
```

The C distortion correction is then determined and corrected:

```
dist_cor DomeHe_ThAr.ms.fits DomeHe_ThAr.dc0.fits
DomeHe_ThAr.dist.txt 0 662 30 2 1 2 2
mdist_cor_sp DomeHe_ThAr.dc0.fits 3 300 3 DomeHe_ThAr.dc1.fits
DomeHe_ThAr.dist.fits 17 1011 1 2 5 3
```

The distortion correction determined using the He lamp exposure can be applied to the science data:

```
mdist_cor_external IRAS18131_A.ms.fits DomeHe_ThAr.dist.txt
DomeHe_ThAr.dist.fits IRAS18131_A.dc1.fits 0 1.353
```

Once transformed to a common reference, it is possible to find the wavelength coordinate solution:

```
disp_cor.pl DomeHe_ThAr.dc1.fits 3700 1.6 3 180 3 4
DomeHe_ThAr.disp_cor.fits DomeHe_ThAr.disp.txt 1
```

[7] Data provided by K. Kehring.

and apply it to the data:

```
disp_cor_external.pl IRAS18131_A.dc1.fits 3700 1.6
IRAS18131_A.disp_cor.fits DomeHe_ThAr.disp.txt 0
```

The differences in the fibre-to-fibre transmission can be determined using dome-flat exposures, and correct the science frames from them:

```
fiber_flat.pl mDomeFF_Cont_IRAS18131_A.disp_cor.fits
fiber_flat_IRAS18131_A.fitsimarith.pl IRAS18131_A.disp_cor.fits /
fiber_flat_IRAS18131_A.fits
IRAS18131_A.fc.fits
```

REFERENCES

Allington-Smith, J., Content, R. (1998), *PASP*, **110**, 1216

Arribas, S., Mediavilla, E., Watson, F. *et al.* (1998), *Fiber Optics in Astronomy III*, **152**, 149

Arribas, S., Mediavilla, E., García-Lorenzo, B., del Burgo, C., Fuensalida, J.J. (1999), *A&AS*, **136**, 189

Becker, T. (2001), 3D-Spektroskopie hintergrundkontaminierter Einzelobjekte in Galaxien der lokalen Gruppe, PhD Thesis, Potsdam University

Benn, C., Ellison, S. (1998), *La Palma night-sky brightness*, ING technical note 115

Fabrika, S., Sholukhova, O., Becker, T., Afanasiev, V., Roth, M., Sánchez, S.F. (2005), *A&A*, **437**, 217

Filippenko, A.V. (1982), *PASP*, **94**, 715

García-Lorenzo, B., Acosta-Pulido, J.A., Megias-Fernández, E. (2002), *Galaxies: the Third Dimension*, 282, 501

García-Lorenzo, B., Sánchez, S.F., Mediavilla, E. González-Serrano, J.I., Christensen, L. (2005), *ApJ*, **621**, 146

Glazebrook *et al.* (2001), http://pdl.perl.org/

Kelz, A., Verheijen, M.A.W., Roth, M.M., Bauer, S.M., Becker, T., Paschke, J., Popow, E., Sánchez, S.F., Laux, U. (2006), *PASP*, **118**, 129

Kissler-Patig, M., Copin, Y., Ferruit, P., Pécontal-Rousset, A., Roth, M.M. (2004), *AN*, **325**, 159

Mink, D.J., Kurtz, M.J. (2001), Astronomical Data Analysis Software and Systems X. *ASPCS*, **238**, 491

Pearson, T. (1995), http://www.astro.caltech.edu/~tjp/pgplot/

Pécontal-Rousset, A., Copin, Y., Ferruit, P. (2004), *AN*, **325**, 163

Peng, C.Y., Ho, L.C., Impey, C.D., Rix, H. (2002), *AJ*, **124**, 266

Roth, M.M., Becker, T., Kelz, A., Schmoll, J. (2004), *ApJ*, **603**, 531

Roth, M.M., Kelz, A., Fechner, T. *et al.* (2005), *PASP*, **117**, 620

Sánchez, S.F. (2004), *AN*, **325**, 167

Sánchez, S.F. (2006), *AN*, **327**, 850

Sánchez, S.F. (2008), *Euro3D Visualization Tool: User Guide for E3D*, v1.3.7. Research Training Network. Available online at http://www.aip.de/Euro3D/E3D/#Docu (July 2009)

Sánchez, S.F., Cardiel, N., in preparation

Sánchez S.F., García-Lorenzo, B., Mediavilla, E. *et al.* (2004), *ApJ*, **615**, 156

Sánchez S.F., García-Lorenzo, B., Jahnke, K. *et al.* (2006a), *NewAR*, **49**, 501

Sánchez, S.F., Cardiel, N., Verheijen, M., Benitez, N. (2006b), *Proceedings of the ESO and Euro3D workshop 'Science Perspectives for 3D Spectroscopy'*, Garching (Germany), October 10–14 (2005) (astro-ph/0512550)

Walsh, J.R., Roth, M.M. (2002), *Msngr*, **109**, 54

Wisotzki, L., Becker, T., Christensen, L., Helms, A., Jahnke, K., Kelz, A., Roth, M.M., Sánchez, S.F. (2003), *A&A*, **408**, 455

Printed in the United States
By Bookmasters

Printed in the United States
By Bookmasters